IET ENERGY ENGINEERING 211

Blockchain Technology for Smart Grids

Other volumes in this series:

Blockchain Technology for Smart Grids

Implementation, management and security

Edited by
Gururaj H L, Ravi Kumar V, Francesco Flammini, Hong
Lin, Goutham B, Sunil Kumar B R and C Sivapragash

The Institution of Engineering and Technology

Published by The Institution of Engineering and Technology, London, United Kingdom

The Institution of Engineering and Technology is registered as a Charity in England & Wales (no. 211014) and Scotland (no. SC038698).

First published 2022

The Institution of Engineering and Technology
Futures Place
Kings Way, Stevenage
Hertfordshire, SG1 2UA

www.theiet.org

British Library Cataloguing in Publication Data
A catalogue record for this product is available from the British Library

ISBN 978-1-83953-535-2 (Hardback)
ISBN 978-1-83953-536-9 (PDF)

Typeset in India by Exeter Premedia Services Private Limited

Contents

5   Blockchain-based secured IoT-enabled smart grid system   115
Sivasankar P, Rathy G A, John Deva Prasanna D S,
Karthikeyan Perumal, Gunasekaran K, and Sivapragash C

6   Deployment of IoT-based sensor data management in smart grids   145
Goutham B, Sunil Kumar B R, Gururaj H L, Ravikumar V,
and Francesco Flammini

**11 Development of novel cryptocurrencies for IoT—blockchain-enabled
    smart grid platforms   273**
Anil Kumar, Gururaj H L, Sunil Kumar B R, Goutham B, and Ravikumar V

About the Editors

Gururaj H. L. is an associate professor at Vidyavardhaka College of Engineering, Mysuru, India. He is a professional member of ACM, founder of Wireless Internetworking Group (WiNG), senior member of IEEE and lifetime member of ISTE and CSI, and board member of the *International Journal of Block chains and Cryptocurrencies (IJBC)*. His research interests include Blockchain technology, wireless sensor networks, machine learning and others. He has published over 75 research papers in international journals.

Ravi Kumar V. is head of the Department in Department of Computer Science and Engineering, Vidyavardhaka College of Engineering, Mysuru, India. A professional member of ACM and lifetime member of ISTE, IETE and CSI, and board member of the *International Journal of Block chains and Cryptocurrencies (IJBC)*, his research interests include Blockchain technology, data mining and machine learning.

Francesco Flammini is a professor of Computer Science with a focus on Cyber-Physical Systems at Mälardalen University, Sweden. He has led the Cyber-Physical Systems (CPS) research and education area within the Smarter Systems complete knowledge environment. An IEEE senior member and the chair of the IEEE SMC Technical Committee on Homeland Security, he is currently a steering committee member and conference coordinator for the IEEE FDC Initiative on Symbiotic Autonomous Systems.

Hong Lin is a professor in computer science at University of Houston-Downtown, USA. He has worked on large-scale computational biology at Purdue University, active networks at the National Research Council Canada, and network security at Nokia, Inc. He established the Grid Computing Lab at UHD, and has published more than 100 research papers in various reputed international journals and conferences.

Goutham B. is an assistant professor at Vidyavardhaka College of Engineering Mysuru, Karnataka, India. He is the faculty member of the Wireless Internetworking Group (WiNG) and an RPA global certified member. His areas of interest include smart grids, cyber security, microgrids, renewable energy sources and electrical machines.

Sunil Kumar B. R. is an assistant professor at Vidyavardhaka College of Engineering, Mysuru, India. He is the faculty member of the Wireless Internetworking Group (WiNG) and lifetime member of ISCA and IAENG. His research interests include

Blockchain technology, wireless senor network, ad-hoc networks, IoT and machine learning.

C. Sivapragash is a professor at Swarnandhra College of Engineering and Technology, India. His research focuses on computer networks, security, reliability and distributed computing. A current project is 'Trace Based Real Time Service Selection and Composition for Secure Cloud Computing in Smart Grids'.

Chapter 1

Blockchain: a new era of technology

Sunil Kumar B R[1], Gururaj H L[1], Goutham B[1],
Ravikumar V[1], and Hong Lin[2]

Many industries, including finance, medical, manufacturing, and education, utilise blockchain applications to benefit from the technology's unique set of properties. Trust ability, collaboration, organisation, identification, credibility, and transparency are all advantages of blockchain technology (BT). This chapter presents the fundamentals of BT and its properties and working procedure.

1.1 Introduction

Stuart Haber and W. Scott Stornetta invented blockchain in 1991. It was designed to timestamp digital documents so that they could not be backdated or tampered with, but it sat unused until Satoshi Nakamoto (false identity) [1] used it to create a digital cryptocurrency in 2009.

Blockchain is a public ledger that is distributed [2] and decentralised. It is a time-stamped collection of immutable data records controlled by a group of computers that are not owned by a single company. Cryptographic concepts are used to protect and link each of these blocks of data, as shown in Figure 1.1.

1.1.1 Properties of blockchain

Figure 1.2 depicts the block in the blockchain along with its properties.

Data – Data stored in the block of the blockchain depends on the type of the blockchain. If you consider Bitcoin blockchain as an example, it stores the details about the transaction, such as sender, receiver, amount, and timestamp. Any type of data can be stored in the block, for example, voting, E-notary and collecting taxes. Once the data have been recorded inside the blockchain, it is extremely difficult to modify it.

[1]Vidyavardhaka College of Engineering, Mysore, India
[2]University of Houston-Downtown, Houston, TX, USA

Figure 1.1   Blockchain

Hash – A hashing technique can be used to make data immutable. When a block is generated, its hash is determined, and if something inside the block changes, the hash will change as well.

Hash of previous – To tie the blocks of data together in blockchain, hash of the previous block is recorded in a new block when it is created, thus creating a chain of blocks.

Metadata – It includes information such as timestamp and block number.

1.1.2   How blockchain works

When the block is created before adding it to the chain, it has to go through several stages. Figure 1.3 shows the workflow of the blockchain.

Step 1: A transaction is requested.

Step 2: Block is created for that transaction.

Step 3: Block is distributed to all the peers in network.

Step 4: Each node in the network will validate the transaction.

Step 5: Node receives a reward for proof of work.

Step 6: Block will be added to the blockchain.

Step 7: Update is sent across the network.

Step 8: Transaction is completed.

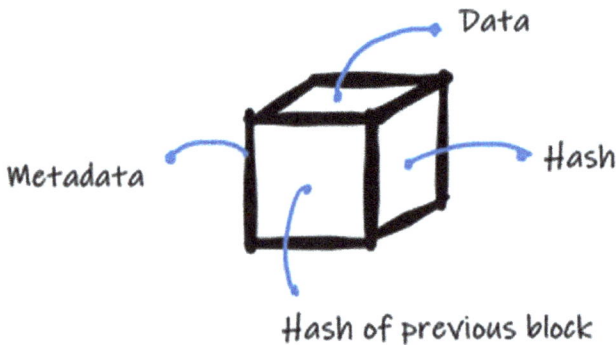

Figure 1.2   Properties of block in blockchain

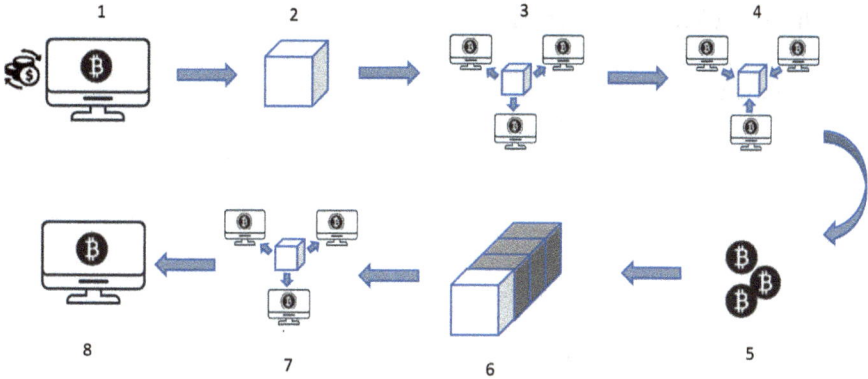

Figure 1.3   Workflow of the blockchain

1.2   Components of blockchain

Blockchain is built with different types of components [3], each of which serves a specific purpose. Figure 1.4 shows the components of blockchain.

1.2.1   Distributed ledger: historical record which is immutable

All blockchain transactions will be recorded in an immutable ledger and distributed across the network.

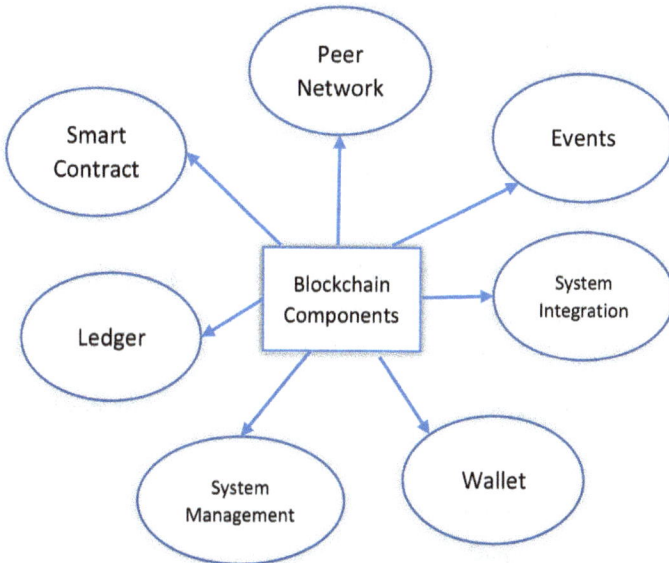

Figure 1.4   Components of blockchain

There are three types of ledgers:

- Single entry – One side entry in either credit or debit column
- Double entry – Tracks both debits and credits
- Triple entry – It is an enhancement of double entry system where all transaction entries are approved and secured by cryptography.

Blockchain uses a triple entry ledger, which includes debit, credit, and link to previous blocks.

1.2.2   Peer-to-peer network: maintains distributed ledger

A peer network stores, updates, and maintains the ledger. All nodes keep their own copy of the blockchain ledger. It is the responsibility of the entire network to reach an agreement called consensus on the contents of each ledger update. This assures that each individual copy of the ledger is identical, without the need for a centralised ledger copy.

1.2.3   Wallet: stores user credentials

The user's wallet on blockchain maintains their credentials and keeps the track of digital assets linked to their address. User credentials and any other information linked with their account are stored in the wallet.

1.2.4   Smart contract: automates the transaction

The first blockchains had minimal configuration and were meant to merely enable financial transactions to be completed and recorded in a historical ledger. Ever since, blockchains have grown into fully working distributed computers in some cases. Smart contracts are programs that are executed when predefined criteria stored on a blockchain are satisfied. They are usually used to automate the execution of an agreement so that all parties may be confident of the conclusion right away, without the need for any intermediaries or time-wasting steps.

1.2.5   Events: blockchain update notifications

Events update the blockchain ledger and the status of the peer network. The generation and distribution of a new transaction over the peer network, as well as the insertion of a new block to the blockchain, are examples of events. Notifications from smart contracts on blockchains that enable such contracts may also be included in events.

1.2.6   Systems management: manages components of blockchain

In a field that is continuously developing, the blockchain is meant to be a long-lasting solution. The capacity to create, change, and monitor blockchain components to fit the demands of its users is provided by systems management.

1.2.7 Systems integration: integrates external system with blockchain

As blockchain technology (BT) has advanced and grown more useful, it has become increasingly common to link blockchains with other systems, most often via smart contracts. Systems integration is also used to emphasise this capacity, even though it is not a particular component of the blockchain [4].

1.3 Blockchain actors

Many actors play a vital role in building a blockchain. Figure 1.5 shows these actors.

1.3.1 Blockchain architect

The architecture and design of the blockchain are the responsibility of the architect. The blockchain architect will plan how the blockchain system will be constructed. He or she will decide what data to be retained, what transactions and business logic must be integrated in the network, and so on.

1.3.2 Blockchain network operator

The blockchain operator oversees recording, updating, and maintenance of the blockchain ledger. An operator can join to create a peer network once the block-chain solution is devised and built. The operator's job is to create and manage peers in the network.

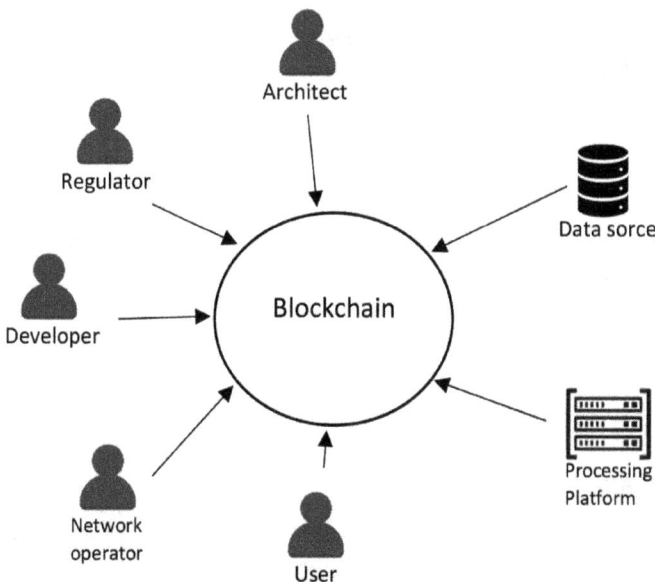

Figure 1.5   Blockchain actors

1.3.3   Blockchain developer

The blockchain developer is responsible for developing smart contracts that operate on the blockchain. The development of smart contract-supporting blockchains has considerably increased the usefulness of the blockchain. To enhance the blockchain's capabilities, programmers create and upload smart contracts. You may also have front-end developers who create applications that access the blockchain in addition to smart contracts.

1.3.4   Processing platforms

Data are processed by an external system. Smart contracts operate on the blockchain, which implies that to stay up with the current state of the network, each peer network member must run the code. If smart contracts often need a considerable amount of processing power to complete, devices outside the peer network could be used to supplement the network's processing capacity.

1.3.5   Blockchain regulator

Many businesses are bound by edicts that regulate how their data are handled and processed. Because of their role within the organisation, a regulator may have greater visibility into the historical record for blockchain systems.

1.3.6   Data storage

Traditional databases for off-chain data storage are represented by data storage. The blockchain provides distributed, immutable storage with built-in integrity checks, but the typical block size and block rate limit its capacity. It is customary to keep data off-chain and a hash of the data on-chain to offer integrity checking for big volumes of data. This ensures that data are not tampered with while also preventing the blockchain from becoming bloated.

1.3.7   Blockchain user

A blockchain user is a business user who is part of a business network. The user is exposed to a blockchain solution in practice. They have no idea what blockchain is. A blockchain user is someone who will utilise the blockchain to perform business transactions. Usually, this involves the usage of blockchain-based software as a backend storage solution. Users rarely interact with the blockchain.

1.4   Types of blockchain

1.4.1   Public blockchain

This is where Bitcoin and other cryptocurrencies got their start and where distributed ledger technology (DLT) gained traction [5]. It overcomes the disadvantages of centralisation, such as lack of security and transparency. Rather than being kept in a single location, information is spread via a peer-to-peer network. It requires some form of data verification due to its decentralised nature. A consensus algorithm [6] is a way for blockchain users to agree on the ledger's present state. Two typical consensus approaches are proof of work (PoW) and proof of stake (PoS).

Anyone with an Internet connection may join a blockchain platform and become an authorized node, allowing public blockchain to be open-source and unrestricted. These users have access to both current and historical data and also the capacity to perform mining operations, which are sophisticated computations necessary to validate and add transaction history.

1.4.1.1 Advantages

• Maximum security

Each public blockchain platform has been built to be as secure as possible. Because shared networks are common targets for Internet hackers, public blockchains strive to maintain a high degree of security processes.

• Complete power to user

Users are bound by a variety of protocols, regulations, and limitations in every other network chain. On a public blockchain, the reality is completely different. Users have total freedom to work with the network and share their ideas because there is no centralised entity supervising it.

• Open for everyone

The public blockchain is an open environment; anybody, regardless of location or device type, may log into the system because all you need is a consistent and trustworthy Internet connection. Because it is available to anyone, you may utilise it to reap the benefits of BT in a totally safe and trustworthy environment.

• Immutability

Immutability is a feature of the public blockchain. This basically means that once blocks are generated and placed into the chain, they cannot be changed, modified, or deleted. Anyone seeking to manipulate the network will be unable to do it. If a user wishes to alter the block, they must start a new chain from the beginning.

• Anonymous nature

The anonymity of the public blockchain is one of the reasons it has attracted so many supporters. Yes, it is a safe and secure open platform on which you can conduct business properly and successfully, but you are not required to reveal your genuine identity or name to participate. No one can monitor your actions on the network if your identity is secured.

- No regulations required

In terms of network server utilisation, there are no precise restrictions for the public blockchain. You have no restrictions when it comes to creating this platform. This is especially useful for consumer platforms where the blockchain is used for more than just internal networking.

- Truly decentralised

True decentralisation is one of the big benefits for any public blockchain firms. In the private blockchain domain, this feature is generally missing. The network is not maintained by a single central platform; rather, it is a distributed system in which each user owns a copy of the ledger.

- Complete transparency

In any shared network system, nothing beats complete transparency. The policy is straightforward: public blockchain enterprises will make your ledger completely accessible to others if you have a copy. You will have complete access to the ledger as a user. This eliminates any discrepancies or grey areas in the network and also the risk of corruption.

1.4.1.2   Disadvantages

Public blockchains have their own set of advantages and disadvantages. Because they are unregulated, they are not suitable for use in any internal system, for example. To put it another way, they are challenging to collaborate with on projects with strict deadlines. Furthermore, due of their anonymity, they are vulnerable to infiltration by anyone wanting to exploit the network.

1.4.1.3   Use cases

- Ethereum network

Ethereum is a decentralised public blockchain network that focuses on executing decentralised application programming code. Simply put, it is a global platform for exchanging information that cannot be altered or destroyed.

- Bitcoin blockchain

In late 2008, during the financial crisis, a pioneering paper titled 'Bitcoin: A peer-to-peer electronic money system' appeared on a little-known online forum. It was written by a mysterious individual known only as Satoshi Nakamoto, a pseudonym for the author's real name [7]. Banks and governments, Satoshi said, had too much power, which they exploited. Satoshi envisioned Bitcoin to fix this: a cryptocurrency that was unregulated and ungoverned by central banks or governments and that you could transfer for free anywhere in the world, with no one in control.

Bitcoin has expanded to a network of roughly 10,000 'nodes' or participants who utilise the PoW technique to authenticate transactions and mine Bitcoin since its inception in 2009.

1.4.2 Private blockchain

A private blockchain is often intended to facilitate internal networking among a small group of people. In terms of management, it varies from its public equivalent. It is often administered by a centralised system overseen by a network administrator. Users must have authorisation to join the network to use this private blockchain.

The basic concept is that the blockchain is controlled by a central authority. This implies that the program will be administered via third-party systems. Furthermore, only individuals with blockchain access and who are participating in the transaction will be able to see what is really taking place. Any other blockchain participant would be unable to see these private transactions.

1.4.2.1 Advantages

- Fast speed and higher efficiency

Public blockchains enable anybody to join the network; however, this reduces the network's efficiency and speed. A private blockchain, on the contrary, utilises as many resources as necessary and is hence quicker and more efficient.

- Empowering organisations

Rather than empowering individuals, private blockchains are typically used to empower companies. To support its procedures, any organisation or business needs a robust network. There are fewer members in private blockchains as the network is smaller at first. This allows them to better manage their environment, resulting in minimal downtime and optimum uptime.

- Compliance

In every sector, compliance is certainly critical. Any technology that does not adhere to stringent compliance rules will eventually fail. To make transactions as smooth and easy as possible, private blockchains adhere to and include all compliance standards within their ecosystem.

● Scalability

You can scale your private blockchain with the help of experienced blockchain engineers to meet your needs. A centralised system facilitates quicker procedures and streamlines decision-making as just a few nodes may govern the network.

● Complete privacy

For data confidentiality concerns, network sharing at the business level frequently demands a higher level of privacy. A private blockchain is the best option if this is one of your prerequisites.

● Completely legal activity

Illegal acts are securely maintained on private blockchains. This is because access to this network is restricted, and acquiring access involves a long authentication procedure. Consequently, any unwanted network access will be filtered out. As only verified users have access to the network, the likelihood of illegal behaviour is reduced.

● Speed

Because just a few users are allowed to join, private blockchains may handle more transactions per second. This cuts down the time to achieve an agreement.

● Balanced and stable

Because only a few users have access to specific transactions, private blockchains are unquestionably a more reliable network solution. Furthermore, because nodes are assigned to each user group, they ensure excellent network stability for users. Private blockchains have fewer nodes than public blockchains, which increases speed and makes transactions easier.

1.4.2.2   Disadvantage

One downside of private blockchains is the contentious allegation that they are not valid blockchains despite the basic concept of blockchain, which is decentralisation. It is also more difficult to get complete trust in the information as centralised nodes choose what is authentic. The limited number of nodes may also contribute to the lack of security. The consensus process may be jeopardised if a few nodes are condemned.

In addition, the source code for private blockchains is typically proprietary and opaque. Users will be unable to independently examine or validate it, thus resulting in security breach. On a private blockchain, there is no anonymity.

1.4.2.3 Use cases

• R3 Corda

Corda is a next-generation blockchain platform that provides privacy, scalability, and security to highly regulated institutions, making it the DLT platform of choice for financial services and beyond.

It operates in an unusual manner in that it is based on a 'state object', with consensus created at the transaction level rather than the full ledger level. As a result, transactions no longer need to be approved by every node in the network.

• Hyperledger

Hyperledger is a software that lets you develop your own blockchain-based services. Its main goal is to foster trust and accountability among the company's shareholders. This blockchain adheres to a set of guidelines, and each unit adheres to a set of guidelines as well. Permissioned transactions are promoted, with each user having the necessary permissions and verifications before being allowed to join the network.

1.4.3 Hybrid blockchain

A 'hybrid blockchain' is a system that combines the greatest features of both private and public blockchains. A hybrid blockchain, in an ideal world, would allow for both regulated and unfettered access.

The hybrid blockchain architecture varies from standard blockchain designs in that it is not available to the public while yet providing blockchain characteristics, such as integrity, transparency, and security.

The hybrid blockchain design, like any other blockchain architecture, is totally adaptable. Members of the hybrid blockchain can decide who may join the network and which transactions are made public. This strategy blends the best of both worlds and ensures that a company's stakeholders may effectively communicate.

Even if transactions are not made public, they may be verified if necessary. Every transaction on the hybrid blockchain network may be kept private while yet being open for verification when needed. Because blockchain is being used, the most important part of it is at work: immutability. It guarantees that each transaction is written just once and that it cannot be altered afterwards.

Once a user has been granted access to the hybrid blockchain platform, he may fully participate in the operations of the blockchain. He has the same ability to do transactions, review them, and even append or change them. However, one element that is kept hidden from other players is the identity of the users. This is done to safeguard the privacy of the user.

1.4.3.1   Advantages

• Works in a closed ecosystem

This implies that while using BT, businesses and organisations do not have to worry about their data being exposed.

• Protecting from 51% attack

Because hackers cannot gain access to the network to carry out the assault, hybrid blockchain is resistant to a 51% attack [8].

• Low transaction cost

Another benefit of using hybrid blockchain is the low transaction costs. Because only a few nodes are necessary to validate transactions, they will almost likely be affordable. The network's most powerful nodes make it simple to validate the transaction, which on the public blockchain might take hundreds of nodes. Transaction costs as low as 0.01$ per transaction are possible.

• Changes the rules when needed

Change is good for business. The good news about hybrid blockchain is that it eliminates the need for regulatory changes. The nature of the shift, on the contrary, is determined by what the hybrid blockchain is attempting to accomplish. In a hybrid system that handles band registration or user identification for verification reasons, however, do not anticipate updating data or modification of transactions.

• Protecting privacy while still communicating with the outer world

Even if private blockchain is the greatest option for privacy concerns [8], they are, however, limited in their ability to communicate with the outside world. Many businesses may wish to maintain their anonymity, but they must also design their blockchain to allow them to interact with all their shareholders, including the general public.

1.4.3.2   Disadvantage

Because information may be protected, this sort of blockchain is not entirely visible. Upgrading can be difficult, and users have little motivation to engage in or contribute to the network.

1.4.3.3   Use cases

● Enterprise services

Hybrid blockchain may be used to create both open-source and enterprise-grade applications. Enterprises in areas like aviation and supply chain, for example, may utilise hybrid blockchain to automate their services and increase their dependability, trust, and transparency for both workers and end-users.

● Governments

Blockchain has the potential to alter the way governments operate. Even governments are aware of this and have begun to implement blockchain in their governance. The government, for example, might utilise blockchain to conduct elections, build a public identification database, record complicated data, automate acquisitions, and give social/humanitarian aid, among other things.

Hybrid blockchains are required to make all of this viable. It gives the government the control they need while also allowing the public to have access to it. Neither a completely private nor a completely public blockchain will function as they either restrict user access or disclose too much data. The correct control of the blockchain may ensure that the government maintains control while using the blockchain.

1.4.4   Consortium blockchain

A consortium blockchain, which is similar to a hybrid blockchain in that it combines private and public blockchain features, is the fourth form of blockchain. It varies, however, in that it necessitates the collaboration of many organisational members over a distributed network. A consortium blockchain is essentially a private blockchain with limited access to a small number of people, obviating the necessity for a public blockchain.

1.4.4.1   Advantages

● Regulations

Enterprise businesses can use both private and consortium blockchain. As a result, it is important to have regulations in place to keep the firm functioning smoothly. All nodes on this sort of blockchain platform must follow the network's regulations. It enables people to operate in a group setting and boosts efficiency at a faster rate. So, if you wish to collaborate with other businesses, consortium blockchains are an excellent option.

* No criminal activity

Private blockchains are comparable to this one. You cannot be anonymous on a federated or consortium blockchain. In truth, anonymity encourages criminal behaviour. However, because the federated blockchain requires established identities, thieves will be unable to access the system.

1.4.4.2   Disadvantage

The consortium blockchain has a lower level of transparency than the public blockchain. Even if a member node is compromised, it may still be hacked, and the blockchain's own rules can render the network useless.

1.4.4.3   Use cases

* Finance and banking

This one is about asset issuance and trade. KYC is another popular application. A consortium of banks establishes a single database that collects and stores all essential information about creditors. The information is taken from the distributed ledger once a bank requires it to identify and assess someone.

* Logistics

The best way to create a network for all supply chain actors is to form a consortium. It is important for product tracking so you can figure out where it came from and how you got it.

1.5   Transaction flow

1.5.1   Client initiates a transaction

Client A has placed an order for radishes. This request is made to peers A and B, who represent clients A and B, respectively. The request is made to peers A and B as the endorsement policy needs both peers to approve every transaction.

The transaction proposal is then produced as a result. An application that uses a suitable software development kit (SDK) (Node, Java, and Python) generates a transaction proposal by utilising one of the available Application Programming Interface (APIs). The purpose is to call a chaincode function with specific input parameters to read and update the ledger.

The SDK acts as a shim, packaging the transaction proposal in the correct format and generating a unique signature for this transaction proposal using the user's cryptographic credentials.

1.5.2   Endorsing peers verify signature and execute the transaction

Endorsing peers ensure that (1) the transaction proposal is well-formed; (2) it has not previously been submitted (replay-attack protection); (3) the signature (using the Membership Service Provider (MSP)) is valid; and (4) the submitter (in this case, client A) is properly authorised to perform the proposed operation on that channel.

The transaction proposal inputs are sent to the chaincode function by the endorsing peers. The chaincode is then run against the current state database, producing transaction results that include a response value, read set, and write set. It is not necessary to update the ledger at this time. The SDK receives this data collection, together with the signature of the endorsing peer and parses it for use by the application as a 'proposal response'.

1.5.3   Proposal responses are inspected

The program examines the suggested answers to verify if they match the supporting peer signatures. If the chaincode just searches the ledger, the application will merely validate the query response before submitting the transaction to the ordering service. Before sending the transaction to the ordering service to update the ledger, the client application validates that the declared endorsement policy has been met (i.e. did peers A and B both endorse). Even if an application decides not to inspect answers or transmits an unapproved transaction, peers will enforce and support it during the commit validation step.

1.5.4   Client assembles endorsements into a transaction

Within a 'transaction message', the application 'broadcasts' the transaction proposal and answer to the ordering service. The read/write sets, signatures of endorsing peers, and the channel ID will be included in the transaction. To perform its job, the ordering service does not need to analyse the whole content of a transaction; instead, it simply takes transactions from all channels in the network, organises them chronologically by channel, and creates blocks of transactions per channel.

1.5.5   Transaction is validated and committed

All channel peers get the transaction blocks when they have been 'delivered'. The endorsement policy was followed, and the read set variables' ledger state has not changed as the transaction execution generated the read set. The transactions in the block are classed as either legitimate or invalid.

1.5.6   Ledger updated

Each peer adds a block to the channel's chain for each valid transaction, and the write sets are committed to the current state database. Each peer delivers an event to alert the client application that the transaction (invocation) has been added to the chain immutably and whether it has been confirmed or invalidated.

Symmetric Keys

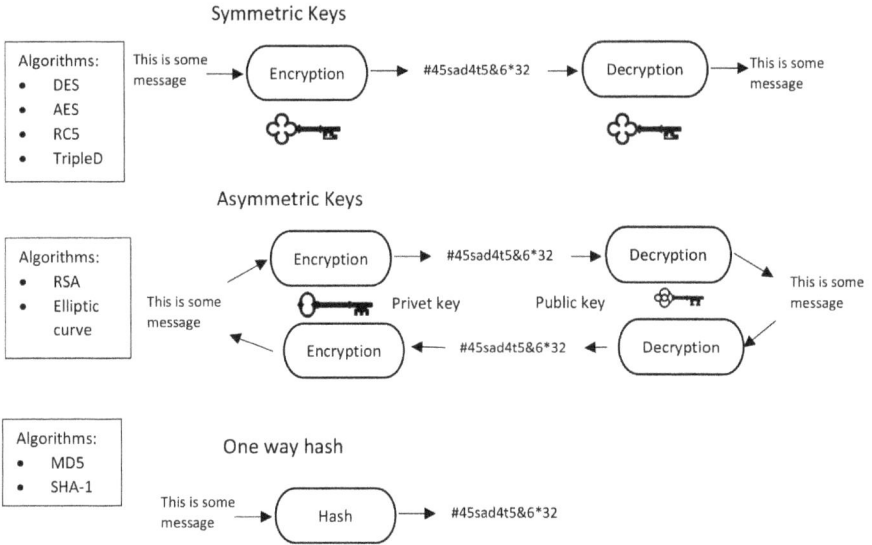

Asymmetric Keys

One way hash

Figure 1.6   Cryptography types

1.6   Cryptography in blockchain

Cryptography is a method of constructing methods and procedures [9] to prevent a third party from having access to or knowledge of data contained in private communications throughout the transmission process. Greek words *Kryptos* and *Graphein* signify 'hidden' and 'to write', respectively.

The following are some of the terminology used in cryptography.

- **Encryption** – It is a process of converting plaintext (plain message) into ciphertext (randomised collection of characters).
- **Decryption** – It converts plaintext from ciphertext.
- **Cipher** – The mathematical function which converts ciphertext from plaintext.
- **Key** – It is a variable that is necessary to deduce the outcome of a cryptographic method.

1.6.1   Cryptography types

To understand cryptography in the context of blockchain, one must first understand the various forms of cryptography. Cryptographic algorithms can be built with symmetric-key cryptography, asymmetric-key cryptography, or hash functions [10], as shown in Figure 1.6.

- **Symmetric-key cryptography** – In this encryption method, we just utilise one key. The same key is used for both encryption and decryption. When using a single common key, the difficulty of securely transmitting the key between the sender and the recipient emerges. It is also known as secret-key cryptography.

- **Asymmetric-key cryptography** – This encryption method makes use of an encryption key and a decryption key, which are referred to as the public and private keys, respectively. This method yields a key pair consisting of a private key and a unique public key generated in the same way. It is sometimes referred to as public-key cryptography.
- **Hash functions** – This encryption technique does not need the usage of keys. It uses a cipher to create a hash value of a certain length from plaintext. Plain text material is nearly impossible to decipher from ciphertext.

1.6.2 Use of cryptography in blockchain

The two types of cryptographic algorithms utilised in blockchains are asymmetric-key algorithms and hash functions. Hash functions are used to ensure that every member of the blockchain views the blockchain in the same way. In blockchains, the SHA-256 hashing algorithm is commonly used as the hash function [11].

1.6.2.1 Hash function

The blockchain achieves the key characteristics from cryptographic hash functions.

- Quickness – In a relatively short length of time, the output may be created. We cannot reverse engineer the input because we just have the output and the hash algorithm.
- Avalanche effect – Tampering the data in small amount might result in a huge difference in the outcome.
- Deterministic – If you use the hash function on any input, you will always get the same result.
- Uniqueness – Each input has a distinct output.

Hash functions can be used to connect blocks and ensure the integrity of the data contained within them. Any changes to the block data may cause inconsistency, leaving the blockchain worthless. To meet this need, the avalanche effect, a property of hash functions, is used.

According to this, even a little change in the hash function's input will result in an entirely different output from the original. As an example, let us look at the results of the SHA-256 hash algorithm.

Input – just hash it
Output–b9e632410fe0acb01a03bcc50338df64b46cd7da4fbc6b558abbfe76fe25acf6
Now let us do some slight changes in input
Input – Just hash it
Here j is changed to J
Output–834f921b13e4f7380b08ce8408bd00ef98a9138fcd8ffdef9c118721c8c31274

You will note a huge difference in the result after changing a single character in the input from lowercase to uppercase. This makes the blockchain data trustworthy and secure; any changes to the block contents will result in a change in hash value, rendering the blockchain invalid and immutable.

1.6.2.2 Asymmetric-key cryptography

In asymmetric-key cryptography, the secret key is generated using a random number process, but the public key is generated using an irreversible approach. Asymmetric encryption offers the benefit of permitting the exchange of distinct public and private keys over untrusted networks.

Digital signatures are commonly used in asymmetric-key cryptography. Because they can be quickly examined and cannot be modified, digital signatures increase procedure dependability. They also have non-repudiation capabilities, making them analogous to physical signatures. Digital signatures may be used to authenticate blockchain data.

The blockchain's underpinnings include hashing, public-private key pairings, and digital signatures. These cryptographic qualities allow blocks to be securely connected by other blocks while also ensuring the blockchain's data stability and immutability.

The public key acts as the person's address in blockchain cryptography. The public key is accessible to everybody, implying that it may be accessed by anybody. The private key is a secret value that is used to get access to address data and undertake any of the operations associated with the 'address' which are often transactions.

Cryptocurrency makes extensive use of digital signatures. They are used on the blockchain for multi-signature contracts and digital wallets as well as to securely sign transactions. Before any action can be taken with these multi-signature contracts and digital wallets, digital signatures from several private keys are necessary.

1.7   Blockchain Merkle tree

BT relies heavily on the Merkle tree [12]. It is a mathematical data structure that works as a summary of all the transactions in a block and is built up of hashes of various data blocks. It also offers content verification over a large dataset in a timely and safe manner. It also assists in data consistency and ingredient verification. Both Bitcoin and Ethereum make use of Merkle trees. Merkle tree is also known as hash tree.

The Merkle tree idea was created by Ralph Merkle. It is simply a data structure tree, with the hash of a data block labelled on every leaf node and the cryptographic hash of its child node labelled on every non-leaf node. Figure 1.7 shows simple Merkle tree. The leaf nodes are the tree's lowest nodes.

By creating a digital fingerprint of the whole collection, a Merkle tree keeps track of all the transactions in a block. It gives the user the option of including or excluding a transaction from a block.

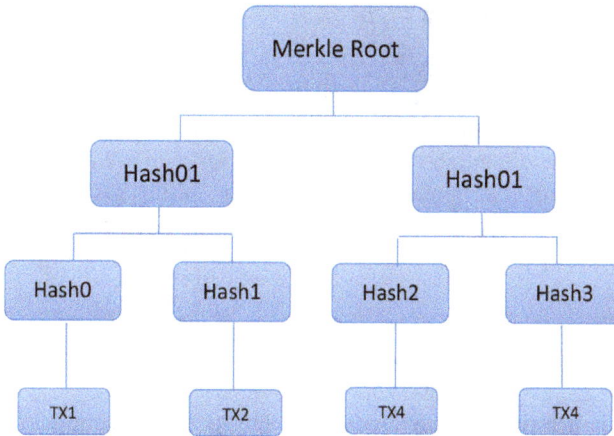

Figure 1.7   Merkle tree

Merkle trees are constructed by hashing pairs of nodes on a regular basis until only one hash remains. This hash is known as the Merkle root or root hash. Merkle trees are built from the ground up.

Every leaf node is a hash of the hashes that came before it, and every non-leaf node is a hash of the hashes that came before it. Merkle trees are binary trees, hence the number of leaf nodes must be odd. If the number of transactions is odd, the final hash will be copied to ensure an even number of leaf nodes.

The Merkle root might be found in the block header. The piece of the Bitcoin block that is hashed during mining is known as the block header. In a Merkle tree, it is composed of the previous block's hash, a nonce, and the root hash of all transactions in the current block. Because the Merkle root is included in the block header, the transaction is protected. Because the root hash contains the hashes of all transactions within the block, these transactions may result in disc space savings.

The Merkle tree verifies that information is correct. If the information or sequence of a single transaction changes, the hash of that transaction will reflect those changes. This change would propagate up the Merkle tree until it reached the Merkle root, causing the Merkle root's value to change and the block to be invalidated. As can be seen, the Merkle tree makes it simple to determine if a transaction belongs in the collection.

1.8   Consensus in blockchain

Consensus methods [13] serve as the foundation of blockchain as they enable all nodes in the network to validate transactions. PoW is Bitcoin's consensus method, and it consumes a lot of energy and time. When compared to Visa and MasterCard, the rate of transaction verification in Bitcoin is quite slow. As a result, a number of consensus-building methods have arisen.

1.8.1   Types of consensus algorithm

1.8.1.1   Proof of work

PoW was one of the first mechanisms of consensus utilised in blockchain applications. It works by computing hash values and verifying transactions until the hash value contains a certain number of trailing zeros. A nonce is an integer that creates a hash with the specified number of trailing zeros. In the hash function, a nonce is a random integer that creates the required number of trailing zeros. Bitcoin and Litecoin, two major cryptocurrencies, both use PoW [14].

1.8.1.1.1   Properties
* PoW is a consensus algorithm for permissionless public ledgers that makes use of the computing power of the node's systems.
* A linear structure is used to represent the blocks. A collection of transactions is represented by each block.
* The cryptographic challenge of obtaining a random integer that leads to hashes with a specific amount of leading zeros is what Bitcoin mining is all about.
* The public and private keys provided to each user are used to validate and sign each transaction.

1.8.1.1.2   Disadvantages
* It is a resource- and power-intensive protocol. When compared with more efficient methods, this way of solving cryptographic difficulties wastes a lot of computer power and electricity.

1.8.1.2   Proof of stake

Ethereum [15], one of the most prominent cryptocurrencies, has opted to use PoS [16] consensus. Let us take a deeper look at this situation. Let us suppose we are miners verifying previous transactions. A person who validates Bitcoin transactions by computing the hash value with a specific number of leading zeros receives the number of Bitcoins authorised.

A validator is picked and a block is assigned in PoS consensus. To begin validating, the miner must set aside a portion of his Bitcoin. If the miner is successful in invalidating the transaction, they will be rewarded the agreed-upon stake plus transaction costs. This is a method of punishing poor behaviour while rewarding good.

1.8.1.2.1   Properties
* Validators are selected depending on the amount of money they have invested in the network.
* The objective is to reduce mining hub concentration and provide all miners a chance to verify.

- It is environmentally beneficial because there is no computational challenge to address.
- No specific hardware is required for mining.

1.8.1.2.2 Disadvantages

- In PoS, an attacker would require more than 50 per cent of the money to take control of the network, but in PoW, an attacker would only need 51 per cent.
- Bribe attacks are one type of potential assault on a PoS consensus-based network. When an attacker reverses the victim's transactions and bribes the miners to confirm them, this is known as transaction reversal.

1.8.1.3 Proof of space

Proof of Space, or PoSpace for short, is a network consensus method that works in the same way as PoW. PoSpace validates transactions utilising disc storage rather than computing resources.

PoSpace uses disc space and compensates miners with the most accessible disc space for each block. The pebbling game is solved using this data structure, which was generated using hard-to-pebble graphs. Pebble vertices in a graph are only possible if all of the parent vertices have been pebbled.

The term 'pebbling' refers to the act of storing the hash values of the parents, while 'removing the pebble' refers to the act of releasing the memory. Plots are created at random to represent all possible solutions to the problem. These plots are saved on discs and solved using a method known as Shabal's algorithm. The miners compare their answers once they have been calculated, and the solution with the best time and space complexity is rewarded with the next block.

Burstcoin is a PoSpace-based decentralised cryptocurrency. The benefit to payment systems is the value proposition.

1.8.1.3.1 Disadvantage

The miners with the most space benefit from this consensus mechanism once again. It is resource-biased, thus miners with little space are unable to fully engage. This is an issue that contradicts the decentralisation notion.

1.9 Conclusion

Blockchain will play a vital role in many sectors as an essential new technology. So, knowing the fundamentals of blockchain is very crucial. This chapter talks about work procedure of BT knowing which reader can understand applicational concepts of blockchain that comes in later chapters easily.

References

[1]   Nakamoto S. 'Bitcoin: A peer-to-peer electronic cash system'. *Whitepaper*. 2009.

[2]   Sharples M., Domingue J. 'The blockchain and kudos: A distributed system for educational record, reputation and reward'. *Proceedings of 11th European Conference on Technology Enhanced Learning (EC-TEL 2015)*; 2015. pp. 490–6.

[3]   Ali Syed T., Alzahrani A., Jan S., Siddiqui M.S., Nadeem A., Alghamdi T. 'A comparative analysis of Blockchain architecture and its applications: Problems and recommendations'. *IEEE Access*. 2019, vol. 7, pp. 176838–69.

[4]   Bonneau J., Miller A., Clark J., Narayanan A., Kroll J.A., Felten E.W. 'SoK: Research perspectives and challenges for bitcoin and cryptocurrencies'. *IEEE Symposium on Security and Privacy (SP)*; 2015. pp. 104–21.

[5]   Barcelo J. 'Bitcoin security and privacy: a study of users experiences'. *KnE Engineering*. 2018, vol. 3(11), pp. 11–28.

[6]   Cachin C., Vukolić M. *Blockchain consensus protocols in the wild [online]*. 2017. Available from arXiv preprint arXiv:1707.01873.

[7]   Narayanan A., Bonneau J., Felten E., Miller A., Goldfeder S. *Bitcoin and Cryptocurrency Technologies: A Comprehensive Introduction*. Princeton, NJ: Princeton University Press; 2016.

[8]   Kosba A., Miller A., Shi E., Wen Z., Papamanthou C. 'Hawk: The blockchain model of cryptography and privacy-preserving smart contracts'. *Proceedings of the 2016 IEEE Symposium on Security and Privacy (SP)*; San Jose, CA; May 2016. pp. 39–58.

[9]   Rawat D.B., Njilla L., Kwiat K., Kamhoua C. 'iShare: Blockchain-based privacy-aware multi-agent information sharing games for cybersecurity'. *Proceedings of the 2018 IEEE International Conference on Computing, Networking and Communications (ICNC)*; Maui, HI; March 2018. pp. 425–31.

[10]   Sharma N., Prabhjot P., Kaur E.H. 'A review of information security using cryptography technique'. *International Journal of Advanced Research in Computer Science*. 2017, vol. 8(Special Issue), pp. 323–6.

[11]   Gupta A., Walia N.K. 'Cryptography algorithms: A review'. *International Journal of Engineering Development and Research*. 2014, vol. 2(2), pp. 1667–72.

[12]   ÖZDEMİR S. 'False data detection in wireless sensor networks via Merkle hash trees'. *Süleyman Demirel Üniversitesi Fen Bilimleri Enstitüsü Dergisi*. 2009, vol. 11(3), pp. 239–45.

[13]   Velliangiri S., Karunya P.K. 'Blockchain technology: Challenges and security issues in consensus algorithm'. *Proceedings of the 2020 IEEE International Conference on Computer Communication and Informatics (ICCCI)*; Nagoya, Japan; June 2020. pp. 1–8.

[14]   Vukolić M. 'The quest for scalable blockchain fabric: Proof-of-work vs. bft replication'. *International Workshop on Open Problems in Network Security*; 2015. pp. 112–25.

[15]   Buterin V. A Next-Generation Smart Contract and Decentralized Application Platform [online]; White Paper. 2014. Available from ethereum.org [Accessed 10 Sep 2020].

[16]   King S., Nadal S. 'PPcoin: Peer-to-peer crypto-currency with proof-of-stake'. *Self-Published Paper*. 2012, vol. 19.

Chapter 2

Integration of blockchain with IoT-enabled sensor networks for smart grids

Francesco Flammini[1], Rakesh K R[2], and Jayanna S S[2]

For the future world, which is going to be completely technology driven, technologies that are resilient, foolproof, transparent, and highly scalable are the requirements, and blockchain is such a technology. With this technology, we can expect the world to be smarter and more lawful. In the coming days, blockchain could be the ideal match for smart applications.

With billions of devices connected to the Internet, Internet of Things (IoT) is making our world simpler, and we can control all these with the click of a button. Not only that but these connected devices collect data via sensors and send it for further processing, which takes thing to whole new level from just managing them to predicting the device health, usage patterns, life of the device, etc. IoT, the next big thing for the world, could transform everything from simple day-to-day work to gigantic industrial applications much more efficient and intelligent.

A wireless sensor network (WSN) comprises a larger number of sensor nodes that detect physical parameters, such as temperature, pressure, and moisture. WSN will be the leading technology in the field of IoT due to the advanced technical evolution of sensors. IoT offers a different strategy in social services to partner with each other over the Internet. People, on the contrary, will have a better understanding of themselves and their surroundings.

The IoT is expected to alter people's lifestyles. Today, IoT is a promising development to transmit data in wired and wireless communication frames, and it can be used in a variety of fields, such as smart homes, smart water networks, electricity generation, transmission, and utilization, where intelligent monitoring and control can be achieved by the network-embedded devices such as sensors.

Using the smart grid as an example, the goal of the smart grid is to respond to the real-time demand by the consumers and smartly manage electricity production, transmission, and storage so that there is minimal or zero loss of power; all of this

[1]Mälardalen University, Västerås, Sweden
[2]Vidyavardhaka College of Engineering, Mysore, India

is accomplished via sensors, which are key prompters to take smart grid to reach its heights. The use of WSN in the smart grid would help WSNs grow as an industry.

2.1   Introduction

The objectives of this chapter are to provide a brief overview of IoT—its mechanism and significance, wireless sensor networks, blockchain, and integration of all these concepts with respect to smart grid application.

2.2   Internet of Things

IoT is exploding right now, and it is affecting everything from how we travel and shop to how businesses preserve inventory records. First and foremost, what precisely is the IoT? What is the mechanism at work here? Is it really a big deal?

The purpose of IoT is to interface any device to the web and other interconnected gadgets. The IoT is a vast network of interconnected objects and individuals that accumulate and send information about how they are utilized and the environment in which they are found [1]. That incorporates everything beginning from the brilliant microwave oven that cooks our nourishment food for the specific measure of time which we indicate, to wearable wellness gadgets that track our pulse and the quantity of steps we require every day, and then utilize the same data to prescribe practice plans, to self-driving vehicles with complex sensors that recognize objects in their way. Additionally, there are arranged footballs that truly can quantify how far and quick they are hit and save the information in an application for future preparation, as shown in Figure 2.1.

2.2.1   Mechanism behind IoT

An IoT platform connects devices and objects with built-in sensors, which gathers information from multiple devices and performs analytics to communicate the most useful information with apps tailored to individual needs.

These modern IoT systems can evaluate which data are useful and which data may be easily ignored. This information can be utilized to note patterns, make recommendations, and anticipate prospective problems.

2.2.2   Empowering technologies for the IoT

WSN, cloud services, information analytics, embedded frameworks, security procedures and configurations, communication conventions, web applications, Internet technology, and powerful search engines are some of the technologies that enable the IoT [2–4].

WSN: A WSN comprises a larger amount of sensory networks that track environmental and physical characteristics. One of the most prominent wireless technologies used by WSNs is Zig Bee.

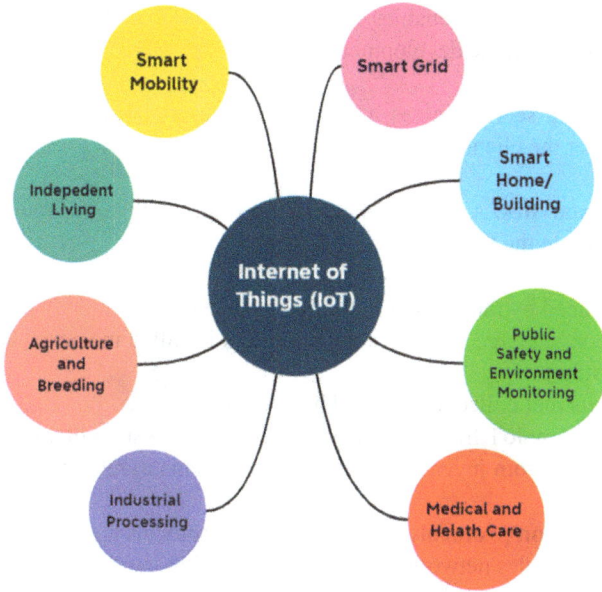

Figure 2.1   Shadow of IoT

The following includes a list of sensor networks (WSNs) that are utilized in IoT systems:

- Nodes collect heat, moisture as well as other information, which is gathered and evaluated in a weather monitoring system.
- Interior air pollution monitoring systems collect information about the quality and concentration of various gases in the indoor environment.
- Soil moisture monitoring systems keep track of soil moisture in a variety of places.
- WSNs are used in surveillance systems to gather information (detection of motion data).
- WSNs are used in smart grids to monitor grids at multiple moments.
- WSNs are used to monitor the condition of structures, such as buildings and bridges, by gathering vibrations from sensors distributed at different positions across the structures.

Cloud technology: One can select from a variety of services while using cloud services.

- Infrastructure as a service: enables clients to quickly facilitate computing and provides virtualized resources on demand. Users can access these resources through virtual machine occurrences and virtual repository.

- Platform as a service: enables user to design and implement applications in the cloud utilizing the development tools, interfaces, software frameworks, and services.
- Software as a service: gives a comprehensive program application or client interface for the client.

Big data analytics: A few occasions of more net worthy information created by the IoT incorporate the following:

- Data produced by the sensors in the IoT systems.
- Information gathered from machine monitors installed in energy and manufacturing systems.
- IoT gadgets that collect fitness and health data.
- Data created by IoT frameworks on vehicle area and surveillance.
- Data received from inventory management systems in retail stores.

Communication protocols: Communication protocols are the backbone of IoT systems, which enables network access and application integration.

- It enables data transfer between devices over a network.
- It defines the route to send packets from source to destination as well as the exchange formats, data encoding algorithms, and device addressing schemes.
- It covers sequential management, traffic shaping, and package restoration.

Embedded systems: It is a framework that combines both software and hardware to perform function. It includes everything from low-cost small gadgets, such as digital watches to advanced cameras, point of deal terminals and other utilities.

2.2.3   Significance of IoT

Across all enterprises, IoT is a basic requirement for clients facing advancement, information-driven improvement and robotization, computerized change, R&D and new applications, plans of action, and revenue sources [2, 3].

So, IoT is an umbrella term with many use cases, technologies, standards, and applications. Moreover, it is part of a bigger reality with even more technologies. The things and data are the starting point and essence of what IoT enables and means. IoT devices and assets are equipped with electronics, such as sensors and actuators, connectivity/communication electronics, and software to capture, filter, and exchange data about themselves, their state, and their environment.

2.3   Wireless sensor networks

A WSN is a collection of dispersed, self-contained sensor devices that track physical or environmental data. A WSN is made up of a network of interconnected sensor nodes that interact and exchange information with each other. These nodes collect

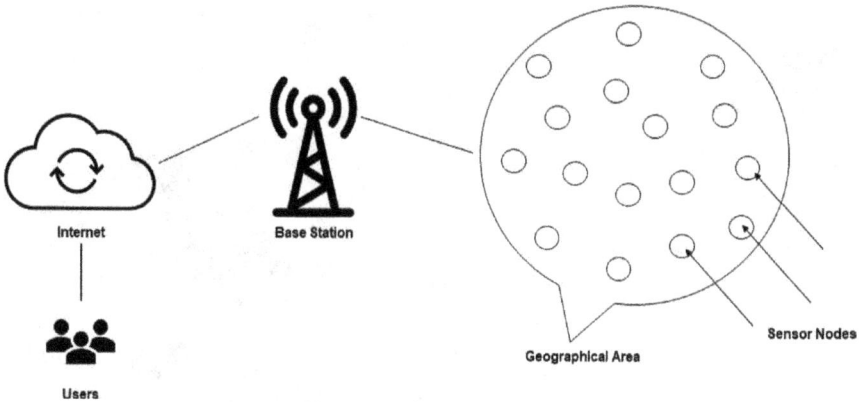

Figure 2.2   Basic architecture of wireless sensor networks

data about the surroundings, which includes temperature, pressure, humidity, and pollution levels, and transmit the sensed information to the base station. The base station functions as a gateway, receiving data from sensor nodes and processing it before providing the updated data to the user through the Internet for further actions [5]. Figure 2.2 shows the basic architecture of WSNs.

WSNs are commonly used for studying ecological parameters, such as pressure, temperature, and vibration factors, as well as tracking human and animal development in timberlands and boundaries, climate and forests checking, and war zone reconnaissance. Sensor nodes communicate using air (wireless) as a communication channel (WLANs) for wireless local area networks.

In the WSNs, the clusters are formed based on the energy of individual sensor nodes; in that cluster, there will be an acquisition node which can send data to the base station or in a multi-hope mechanism. The energy consumed by sensor nodes is related to the number of times the data have been transmitted and the distance of data transmission. The sensor nodes that are at a longer distance from the base station will consume more energy compared to the nodes that are nearer to the base station; this causes energy imbalance in the cluster. The solution to this problem is to use clustering routing mechanism.

2.3.1   Characteristics of WSNs

Sensor networks are widely recognized as one of the most crucial innovations of this century. The ability to instrument and regulate the environment, cities, and factories is enabled by equipment with many sensors onboard, linked via wireless links and the Internet, and deployed in large numbers. Also, networked sensors can be used for surveillance and other tactical purposes applications. Sensor networks for different applications may differ quite a bit, but they share common characteristics.

Generally, sensors are electrical, electronic, or electromechanical devices, although there are other types of sensors. Sensors are transducers, and they convert the input they receive into another, usually electrical, form. They can be direct or

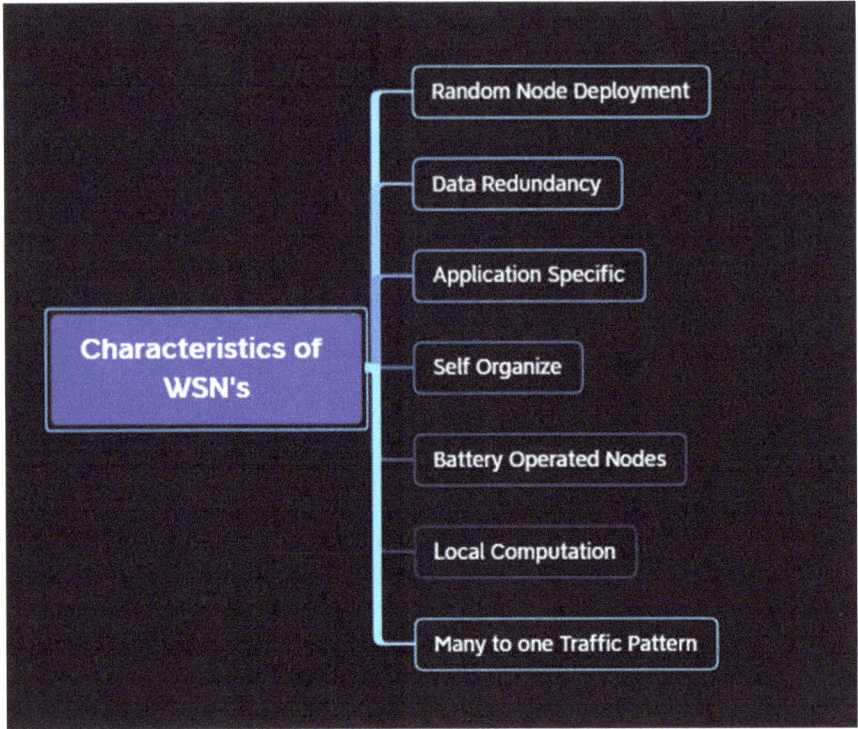

Figure 2.3   Characteristics of wireless sensor networks

paired sensors. Thermometers and electrical meters are examples of direct sensors. To convert analog signals into digital signals, paired sensors use analog-to-digital converters. Medical, industrial, environmental, robotic, and military applications rely on sensors. A growing number of sensors are now being built with micro-electro-mechanic systems, as depicted in Figure 2.3.

The following are the main qualities of a good sensor/transducer:

1. It should be relevant to the property under consideration.
2. It must be unaffected by any other property.
3. It is preferred that the sensor's output signal is proportionate to the value of the measured feature.
4. It should have a long life span.
5. It should not use a lot of energy.

A wireless network is composed of many nodes which use sensing devices (sensors) to monitor various environments and conditions, such as movement, pressure, temperature, noise, acoustic noise, emissions, concentration of oxygen or dioxide, road severity and designs, and many more, at different locations. These devices are usually small and minimal, enabling them to be bulk and distributed in huge volumes. WSNs differ greatly from traditional mobile ad hoc networks in that they frequently have extremely limited resources in terms of energy, storage, computational power, and throughput. A radio transceiver, a tiny microcontroller, and a power supply, which is commonly a battery, are all included in the node.

2.3.2 *WSNs data transmission energy consumption model*

Let us consider the basic radio energy consumption model, which includes a sender and a receiver, and the sender wants to communicate to the destination through a channel [2].

The above process is divided into three steps

1. The energy consumed $E_T(x, n)$ while transmitting n bits of data to a node with of x meters of distance consists of two parts: transmission loss E_{elec_t} and amplifier loss E_{amp} are given by

$$E_T(x, n) = E_{elec\_t} + E_{amp}$$
$$= \begin{cases} n \times E_{elec} + n \times \in_{fs} \times x^2, \ (x < x_0) \\ n \times E_{elec} + n \times \in_{anp} \times x^4, \ (x \geq x_0) \end{cases} \quad (2.1)$$

The above equation denotes the energy consumed E_{elec} for transmitting a unit bit, $\in_{fs}$ denotes data energy consumption of a unit bit in free space, $\in_{anp}$ denotes the data energy consumption of a unit bit in multi-path attenuation mode, and $x_0 = \sqrt{\frac{\in_{fs}}{\in_{anp}}}$ is the critical value for dividing the spatial model.

2. The energy consumed by the receiving node to receive n bit data is given by

$$E_{R(n)} = n \times E_{elec} \quad (2.2)$$

3. The energy consumption of data transfer node for fusing n bit data is

$$E_{DA(n)} = n \times E_{da} \quad (2.3)$$

E_{da} is the energy required to fuse unit bit data.

2.4 Smart grids

Conventional electrical circulation frameworks are being used to carry electrical power generated at a centralized generating station by boosting voltage values and then progressively lowering voltage values to deliver it to end consumers. With the growing importance of sustainable energy in particular, smart power market arrangements based on continuous evaluation are now necessary. The smart grid protects user safety while monitoring, updating, and continuously reliably supplying electricity by combining smart meters and traditional analytics into the power system to enable electronic communication between suppliers and users. One of the

Table 2.1   Conventional grid vs smart grid

Characteristics	Conventional grid	Smart grid
Technology	Electromechanical	Digital
Distribution	One-way Distribution	Two-way distribution
Generation	Centralized	Centralized and distributed
Monitoring	Blind or manual	Self-monitoring
Control	Limited control system	Pervasive control system
Restoration	Manual	Self-healing

key purposes of the smart grid is to provide fulfilling experience in terms of electricity to its customers [2–4].

A smart grid framework will allow digital technologies to be used to defend against attacks, engage customers, optimize power supply and generation, provide high-quality power, accommodate influential ideas, and self-heal. The advantages of using smart grids with respect to conventional frameworks are listed in Table 2.1 below.

2.4.1   Purpose of the smart grid

In the foreseeable future, the consumer profile may alter potentially. Electric vehicles (EVs), which are likely to become commonplace soon, will put a significant strain on the infrastructure. EVs must always be charged; energy must always be used efficiently in conjunction with infrastructure; and power needs should be predicted. As a result, the network will become overburdened [6]. By establishing an agreement among suppliers and purchasers, resources for the smart grid will be used more efficiently. Smart meter reading and system management at specified time intervals, as well as power usage monitoring, will be possible in the smart grid environment.

To summarize, smart grid benefits include increased autonomous control actions, operator support, renewable energy systems, enhanced economic efficiency through creative technologies for diverse types of goods, improved quality of service, situational awareness, and efficiency development, among other benefits.

2.4.2   Grid structure in a traditional Grid

Traditional power distribution framework has been used to deliver electrical energy created at centralized power stations by gradually boosting voltage values and subsequently gradually decreasing voltage values to deliver it to consumer's end.

The traditional grid is an interconnected network of power lines that transfer electricity from generators to consumers. Electricity is generated by power plants, transmitted by high-voltage transmission lines to users and distributed via distribution lines, as shown in Figure 2.4.

The popularity of distributed generation is growing due to advancements in power generation, transmission, distribution, regulation, and control technologies.

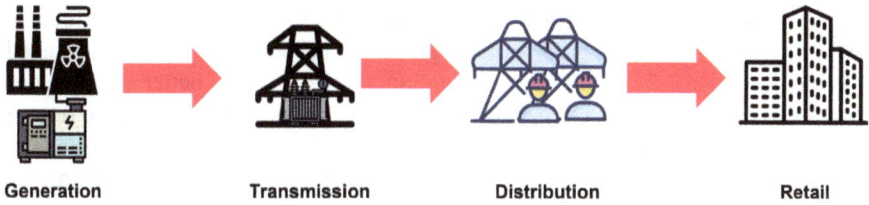

| Generation | Transmission | Distribution | Retail |

Figure 2.4   Traditional electrical distribution systems

The structure of electrical energy generation is depicted as centralized, decentralized, and distributed in Figure 2.5 [7–9].

The world's energy industry has become more appealing and competitive because of this transformation in the electric power business. Switching from a regular grid to a smart grid has three benefits.

1. the growing demand for energy
2. reducing losses and unauthorized uses
3. the capability of current plants to produce and transport goods has been increased.

Smart grids are self-contained systems that allow each kind and scale of production to be incorporated into the grid, reducing manpower requirements while guaranteeing that all customers have access to reliable, safe, and high-quality power.

2.5   IoT-enabled sensor networks for smart grids

The platform that connects every person, enterprise, and municipal infrastructure service is referred to as the "grid." The "smart grid" will be the next phase of electricity grids that were modernized with communication technologies and connections to enable better resource optimization [1].

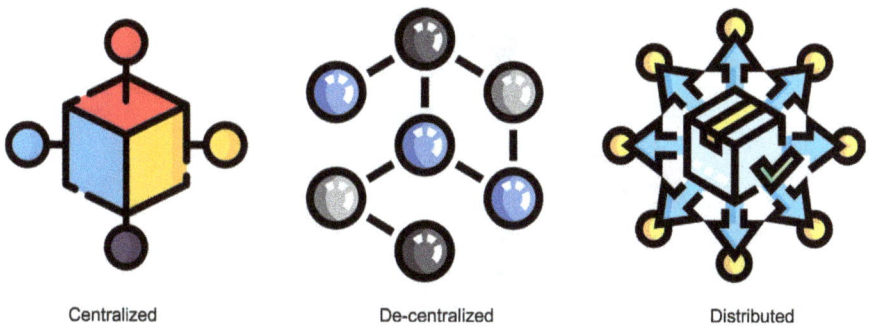

| Centralized | De-centralized | Distributed |

Figure 2.5   Generation of energy in a distributed manner

Sensors, communication modules, gateways, and routers are all part of today's IoT-enabled energy "smart grid" technology. These technologies allow for advanced communication and connectivity, allowing individuals to create better energy judgments, communities to conserve energy and money, and power companies to restore power more efficiently after an outage.

For a variety of reasons, municipalities are increasingly turning to smart grid technologies. These include the need to reduce energy use, improve citizen service, safeguard against calamities, and modernize aging methods that are costly to operate. Wireless, both cellular and radio frequency (RF), has also become more economical and simpler to utilize in smart grid applications because of technological advancements, as shown in Figure 2.6.

2.5.1   Smart grid: in action

Every citizen in the world requires energy from the grid; thus, cities that implement smart grid technology will benefit all residents, civic services, and national infrastructure. A smart grid enables a utility to analyze an entire system in far greater depth than previously feasible. Smart meters, for example, allow the power provider to uncover actual energy usage with precision and accuracy that conventional equipment simply cannot match. This will help to effectively anticipate and respond to demand spikes, lowering the danger of blackouts.

In the case of a blackout, IoT devices deployed in transformers and substations that use cellular and RF technology can automatically redirect electricity. This may result in a speedier, more straightforward solution than dispatching repair personnel in a van every occasion when the power goes out.

Our electric power system's robustness will be strengthened, and it will be better prepared to respond to disasters, such as strong thunderstorms, earthquakes,

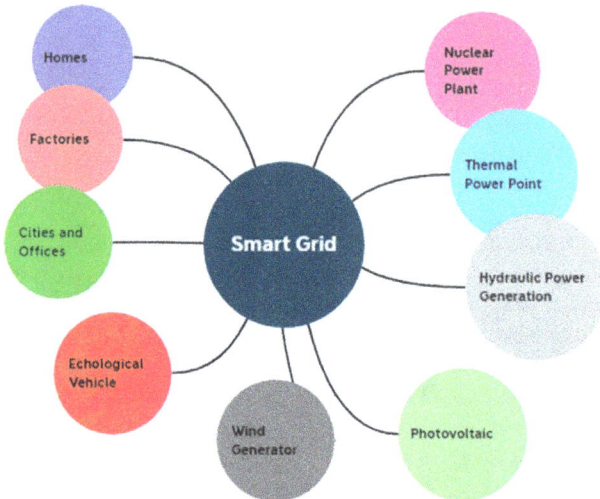

Figure 2.6   Applications of smart grid

big solar flares, and terror attacks, thanks to the smart grid. In case of equipment failure or outages, the smart grid's multiple support capabilities will enable quick rerouting.

The use of IoT in smart grids allows information to be shared throughout all grid components. IoT plays a critical role in the implementation of smart cities with smart grids. Smart energy management is also possible thanks to IoT. IoT takes advantage of smart meter's strong sensing capabilities and greater connectivity features. In large-scale environments, consumption patterns can be studied and exploited to increase energy and billing efficiency. Several researchers are working on ways to increase the privacy and security of smart grids based on IoT.

When putting these systems in place, there are several challenges to overcome. The main issue is that of security and privacy. Safety must be enforced through data collecting, control messaging, equipment monitoring, and notifying. Other factors to consider are privacy, integrity, availability, authentication, nonrepudiation, and access control, as shown in Figure 2.7.

2.5.2   Implementation challenges for IoT-based smart grids

The major challenges that will occur while implementing IoT enables smart grids are as follows:

- Communication and connection stability

In smart grids, the system reaction transfers excess electricity to areas in which there is local power scarcity [10]. The potential of systems to communicate is crucial. To work properly, IoT requires high-speed Internet connectivity. The optimality of power redistribution is reduced when the connection is low. Low latency and quick response are required for optimal communication between the components of a smart grid. Gathering scalable data is problematic due to a lack of tools for real-time analysis.

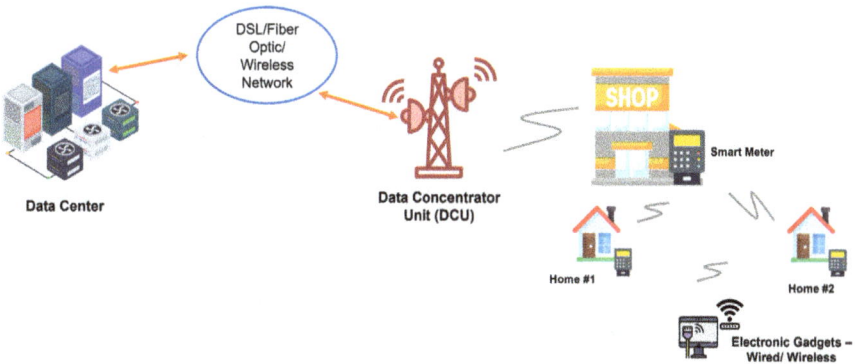

Figure 2.7   Smart grid architecture based on IoT

In large disasters, it is also difficult to collect data and enable activity monitoring technology. Cloud storage is also essential as data will be broadcast in large numbers when the system is deployed in smart cities.

• Budget

A wireless network is easier to set up than a wired network, which takes careful planning and implementation. When establishing smart grids, power outages are unacceptably frequent. This factor could lead to greater implementation costs. Smart grids and the IoT, despite their high implementation costs, save a lot of energy by promptly rerouting power when a power outage is detected. In France, the United States, and other nations, smart meter installations are being carried out in real time, as are several devices connected to the Internet.

In smart grid IoT applications, power distribution causes severe interference. To avoid this, chips and equipment must be engineered to withstand high temperature variations, being anti-vibration and anti-electromagnetic, and be supplied with water and dust-proof systems as well as novel technologies to prolong the module's lifetime.

• Information protection

Several security issues exist in smart grids and the IoT. Internet-based security challenges, cyber threats, resource restrictions, confidentiality, trust management, authorization and authentication, cyberattacks, scalability, confidentiality, and identity spoofing are some of the most common concerns. Security technologies, such as deep packet inspection and information management, can help protect IoT-based smart grids. These solutions can be used to address specific vulnerabilities and safeguard data threads. In wired networks, physical damage and disruption are typical occurrences. Despite the use of robust encryption algorithms, data packets on wireless networks are occasionally captured and deciphered.

In a smart environment, cameras and sensors may catch every action, attracting cybercriminals. As a result, it is critical to create a safe environment and avoid data leakage and misuse. To provide this security, blockchain and similar systems can be utilized. Vulnerabilities and problems may arise because of the system's poor design. Because of the power limits in IoT and smart meters, it is difficult to combine dependable and complicated security techniques [6]. Remote control of smart meters, breach of consumer confidentiality, manipulation of energy transactions and grid destabilization, and data monitoring for fictitious purposes are among the most common risks to information security.

2.5.3   *How smart cities are using smart grid technology to benefit their communities*

Smart city solutions include everything from smart traffic control to smart urban lighting, as well as treatment of water and wastewater management. Traffic light monitors can transmit data to a central body for decision-making. Smart transportation systems, on the other hand, can handle both surface traffic and public transportation, including traffic routing and lighting to decrease or remove overcrowding.

In addition, based on real-time conditions, IoT sensors in streetlights can change the timing and brightness of turn-on and turn-off. It may appear that a few watts are nothing. When thousands or large numbers of bulbs are installed in a city, the cost savings and environmental benefits quickly add up. If a light needs to be serviced, those same sensors can send out a warning. There is no need to wait for an irate client to complain about the streetlights being down.

Furthermore, using a powerful remote access technology, professionals may immediately troubleshoot the problem and assess whether a vehicle should be dispatched. Previously, a truck roll—a pricey proposition compared to a quick software fix or rebooting from a home office monitoring system—was inescapable.

Demand response is enabled by smart meters, which allows homeowners and business owners to access real-time pricing information and modify their energy usage accordingly. For example, in the winter, turn off the air conditioner or lower the thermostat. Smart meters will primarily assist electric car owners [11]. EV users will be able to recharge their cars when electricity is lowest and avoid recharging during peak demand, if possible, thanks to real-time pricing information.

2.5.4   Advantages of smart grid

While there are many benefits to implementing a smart grid, the following three instances show how useful it may be.

- Renewable energy generation is made possible by the smart grid

Electricity is transported from a large, centralized power plant to a vast network of surrounding residences and enterprises in conventional energy grids. At current time, the electric grid is not designed to accept feeds from residential and commercial buildings that generate electricity using photovoltaic panels or wind turbines. A smart grid is one that accepts energy from renewable sources.

Importantly, the smart grid, when used in combination with remotely connected smart meters, can measure how much energy a net-positive firm generates and reimburse it appropriately. Solar panels and equipment can also be monitored using the smart grid.

A smart grid, as previously indicated, may mitigate the consequences of a calamity on a power station, such as a terror attack or catastrophic event, a task made feasible by decentralized energy generation. A city was traditionally powered by a modest number of power plants. Under the traditional model, these services are exposed to risks, such as severe blackouts and energy shortages. In a decentralized architecture, even if the centralized power station is shut down, numerous possible forms, such as wind and solar, can supply the grid's energy. This decentralized system is significantly more difficult to take down, and it can provide a level of reliability that is impossible to attain when a single plant supplies electricity to a whole city.

- Better predictions and billing

The advantages of smart meters are twofold. First, they can collect a massive amount of data via wireless IoT devices, knowledge that companies have never had access to before. Companies can use this information to help forecast when and where electricity consumption will peak.

Second, the smart grid promises more efficient invoicing for consumers. Electricity costs at peak demand were previously averaged across cities and suburban communities. If you use electricity at a high rate, you will now be charged for it. Your cost will also decrease if you turn-off the appliances and preserve electricity. Everyone will be encouraged to utilize electricity responsibly because of this.

- Smart Grids are more resilient than traditional grids.

When a blackout occurs, smart grid technology allows power to be transferred automatically, decreasing the impact on homes and businesses. Smart applications can provide information on the condition of equipment, allowing repairs to be made before it breaks down. Rather than responding to customer calls reporting outages, providers can advise users (through email/social media) when an outage occurs.

2.6   Blockchain for IoT

The onset of blockchain has led to many improvements over different sectors where there are trust issues.

Decentralization: Blockchain, which follows decentralized design [12], is a reassuring way for successfully tackling single point of failure issues, a bottleneck by eliminating the prerequisite for trust in outsiders inside the IoT organization. The configuration of a node in blockchain would not influence the working of the blockchain IoT organization. Information on the blockchain is for the most part held in a distributed organization with multiple hubs, and the framework is profoundly strong to vindictive assaults and disappointments. By decentralizing the administration of and resource access in an application, more noteworthy and more attractive services can be accomplished. The disadvantage of decentralization commonly is lower throughput in transaction; however, in a perfect world, the drawbacks outweigh the further developed security and administration levels they produce, as shown in Figure 2.8.

2.6.1   *Advantages of decentralization*

- **To improve data coordination:** Companies share information with their stakeholders often. These data are usually stored in stakeholder's databases, which would probably be processed only during the migration process or when it needs to be shared further down the hierarchy [13]. When the data are transformed, they create new challenges of loss of data or incorrect data to enter the system.

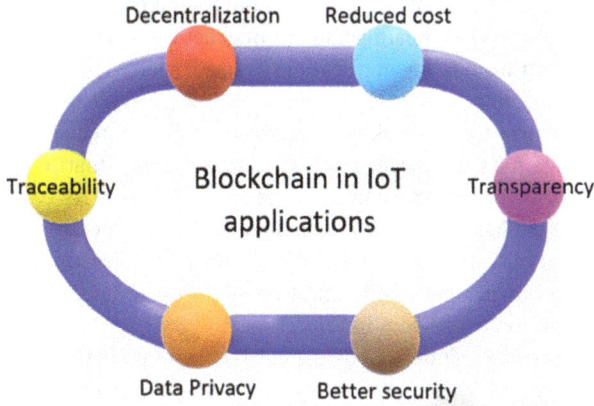

Figure 2.8   Features of blockchain

Utilization of decentralized data sets guarantees, each substance approaches an ongoing, shared perspective on the information.

- **Reduces weak points:** In sectors where there may be too much dependency on specific users, decentralization can decrease flimsy spots in the system. These flaws could lead to systemic problems, such as service failure or inefficiency owing to asset depletion, repeating blackouts, bottlenecks, an absence of adequate impetuses for great assistance, or defilement.

- **Optimizing distribution of resources:** Decentralization will lead to resource distribution optimization, resulting in improved consistency and service performance and also reduced risk of catastrophic failure.

  Even if some of the nodes fall offline, the network's availability or security cannot be jeopardized. Conventional data sets, then again, depend on more than one server and are susceptible to technological failure and cyberattacks. Moreover, blockchain's distributed design furnishes all organization members with equivalent approval consensus to guarantee the precision of IoT information and assure immutability.

- **Improved security:** From numerous perspectives, blockchain is more dependable and secure than other conventional systems. Before the network's nodes can document a transaction, it must first be agreed upon [14]. An encrypted transaction is connected to the preceding transaction once it has been approved. Because information is kept on a series of servers, not on a single server, hacking is nearly impossible. The use of the private/public key approach is of paramount relevance in blockchains for security. To safeguard transactions between nodes, blockchain uses asymmetrical cryptography. The keys are generated using strings and random numbers, making it impossible to distinguish the private key from the public key. The tamper proof data in the blockchain avoids the data breach.

Furthermore, blockchain has the potential to make changes to how personal data are shared to avoid cheating and criminal actions in any sector where sensitive information from many applications is used, such as government, financial services, and healthcare. Furthermore, BCIoT can provide users with trustworthy access control using blockchain-enabled smart contracts, which automatically authorize all IoT device functions. Furthermore, smart contract services provide users with data provenance. This gives data owners control over how their data are exchanged on the blockchain. Users can specify access protocols for self-executing smart contracts on the blockchain, ensuring personal data protection and ownership. With smart contract-based authorization, malicious access may be confirmed and disabled.

- **Better traceability:** Transactions cannot be easily traced back to the starting point in conventional ledger systems like blockchain. As all transactions are recorded in an immutable chain of data blocks, anyone can check the recorded transaction for authenticity and avoid unethical practices. The patient records of a hospital can be accessed by any doctor for medical treatment from anywhere and from any other hospital.

- **Improved transparency:** Data can be exchanged more easily in blockchain as these accounts are accessible to all arranged clients. Blockchain is a type of distributed network where all members share almost identical records instead of individual duplicates in the standard organization. This common archive can only be changed by agreement, implying that everyone needs to agree. After every few transactions or based on the preset of transactions, a similar replica of blockchain information is sent over a wide organization of nodes in the network to check the transactions. Therefore, the blockchain clients have reasonable control over the network to connect, audit, and trace exchange exercises. In the event that one single exchange record is altered, all resulting records should be recalculated, and plotting would be required for the whole organization. Blockchain is more precise, vigorous, and straightforward than traditional networks. The straightforwardness additionally prompts ensuring the validity of the blockchain-based frameworks by decreasing the chance of unapproved information adjustments.

- **Privacy of data:** Blockchain's trustworthy trait and immutability transaction management on the blockchain are incredibly proficient to shield IoT information from change [15]. Blockchain records data blocks in a genuine, with full integrity ensured, through digital signatures and immutable hash chains. Basically, the blockchain lets nodes audit exchanges across the organization, so information rights are preserved.

- **Cost reduction:** Expense saving is one of the fundamental objectives of most organizations. Companies could save a lot of operation costs by using blockchain as the deployment of infrastructure cost for setting up the public blockchain is less, and it does not require any middleman to implement it. Blockchain clients do not have to audit a lot of paperwork to complete a transaction, as each node has to reach an immutable ledger. While blockchain can eliminate the expense of vendor services, it requires immense work for consortium blockchain and private blockchain and in setting up public blockchain some expense for exchange handling transactions has to be paid (in Ethereum blockchain we call it gas).

- **Immutability:** Blockchain stores transaction data permanently for a long time. Transactions are time-stamped subsequently to be verified by the blockchain nodes and after which it is embedded into a block by encoding it using a hashing technique. Hashing systems interface blocks together and develops a sequential chain. Header in each block consistently stores the hash value worth of metadata of the previous block, which makes the chain secure and immutable. Hence, the blocks in the blockchain cannot be edited, changed, or altered once it is approved and is part of the blockchain. The encoded link between any two blocks can withstand any change or alteration. Regardless of whether any transaction takes place, it will be easily recognized.

2.6.2 Challenges of adopting blockchain in IoT applications

The Figure 2.9 comprises of power utilization, execution, and security-related aspects: The high computational force needed to run blockchain calculations has dialed back the headway of these innovation put together applications with respect to asset compelled gadgets [16]. Bitcoin's energy utilization is compared with the household power utilization of Ireland, which IoT gadgets cannot embrace. The whole Bitcoin network assimilates significantly more energy than a few countries,

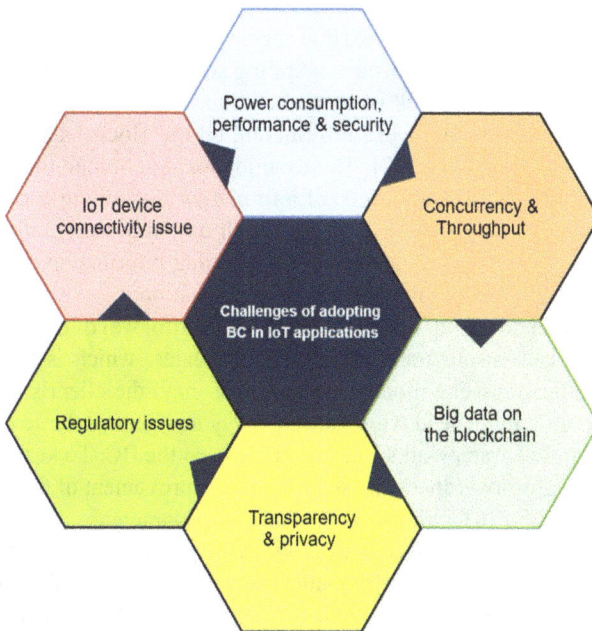

Figure 2.9   Challenges of IoT in blockchain

including Austria and Colombia. Furthermore, specialists have scrutinized the presentation of blockchain to deal with IoT information and have recommended enhancing its focal calculations to build the quantity of affirmed blocks for each second. For example, the end of the BC Proof of Work (PoW) agreement instrument can decrease power utilization and further develop execution. In effect, the PoW is malignant, Sybil assaults and seals the blocks. Thus, the objective is to refine BC cycles to suitably adjust security and productivity.

- **Information simultaneousness and throughput issue:** In IoT frameworks, the IoT gadgets consistently stream information, which leads to high parallelism. The blockchain throughput is limited by its complex cryptographic security convention and agreement instruments. The quick synchronization of new blocks among BC hubs in a chain-organized record requires a higher measure of data transfer capacity, which can further develop BC throughput. Hence, the test is designed to improve blockchain's throughput to address the issue of successive exchanges in IoT frameworks.
- **Availability difficulties of IoT:** The IoT gadgets are needed to be associated with high figuring stockpiling and systems administration assets to impart IoT information to expected partners. The IoT has restricted abilities to associate them with BC innovation to give novel business freedoms to the execution of new applications and administrations in different areas.
- **Handling big data on the blockchain:** In the blockchain organization, each member keeps a nearby duplicate of the total appropriated record. Upon the affirmation of another block, the block is communicated all through the whole distributed organization, and each hub attaches the affirmed block to their nearby record. While this decentralized stockpiling structure further develops productivity, tackles the bottleneck issue, and eliminates the requirement for outsider trust, the administration of IoT information on the blockchain puts weight on the members' extra room [17]. The examination determined that a blockchain hub would require around 730 GB of information stockpiling each year if 1000 members trade a solitary 2 MB picture each day in a blockchain application. Subsequently, the test is to address the expanding information stockpiling prerequisites when blockchain manages IoT information.
- **Difficulties in keeping up with both straightforwardness and security:** Blockchain can ensure transparency of exchanges, which is fundamental in certain applications like money. Be that as it may, the client's secrecy might be unfavorably influenced while putting away and getting the IoT information from certain IoT frameworks, such as eHealth on the BC. To keep a reasonable level of straightforwardness and security, the improvement of financially savvy access control for IoT utilizing blockchain is fundamental.
- **Directing difficulties of BC in IoT:** While a few BC mechanical components including decentralization, permanence, secrecy, and mechanization are promising security answers for different IoT applications, these changes pose different types of administrative challenges. The unchanging nature infers that information is forever distributed in Distributed Ledger Technology (DLT)

on the shared organization and cannot be erased or changed. Also, because of the shortfall of administration [18], records cannot be separated for keeping up with protection prior to distributing them on the BC. Activities coming about because of executing code, for example, shrewd agreements on a DLT, can break the law. Due to the secrecy of the DLT, it is difficult to identify the groups that are performing trades for illegal administrations. While the robotization component of the BC brings many benefits, the entertainers that cause a few practices remembering mistakes for code and jumbling code are equivocal. Current IoT laws and guidelines are becoming obsolete particularly with the appearance of new problematic innovation, for example, blockchain should be reexamined to attempt the DLT.

2.7 Vulnerabilities in IoT

2.7.1 Problems at sensing layer

- **Hub capturing:** IoT applications contain a couple of low-power centers, such as actuators and sensors [18]. These center points are feeble against a grouping attack by the adversary parties. The defaulter might endeavor to get or override the center point of the IoT environment with a noxious center point. The new center point may emit an impression of being the piece of the structure, yet it is obliged by attacker. This may provoke security breach of the absolute IoT system.
- **Malignant code-injection attack:** In this type, the attacker may inject malignant code to the core system. By considering all the factors, the rmware or programming of IoT center points is upgraded real time, allowing attackers to insert malicious code into a part. By using such malicious code, the aggressors may change the centers to play out some incidental limits or even endeavor to get to the absolute IoT environment.
- **Bogus data-injection attacks:** When the assaulter obtains the center, he or she might use it to implant wrong information in the IoT environment. This might incite fake results and might collapse the IoT environment. This procedure could be used to launch a DDoS assault.
- **Interference and listening in:** IoT applications comprise different hubs conveyed in open conditions. Accordingly, such IoT applications are presented to snoops. The aggressors might listen in and catch the information during various stages, such as information transmission or validation.
- **Lack of sleep attack:** In attacks like these, the adversaries endeavor to drain the battery of the energy-controlled IoT end devices. This leads to refusal from the centers of organization in the IoT environment in view of a chargeless battery. The edge devices with malignant code may be in a loop extending the power usage.
- **Side-channel attacks (SCA):** The distinctive side-channel attacks are known to cause spilling of important data, along with the nodes being attacked. The

processors with microarchitectures, electromagnetic transmission, and their capacity use uncover delicate data to fraudsters. Laser-based attacks, attacks with precise timing, power use, and electromagnetic attacks may be targeted by SCA. Present day chips try various ways to defy these attacks while doing the cryptographic actions.

- **Booting attack:** The end devices are helpless against various attacks while on the boot cycle. This is because the inbuilt safety efforts are not available by then, in the instance. The main devices may be attacked at the time of restarting the device. As end contraptions are ordinarily low controlled and now and again move via rest cycles, it is like a way to secure the boot configuration in these systems, as shown in Figure 2.10.

2.7.2   Problems of security in network layer

Here, the basic drawback in association layer is to impart the data obtained from identifying the layer to computational system for readiness. The following are the security risks present at the association layer.

- **Phishing site attack:** Phishing attacks frequently insinuate attacks in which a few IoT nodes can be focused around by the assailant's inconsequential work. The attackers hope that a few devices nearby will be transformed into a delayed attack. It is conceivable to encounter phishing areas among the customers visiting webpage pages on the Internet. At the point when the customer's record and private key are stolen/lost, the entire IoT data consumed by the customer turns powerless against advanced assaults. The association layer is significantly in danger against phishing attacks.
- **Access attack:** Attacks of this form are furthermore suggested as state-of-the-art tenacious risk. This kind of attack occurs when an unauthorized user or attacked gains access to an IoT group and remains undiscovered in the environment for a lengthy period of time. The objective here is to capture huge data or information, not to hurt the association. Because the IoT environment reliably receives and moves large amounts of data, they are consequently significantly helpless against such attacks.

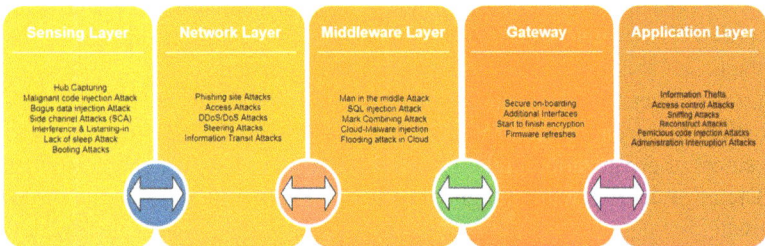

Figure 2.10   Vulnerabilities in IoT

- **DoS (Denail of Service)/DDoS attack:** In similar attacks, the attacker bombards the victims with endless unfortunate requests. This hampered the people's ability to be goal specialists, subsequently disrupting the organization's authentic customers. In the event that there are different ways taken to persuade the objective node, this is named as Distributed Denial of Service (DDoS) or scattered refusal of organization attack. Attacks of these sort are no explicit to IoT applications, yet because of unpredictability and heterogeneity of IoT associations, such attacks are more on the association layer of the IoT. The end nodes in IoT are set in an immovable fashion because of which these spots become the prime point for attackers to start with DDoS attacks.

- **Information movement attacks:** Exchange and accumulation of enormous data happen in IoT. Information is critical, making it an attractive choice for software engineers and various adversaries. Data that are kept close to users or in the cloud are at risk of being hacked, but the data that are coming in or moving from one location to the next are essentially more vulnerable to advanced attacks. A huge amount of data are exchanged between cloud, actuators, and sensors in an IoT environment. IoT environment is basically inevitable or guard-less against data breaks because of the huge information exchange that happens in the ecosystem.

- **Steering attacks:** Here, the attacked node tries to lead the data exchange inside the ecosystem in a malicious manner. Sinkhole attacks are an exceptional kind of guiding attack in which a foe plugs a counterfeit most concise guiding path and attracts center points to travel via traffic. A worm-opening attack is another attack which can turn into a veritable security risk at whatever point along with various attacks, for instance, sinkhole attacks. A warm-opening is an out of band relationship between two center points for speedy movement of the package. An attacker can make a warm-opening between a compromised center and a device on the web and endeavor to avoid the crucial security shows in an IoT application.

2.7.3   Security problems at middleware layer

In IoT, the purpose of the middleware is to create a reflection layer in between the association layer and the application layer. In addition, middleware will provide fantastic preparing and limit capacities. This layer offers APIs to satisfy the solicitations of the application layer. Middleware layer brings together subject matter experts, tireless data stores, lining structures, AI, etc. In IoT ecosystem, the middleware is considered to be defenseless against various attacks. Most of the middleware can be corrupted by the attacker in the IoT ecosystem. The cloud and database security are two more important security features in middleware. The following are examples of predicted attacks in the middleware.

- **Man-in-the-middle attack (MQTT):** The MQTT convention utilizes a distributed buy-in model of correspondence among customers and supporters, with the MQTT dealer serving as a viable mediator. This aides in decoupling the distribution and the buying in customers from one another, and messages can be sent without the information on the objective. In the event that the assailant can handle the expedite and turn into a man-in-the-center, then, at that point, he/she can oversee all correspondence with no information on the customers.

- **SQL injection attack:** Middleware is a medium which is vulnerable to SQL injection (SQLi) attacks. Aggressors here can embed malicious SQL declarations in a software/program. Th aggressors can get personal information of any customer and can even change values in the informational index. Open Web Application Security Project has recorded SQLi as a top risk to web security.
- **Mark combining attack:** In the web organizations used in the middleware, XML markup are utilized. In an imprint wrapping attack, the attacker breaks the imprint computation and can execute exercises or adjust sneaked around messages by exploiting shortcomings in simple object access protocol.
- **Cloud malware injection:** Here, the attacker attempts to take control, add malicious code, and take over Virtual Machine (VM) in the cloud by portraying to offer help by virtual machine event or lucrative code for module. Consequently, the attacker can get permission to help sales of the setback's organization and can obtain sensitive/personal data that can then be changed by the event.
- **Flood attack in cloud:** This attack is similar to a DOS and it impacts the Quality of Service (QoS). For compromising cloud environment, the aggressors unendingly send different sales to help. These attacks can tremendously influence cloud structures by extending the stack on the cloud laborers.

2.7.4   Security problems at gateways

Gateway is an expansive layer that plays a significant part in associating different individuals, gadgets, cloud administrations, and things. Gateways likewise help in giving equipment and programming solutions for IoT gadgets. Gateways are utilized for unscrambling and encoding IoT data and deciphering data to ensure correspondence between various layers. IoT frameworks today are heterogeneous including ZigBee, Z-Wave, LoraWan, and TCP/IP stacks with numerous entryways in the middle. A portion of the security challenges for IoT entryway is discussed below.

- **Secure On-boarding:** When another gadget or sensor is introduced in an IoT framework, ensure encryption keys. Key exchange happens through gateways and is communicated to all the new devices. The gateways are defenseless to man-in-the-middle assaults and snooping to find the encryption keys, particularly while new devices are registered with the environment.
- **Additional interfaces:** Whenever new devices are being registered, it is important to minimize the attack area. Just the vital connections and conventions ought to be executed by an IoT gateway maker. A portion of the functionalities and administrations ought to be confined for end-clients to stay away from indirect access confirmation or data break.

- **Start to finish encryption:** The start to finish application layer security is needed to guarantee the privacy of the information. The system ought not to let anybody other than the novel beneficiary to decode the scrambled messages. Despite the fact that Z-wave and Zigbee conventions support encryption, this is not start to finish encryption, on the grounds that, to decipher the data starting with one convention then onto the next, the doors are needed to unscramble and re-scramble the messages. This decoding at the passage level makes the information powerless to information breaks.
- **Firmware refreshes:** Most IoT gadgets are asset-based, so they do not have a user interface (UI) or the calculation ability to download and introduce firmware refreshes. Firmware download is carried out, and to apply it for firmware upgrade, its usually taken care by gateways. To ensure a safe firmware upgrade, the latest firmware is downloaded and verified before a firmware refresh is done.

2.7.5   Security problems at application layer

The application layer directly oversees and offers help to the end customers. IoT applications, such as smart homes, smart meters, astute metropolitan networks, and insightful organizations, lie in this layer. Similar to data theft and assurance issues, explicit security is ensured in this layer. Different applications find this more explicit in terms of security. In between the association layer and application layer, different IoT applications are available in the form of sublayers, named as an application support layer or middleware layer.

Significant security issues experienced by the application layer are examined below.

- **Information thefts:** IoT applications manage a part of basic and private information. Information on the move is much more powerless against assaults than information that is stationary, and in an IoT environment, data generation will be huge as the number of nodes increases. If data security is of the major concern, then clients will not find IoT environment safe enough to share personal information, hence security aspects here is of great importance. Information encryption, information disengagement, client and organization confirmation, protection the board, and so forth are a portion of the methods and conventions being utilized to get IoT applications against information robberies.
- **Access control attack:** Access control is approval system that permits just real clients or cycles to get to the information or record. Access-control assault is a basic assault in IoT systems on the grounds that once the entrance is compromised, then, at that point the total IoT application becomes powerless against assaults.
- **Administration-interruption attacks**: These assaults are also called as unlawful interference assaults or DDoS assaults in existing writing. There have been different cases of such assaults on IoT environment. Such assaults deny real

clients from utilizing the administrations of IoT environment by artificially making the workers or organization too occupied to even consider reacting.

- **Pernicious-code injection attacks:** Attackers for the most part go for the least demanding or easiest strategy they can use to break into a framework or organization. On the off chance that the framework is helpless against malevolent contents and confusion due to insufficient code checks, for the most part, aggressors use cross-site scripting (XSS) (cross-webpage scripting) to infuse some malevolent content into a generally confided in site. An effective XSS assault can bring about the capturing of an IoT environment and can incapacitate the IoT framework.

- **Sniffing attacks:** The assailants might utilize sniffer systems to screen the organization traffic in IoT systems. This might permit the aggressor to access private client information in case there are insufficient security conventions carried out to forestall it.

- **Reconstruct attacks:** If the programming cycle is not secured, then, at that point the aggressors can attempt to reinvent the IoT protests distantly. This might prompt the capturing of the IoT network.

2.8   Scalability in IoT

Scalability is one of the most important feature in any technology. There is significant research involved in this field. Self-aware computing is a new approach for applications and systems that synthesize information about the state of the IoT device and the environment of the device and use this information to predict their behavior. Self-aware device systems can be used to employ advanced levels of autonomous actions to make real-time adaption and to deal with complicated functionalities [19].

- **Self-awareness of the node:** On node level, self-awareness is all about viewpoints that are straightforwardly identified with the singular system/device and the environment around it. This can go from choosing unique calculations dependent on changing environmental conditions (e.g. lighting) or changing objectives and imperatives (e.g. low battery) to investigate and apply diverse social ways to deal with to survive quickly unfurling circumstances (e.g. increment/lessen asset usage).

- **Self-awareness of the network:** When an IoT gadget starts to collaborate with others all together to overcome the difficulties, monitoring the state, capacities also, objectives of different gadgets in the environment permits every node to settle on significant choices about the others intended activities. Exclusively by having this information, the most ideal result can be accomplished. Network-level self-awareness is taking care of this while being decoupled from the node-level self-awareness with an end goal to isolate concerns—focusing on data from different devices in the network instead local information.

To examine and assess the methodology of node and network awareness to visual mobile sensor networks and the instance of covering moving objects with a predefined number of cameras. Here, a set of cameras $Cm = \{cm1, cm2..., cmn\}$ has to cover each moving object of interest $Ob = \{ob1, ob2,..., obm\}$ with at least r cameras where the object's interest can change during runtime:

$$imp(obj, t) = \begin{cases} 1; & \text{if oj is of interest} \\ 0; & \text{otherwise} \end{cases} \tag{2.4}$$

An object ob_j is r-covered by the n cameras if

$$cov(oj, ci, t) = \begin{cases} 1; & \text{if oj within FOV of ci} \\ 0; & \text{otherwise} \end{cases} \tag{2.5}$$

$$rcov(oa, r, t) = \begin{cases} 1; & \text{if } \sum_{i=1}^{n} cov(oj, ci, t) \geq r \\ 0; & \text{otherwise} \end{cases} \tag{2.6}$$

Finally, we have to increase the times whereas many objects as possible are being covered by at least k cameras in parallel. For a finite time, horizon T this can be formulated as

$$Performance = \frac{\sum_{t=1}^{T} \sum_{j=1}^{m} rcov(oj, k, t)}{\sum_{t=1}^{T} \sum_{j=1}^{m} imp(oj, t)} \tag{2.7}$$

2.9 Conclusion

This chapter emphasizes on the IoT, its mechanism, and significance with respect to the current trends. Sensors will be the medium to exchange information from one part of the location to different locations. Smart grids play a major role compared to conventional grid were the energy usage and controlling are critical. The later section of this chapter gives the complete picture of how these IoT-enabled networks will be the part of smart grids, its implementation challenges, and benefits.

The role of blockchain in IoT is discussed. Various advantages and disadvantages or challenges faced in adopting blockchain in IoT are considered. The overall vulnerabilities of IoT environment is discussed in detail here.

References

[1] Mocrii D., Chen Y., Musilek P. 'IoT-based smart homes: A review of system architecture, software, communications, privacy and security'. *Internet of Things*. 2018, vol. 1-2(6), pp. 81–98.

[2]   Liu X., Wu J. 'A method for energy balance and data transmission optimal routing in wireless sensor networks'. *Sensors*. 2019, vol. 19(13), p. 3017.

[3]   Mugunthan S.R., Vijayakumar D.T. 'Review on IoT based smart grid architecture implementation'. *Journal of Electrical Engineering and Automation*. 2019, vol. 01, pp. 12–20.

[4]   Bhuiyan., Alam M.Z., Zaman M. 'Protected bidding against compromised information injection in IoT-based smart grid' in Springer C. (ed.). *International Conference on Smart Grid and Internet of Things*; 2018. pp. 78–84.

[5]   Obaidat M., Misra S. *Principles of Wireless Sensor Networks*. Cambridge: Cambridge University Press; 2014.

[6]   Reka S., Dragicevic T. 'Future effectual role of energy delivery: A comprehensive review of Internet of things and smart grid'. *Renewable and Sustainable Energy Reviews*. 2018, vol. 91, pp. 90–108.

[7]   Yun M., Yuxin B. 'Research on the architecture and key technology of Internet of Things (IoT) applied on smart grid'. *IEEE International Conference on Advances in Energy Engineering*; 2010. pp. 69–72.

[8]   Shahinzadeh H., Moradi J., Gharehpetian G.B., Nafisi H., Abedi M. 'IoT Architecture for Smart Grids'. *IEEE International Conference on Protection and Automation of Power System (IPAPS)*; 2019. pp. 22–30.

[9]   Lombardi F., Aniello L., Angelis S.D., Margheri A., Sassone V. *A Blockchain-based Infrastructure for Reliable and Cost-Effective IoT-aided Smart Grids*. IET; 2018. pp. 42–6.

[10]   Hadjioannou V., Mavromoustakis C.X., Mastorakis G. 'Security in smart grids and smart spaces for smooth IoT deployment in 5G' in Springer C. (ed.). *Internet of Things (IoT) in 5G Mobile Technologies*; 2016. pp. 371–97.

[11]   Ghasempour A. 'Optimum number of aggregators based on power consumption, cost, and network lifetime in advanced metering infrastructure architecture for Smart Grid Internet of Things'. *2016 13th IEEE Annual Consumer Communications & Networking Conference (CCNC)*; 2016. pp. 295–6.

[12]   'IEEE draft standard for framework of Blockchain-based Internet of things (IoT) data management'. *P2144.1/d2*. 2020, pp. 1–18. 16 July 2020.

[13]   Bhattacharjee S.B., Sengupta S. 'Blockchain-based secure and reliable manufacturing system'. *International Conferences on Internet of Things (iThings) and IEEE Green Computing and Communications (GreenCom) and IEEE Cyber, Physical and Social Computing (CPSCom) and IEEE Smart Data (SmartData) and IEEE Congress on Cybermatics (Cybermatics)*; 2020. pp. 228–33.

[14]   Putz., Pernul G. 'Detecting blockchain security threats'. *2020 IEEE International Conference on Blockchain (Blockchain)*; 2020. pp. 313–20.

[15]   Piao C., Hao Y., Yan J., Jiang X. 'Privacy preserving in blockchain-based government data sharing: A Service-On-Chain (soc) approach'. *Information Processing & Management*. 2021, vol. 58(5), p. 102651.

[16]   Al-Khazaali A.A.T., Kurnaz S. 'Study of integration of block chain and Internet of things (IoT): an opportunity, challenges, and applications as medical sector and healthcare'. *Applied Nanoscience*. 2021, vol. 13(2).

[17]   Hassani H., Huang X., Silva E. 'Big-Crypto: Big data, Blockchain and Cryptocurrency'. *Big Data and Cognitive Computing*. 2018, vol. 2(4), p. 34.

[18]   Tawalbeh Lo'ai., Muheidat F., Tawalbeh M., Quwaider M., Tawalbeh L., Privacy I. 'IoT privacy and security: Challenges and solutions'. *Applied sciences*. 2020, vol. 10(12), p. 4102.

[19]   Esterle L., Rinner B. 'An architecture for self-aware IOT applications'. *2018 IEEE International Conference on Acoustics, Speech and Signal Processing (ICASSP)*; 2018. pp. 6588–92.

Chapter 3

Blockchain technology as a solution to address security and privacy issues in IoT

D S Krishna Prasad[1], H R Prasanna Kumar[2], and Anand Nayyar[3]

Nowadays, smart applications are becoming part of small-scale to large-scale industries. Smart applications are making farming, tourism and hospitality, energy sector healthcare, supply chain, and so on smarter than ever. Industrial Internet of Things (IIoT), which is also called Industry 4.0, is the new evolution in the industrial sector that includes automation and smart applications. Internet of Things (IoT) is one of the major pillars of Industry 4.0. IoT devices that collect real-time data from the environment and exchange data among themselves. However, the security and privacy of IoT devices are a major concern as data transmission happens over public networks. To handle the issues, blockchain technology (BCT) can be one of the solutions. The blockchain proved its potential in enhancing the security and privacy of Internet data with various applications. In this regard, the chapter provides an overview of the idea of incorporating blockchain in IoT to overcome the security issues of IoT systems. The main objective of this chapter is to how blockchain features overcome the security and privacy threat in IoT system. Initially, the chapter focuses on several security and privacy concerns in IoT. We associate the features of the BCT with the challenges faced in IoT with respect to security and privacy issues. Some of the applications of blockchain in IoT have also been discussed that primarily target security and privacy issues of IoT. The chapter also focuses on challenges and future works in blockchain integrated with IoT infrastructure.

[1]Canara Engineering College, Mangaluru, India
[2]PESITM, Shimoga, India
[3]Duy Tan University, Da Nang, Vietnam

3.1   Introduction

The IoT refers to a network of all types of devices over the Internet that shares data with each other. The tremendous growth in IoT is resulting in advancements in Information and Communication Technology business. The popularity of IoT has increased rapidly, as these technologies are used for various purposes. The application of IoT includes healthcare, transportation, supply chain management, education, and other business developments. IoT has the potential to enable a new business paradigm. IoT has become part of our day-to-day life. It is assessed that, by 2025, there will be more than 30 billion IoT devices [1].

As billions of devices across the world are being connected to the growing network size, security and privacy of sensitive data are still a major concern. Industries across the world are finding new technology that promises to improve overall online security. Protecting personal information and transaction details identified endeavors, and their participants are some of the main objectives of IIoT [2].

BCT is one of the prevalent technologies of the decade. Technology is playing an important role in both research and industry for its potential in addressing various problems in the growing technological world. Decentralized, immutable, transparency, autonomy, and anonymity features made BCT unique by its nature [3]. The blockchain is considering the development that will deal with the cybersecurity risk management for IoT. Therefore, BCT has the possibility to build an innovative cybersecurity system, providing the assurance of confidentiality of sensitive information. Blockchain is replacing neither the technology nor the system; however, it can resolve some issues and offer some solutions to built-in challenges, as shown in Figure 3.1.

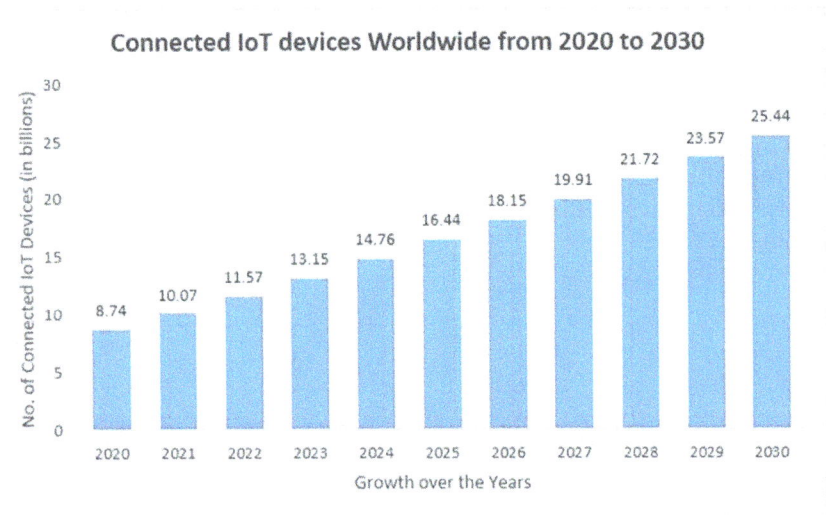

Figure 3.1   Growth of connected IoT devices from 2020 to 2030 [1]

Blockchain is a Peer-to-Peer (P2P), distributed, and immutable ledger that is controlled by a group of peers [4]. Transaction data (i.e. block) are secured and connected (i.e. chain) to each other using cryptographic principles. BCT is distinguished by several fundamental characteristics, such as decentralization, persistence, anonymity, and auditability.

The consensus algorithms of blockchain make transactions transparent and verify the validity of a new block that is new to the chain. Even though the initial focus of BCT was on cryptocurrencies, it is evident that BCT has more potential in transforming traditional business models into smart and transparent models. This will enhance the work process in manufacturing, agriculture, healthcare, supply chains, and logistics. With smart contracts, execution processes, regulatory workflow automation, information monitoring and reporting, tracking of compliances, and approval processes are getting easier.

Manufacturing companies require defining and shaping their core value drivers enabled by the digital platform to succeed in the next industrial revolution. Smart applications will drive the operational efficiencies in Industry 4.0. It also helps to grow opportunities through creative innovations and address solutions, which will increase customer value. They will finally lead to complete new business models and service providers that are enabled through digital technology.

It is evident that platforms like the IoT, robotics, smart sensors, 3D printing, and augmented reality are emerging technologies that drive Industry 4.0. Blockchain can become more effective by integrating Industry 4.0 with cyber and physical systems. BCT becomes more impacted by its secure and interconnected nature on industrial and business models. Blockchain has the potential to become one of the core technologies that can act as a common base for Industry 4.0. Core value drivers, such as a smart factory, and smart supply chain systems, use the blockchain for their data management. Modern technology enablers, such as cyber security systems, cloud computing, IoT, and additive manufacturing, are built upon BCT in Industry 4.0. The core drivers and technology enablers for Industry 4.0 are depicted in Figure 3.2.

In the next section, we present major security and privacy issues in IoT, which is the motivation for this chapter. In Section 3.3, brief introduction of BCT that includes core components, key features of blockchain that understand how blockchain is suitable for IoT systems' security and privacy issue. The same section also presents the integration of blockchain and IoT-layered architecture followed by some of the application areas that benefitted from a combination of these two technologies. Section 3.4 focuses on major open issues and research areas on this topic. Finally, we conclude the overall discussion.

3.2   Security and privacy concerns in the IoT

The growth of any digital technology will result in new limitations. This is true with the IoT network too. The cyber attackers search for limitations of the technology and take the advantage of the same for their benefits. IoT networks or devices have some

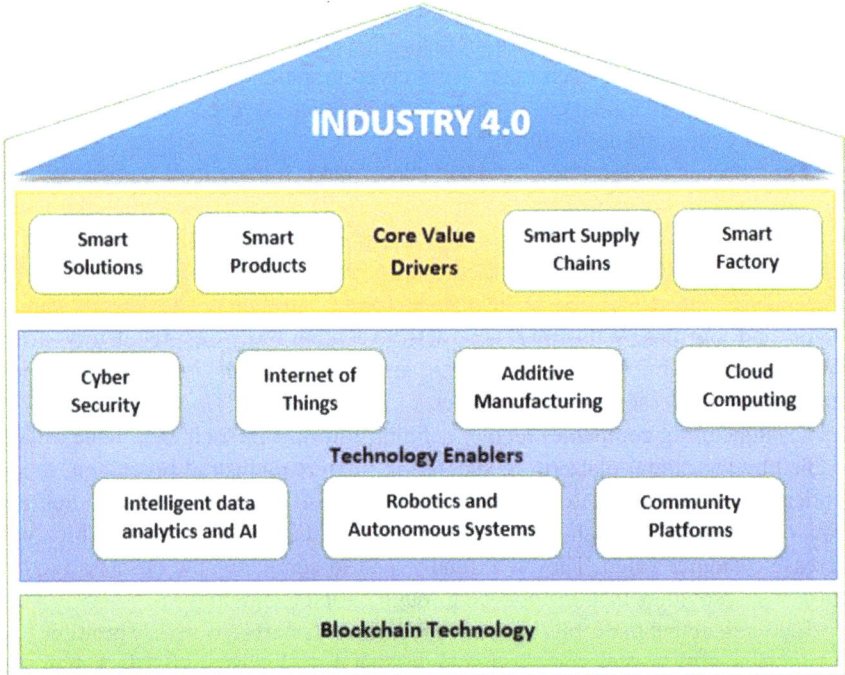

Figure 3.2   Blockchain technology as the backbone for Industry 4.0

limitations that make them more prone to security and privacy attacks. We discuss some of the major limitations here.

3.2.1   Limitations of IoT network

All the time hackers find the vulnerabilities and limitations of the technology before they actually attack the system. Likewise, there are some technical limitations in IoT. Some of the major limitations are listed below, which make IoT devices more susceptible to security attacks.

- Limited device capabilities: IoT devices are heterogeneous in nature. Sometimes, it may include small sensors deployed on the fields to sense various parameters. These devices have limited memory and computational power. Due to this, a traditional network with multifactor security layers and complex algorithms and protocols may not suit a real-time IoT system.
- Use of less secure wireless media: IoT devices may use both wired and wireless networks for data transmission. However, most of the devices nowadays are standalone wireless devices. These IoT's Ad-Hoc devices use insecure wireless communication media, such as radiofrequency identification (RFID), low-power wide area networks, ZigBee, Bluetooth, and 802.11a/b/n/g/p. Compared

to wired media, wireless media is more vulnerable to security and privacy attacks.

- Heterogeneous applications: Devices in IoT may include various manufacturers. They may have application-oriented functionality and uncommon operating systems. It is challenging to develop standard security protocols with the heterogeneity in the information types and transmission formats.
- Vulnerable environments: IoT devices tend to be deployed in unprotected environments. This makes it easier for attackers to access, compared to firewalled networks [5].
- Centralized architecture: Traditional IoT system uses centralized architecture. The centralized architecture of IoT systems can have a harmful effect on security. In a centralized architecture, data from each device and sensors of the IoT network will be communicated to a base station. Each device may interact with a common database. This type of system is advantageous for the attacker to concentrate on a single system.

3.2.2 Security and privacy attacks of IoT

After the introduction of the IoT system, major applications transformed into IoT-equipped smart applications. The introduction of IoT carries various benefits, as it will positively influence the way devices to carry out data transmission and potentially change the world. However, with the benefits comes risk. Cybercriminals find more security exploits through these connected devices. Security and privacy measures are still a need of the hour.

3.3.2.1 Security attacks

The layered architecture for IoT is evolved from its three-layered architecture to four- and five-layered architecture. The physical, network, and application layers are considered three layers of basic-layered architecture. The security issues are present in each of these layers. Layer-wise security issues are shown in Figure 3.3.

3.3.2.2 Perception layer attacks

This layer is also called the physical layer or sensor layer [6]. In this layer, smart devices, which may have a sensor, actuators, and a microcontroller, are connected to the network. These devices may use various wireless technologies, such as Bluetooth, GPRS, Wi-Fi, and ZigBee technology, to get connected to each other [7]. Some of the most frequent security attacks at this layer are described below:

- Node capture attacks: An adversary takes over the control of the IoT node by a physical attack, e.g. modifying an integrated circuit board to exploit stored data or access data from the network [8]. The attacker can get access to the network by adding a new device or replacing a genuine device. This type of attack makes it difficult to identify false devices in the IoT network.

Application Layer	•Phishing Attack •Cross Site Scripting •Malicious Code Attack
Network layer	•Spoofing •Man in the Middle attack •Sinkhole Attack •DoS Attack
Perception layer	•Node Capture Attack •Replay Attack •Side Channel Attack •False Data Injection

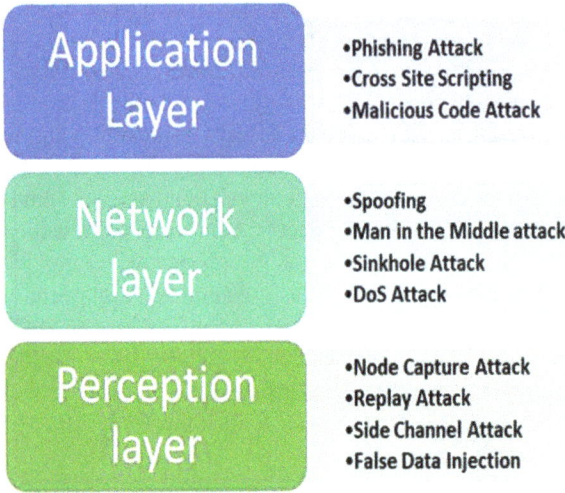

Figure 3.3   Layer wise security attacks in IoT

- Replay attacks: In this attack, the adversary eavesdrops on the network communication and then delays or resends (replay) as the attacker wants. The attacker tries to become a legitimate member of the network and then access the required data from the other devices.
- Side-channel attacks: The intruder attempts to develop access to the device information resembling power consumption and processing time from the IoT devices. In this attack, the attacker tries to get the cryptographic key to get the plaintext from the ciphertext.
- Eavesdropping: It is a type of passive attack. The eavesdropper takes the advantage of an insecure wireless communication medium to access and read the transaction data. Passive attacks are more difficult to identify than active attacks.
- False data injection: Various types of IoT devices, such as, CCTV, sensors, and home appliances, will be deployed all over the open environments. These devices will be susceptible to false data injection attacks, where an attacker will try to access the device and inject false data. As a result, the devices that are associated with the victim device will receive the wrong data.

3.3.2.3   Network layer attacks

The network layer is an intermediator between the perception layer and the application layer. It collects data from the lower layer and transmits the same to the upper layer. This layer takes the responsibility of transmitting the data within and across the IoT network. The network layer of the IoT architecture is one of the most sensitive layers, which is more prone to security attacks. Common security attacks associated with the network layer are described below:

- Spoofing: In spoofing, the attacker tries to access the network by exploiting vulnerable points. The attacker acts as a legal node and tries to access sensitive information. Spoofing is sometimes used to mislead the routing packets in the IoT network. Bluetooth spoofing is the latest cyberattack targeting IoT devices.
- Man in the Middle (MitM) attack: In an MitM attack, the attacker attempts to intercept communication between two devices. They access the data packets from the network during transit. Captured data then forwarded to the other node. Neighboring nodes falsely think that the data received from a legitimate node. A simple MitM scenario can be an attacker sending a false temperature resulting in a piece of machinery overheating. This can end up with serious issues, such as defective product generation or damage to the machinery. There are two types of MitM attacks: passive and active attacks. Passive is more difficult to detect than active MitM attack.
- Sinkhole attacks: This attack is considered as routing attack in IoT applications. In this, the attacker attracts the adjacent nodes with forged routing information. The attacker makes all the data to be transmitted through some promised node. It creates high network traffic and collapses the data communication. Along with traffic analysis, other attacks such as selective forwarding or Denial of Service (DoS) can be combined with the sinkhole attack.
- DoS attacks: In a DoS attack, the adversary makes a server unavailable to its intended clients. The attacker interrupts the server by sending a large number of service requests. As a result, the server will be unable to serve the request from legitimate clients. In a scenario in IoT, an attacker can send a large number of authentication requests to reduce sensor energy and thus stop sensors from collecting and transmitting data.

3.3.2.4 Application layer attacks

The application layer is the topmost layer of IoT architecture. It defines all services and protocols that are used to transmit the data at the application level. It provides an interface between users and the IoT systems. Services depend on sensors that collect real data from the environment. In summary, the application layer provides overall services requested by the application. Following are some of the major security attacks associated with the application layer of IoT:

- Phishing attacks: Phishing attacks result in dangerous data breach in IoT [9]. In a phishing attack, the attacker tries to get a user's sensitive data through spam email, websites, blogs, fraudulent messages, social media, mobile applications, and forms [10]. Phishing attacks are performed using IoT devices including smart TV, smart refrigerators, etc. [11].
- Cross-site scripting: In this attack, the adversary injects malware scripts like javascript into the code of a third-party application or website. It is also called an injection attack. With these applications, the attacker gets the user credentials and uses them for personal benefit.

- Malware attack: It is malicious software projected to cause damage to the system by executing malware. Once the attacker succeeded in injecting malware, the victim devices can be used for distributed denial of service attacks.

3.3.2.5   Privacy attacks

IoT devices collect a huge amount of raw real-time data from the surrounding environment. These data may be low-to-high sensitive data. Maintaining the secrecy of sensitive data is one of the main responsibilities of an IoT network system. Preventing privacy attacks is also one of the challenges in addition to security threat measures. These privacy attacks may include sniffing, de-anonymization, and inference attacks [12]. The impact of these attacks is on the confidentiality and privacy of the IoT data. The following section discusses some of the well-known privacy attacks on IoT:

- MiTM: Based on the impact of the attack, it is categorized into active MiTM attacks and passive MiTM attacks. A passive MiTM attacker silently reads or observes the data without any modification. Passive MiTM attacks are difficult to identify. These attackers may use the sensitive information collected during data transmission for their financial benefits.
- Data privacy: The data privacy attack is also classified into active and passive data privacy attacks. Data leakage, data fiddling, identity stealing, and re-identification are some of the privacy attacks related to data privacy [13]. The re-identification attacks are sometimes called inference attacks. The gathering of information is the main goal of data privacy attack.

Apart from security and privacy disputes in IoT, the other two aspects also play an important role. They are trust management and authentication. These two parameters have a significant functions in IoT in enhancing reliability, privacy, and security.

Trust management: In IoT, data are communicated between a large number of interconnected devices. Trust management plays an important role in securing all the participating IoT devices. A robust trust management model is need for an hour for ensuring trust among all the IoT devices.

Authentication and authorization: IoT network comprises smart devices like sensors, actuators, and some connecting devices, which are connected through the Internet. The authentication and authorization process ensures that only intended devices have the permission and access to resources in the network. The authentication process identifies the legitimate device, and the authorization process permits the authentic device. Public key infrastructure is one of the techniques used by many organizations to authenticate the legitimacy of an IoT device. Registering the devices with the authentication certificate is another popular way of device authentication. Sometimes, communication protocols too provide strong IoT authentication and ensure secure data transmission between authentic devices. The best example is the

Message Queuing Telemetry Transport protocol. This protocol securely connects devices with the server that stores digital identity and authentication certificates.

3.3　Blockchain technology concepts

Before really looking into the benefits of BCT in securing IoT systems, it is good enough to discuss fundamentals and some of the key features of the technology. **Blockchain:** Blockchain is a distributed digital ledger formed by a sequence of blocks. The distributed ledger records all the valid transactions on a network. A new valid transaction is time-stamped and stored in a new block along with required data for processing. Each block in the chain is identified by its cryptographic hash value. The addition of a valid new block is attached to a previous block by storing the hash value. This forms a linear sequence of chains which is called "blockchain." In other words, more than one connected blocks forming a connected sequence of block are called the blockchain. Figure 3.4 shows the structure of a blockchain. A block in a blockchain contains a hash value of the previous block, time-stamp, and other information for processing. The data section contains the transaction data of the block.

3.3.1　Blockchain types

There are four types of blockchain in the market at present, namely, public, private, consortium, and hybrid blockchain.

- Public blockchains are publicly accessible through the Internet. Anybody can create an account and join the blockchain network. Every node has an equal opportunity to become the miners or transaction validators based upon the power of computation. The first cryptocurrency called "Bitcoin" was built on a

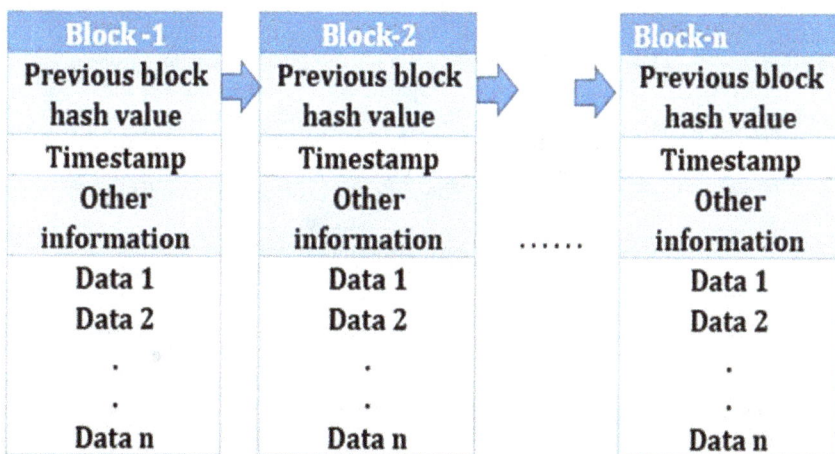

Block -1	Block-2	Block-n
Previous block hash value	Previous block hash value	Previous block hash value
Timestamp	Timestamp	Timestamp
Other information	Other information	Other information
Data 1	Data 1	Data 1
Data 2	Data 2	Data 2
.	.	.
.	.	.
Data n	Data n	Data n

Figure 3.4　Simple structure of a blockchain

public blockchain [4]. Scalability is the key challenge of a public blockchain. The network slows down with more nodes in the network.

- Private blockchain: As the name indicates, it is restricted to only permissioned users. It is also called permissioned blockchain. Normally, these blockchains are used inside an organization for internal use cases. As it is local to one organization, the number of nodes will be less. This results in faster transactions. Multichain is an example of a private blockchain. A private blockchain is not fully decentralized because the consensus process is managed by a single entity rather than a group or all the entities. So, achieving trust among the multiple private blockchain is difficult.
- Consortium blockchain: It is considered a semi-private blockchain. Here, instead of a single entity, a predefined group of nodes from different organizations manages the consensus procedures and validates the transaction. This builds trust between the organizations. It is sometimes called a federated blockchain. Hyperledger and Corda use consortium blockchain.
- Hybrid blockchain: It is a combination of public and private BCT. It offers the benefits of both public and private blockchain. It provides privacy and security benefits of permissioned blockchain and transparency benefits of a public blockchain. It provides the flexibility to choose the level of transparency for different data sets. IBM (International Business Machines Corporation) food trust is an example of a hybrid blockchain, The comparison between public, private and consortium blockchain [14] is shown in Table 3.1.

Table 3.1   Comparison between blockchain types

Comparison criteria	Public blockchain	Private blockchain	Consortium blockchain
Architecture type	Fully decentralized	Partial decentralized	Partial decentralized
Openness	Open to all	Open to one specific organization	Open to a group of organization
Read/Write access	All have equal access to read and write	Read access can be given to public or organization but write access is restricted to organization only	Read access is open to all. Write access is restricted for selected group of organization
Anonymity level	Maintains high privacy	Low privacy level	Low privacy level
Speed of transaction	Slow compared to other types	Faster compared other types	Faster compared to public and slower than private
Consensus process	Permissionless	Permissioned	Permissioned
Immutability	Highly tamper proof	Data can be tampered	Data can be tampered

3.3.2   Core components of a blockchain

Blockchain technology comprises some of the core components that collectively drive the technology. The following are the key components of the blockchain concept:

Smart contracts: Smart contracts are a set of code that contains a set of transaction protocols, which acts as an agreement between unknown parties. They execute automatically when predetermined terms and conditions are met. A smart contract includes an executable set of codes, functions, and state variables. The functions contain various parameters required during the transaction and change based upon the set of code execution. High-level languages, such as Solidity (similar to JavaScript and C++) or Python, are used to implement smart contracts. Solidity is one of the language-specific compilers that compile source code to byte code. The compiled contracts are loaded to the blockchain network. The uploaded contract gets unique addresses by the blockchain network. The set of codes in the contract will be triggered whenever any user within the network sends the transaction details. To validate the new block, each node in the network executes the smart contract. Figure 3.5 shows the simple structure of a smart contract.

3.3.2.1   Merkle tree

A Merkle tree [15] is a data structure in the form of a complete binary tree. Merkle tree in the blockchain is used to store the summary of all transactions securely and efficiently. In the blockchain, each transaction is hashed, which is called "leaves." Then, each pair of hashed transactions are concatenated and hashed together, which are called "branches." The branches are then paired and hashed together until there is one hash, which is called the "root." If there is an odd number of transactions, a copy of the transaction is considered and its hash is concatenated with itself. In a Merkle tree, leaf value can be validated with the publicly known root value.

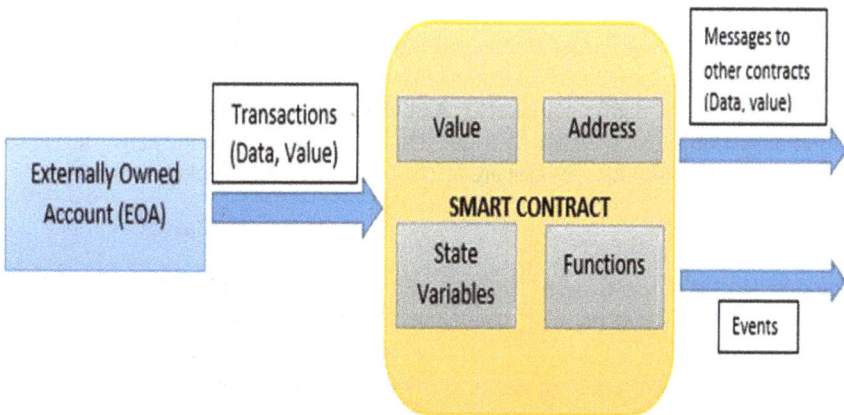

Figure 3.5   Sample structure of smart contract

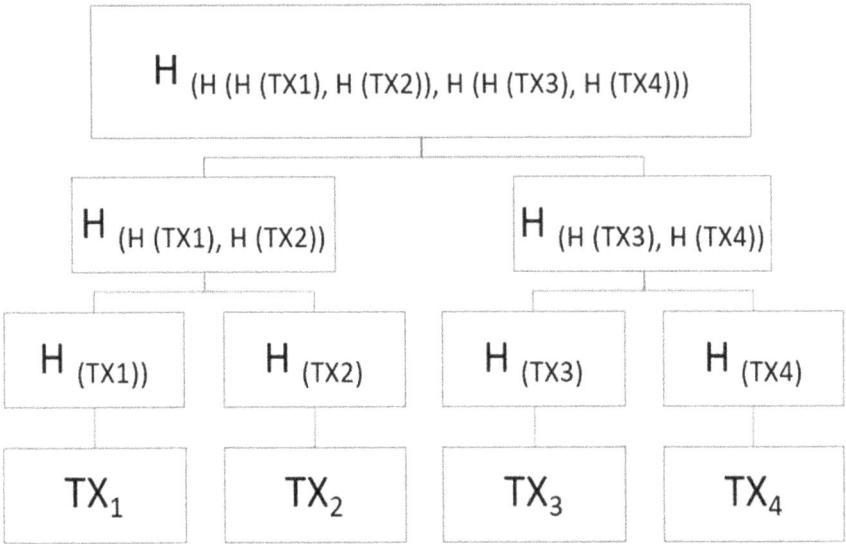

Figure 3.6   Sample structure of Merkle tree

A Merkle tree is also known as a binary hash tree. The sample of the Merkle tree is shown in Figure 3.5. In the figure, TX1, TX2, TX3, and TX4 are the transaction numbers. The function $H(x)$ indicates the hashing function on a particular transaction of concatenated hash values. Hashing is done until the root node as indicated in Figure 3.6.

3.3.2.2   Node

A node is a user in the form computer in a blockchain network. A node is a computer or any smart device that stores a copy of the distributed ledger and acts as a communication point in a blockchain network. It also performs network functions, as well as verifying and validating new blocks. As blockchain is distributed network, every valid transaction is recorded and distributed to each device within the network. These devices are called as nodes. Nodes are also called as peers in blockchain network.

3.3.2.3   Blocks

The details of a transaction are wrapped into a unit called a block in a blockchain network. Each block in the blockchain can be considered as a page in a ledger. A block contains information about a valid transaction that happened in the blockchain network. A block contains two sections: the header and body section. The header contains a hash value of the preceding block, hash of the Merkle tree, time stamp, and other transaction-related details. The body of a block contains all transactions that are confirmed with the block.

3.3.2.4   Transaction

In general, a blockchain transaction is an event that changes the state of data within all nodes. For example, in a Bitcoin, a transaction can be a transfer of 10 Bitcoins between two nodes. A valid transaction in the blockchain will affect the digital ledger entry in all nodes.

3.3.2.5   Miners

A node or a group of nodes that verifies the valid transaction are called miners. The miners validate the transaction based upon the consensus algorithm. The number of nodes chosen depends upon the type of blockchain used.

3.3.2.6   Consensus algorithm

The consensus algorithm is a set of rules through which all the nodes reach a common agreement about the present status of the digital ledger. The main goal of the algorithm is to build trust between unknown peers in the blockchain network. There are different consensus algorithms, such as proof-of-work (PoW), proof-of-stack (PoS), practical Byzantine fault tolerance (pBFT), and proof-of-capacity (PoC). The details of these algorithms are not in the purview of this chapter.

3.3.2.7   Proof of elapsed time

Proof of elapsed time (PoET) is developed by Intel Corporation. It follows the lottery system to ensure that the chance of winning is equal across the participants. It uses a permissioned blockchain network. The algorithm generates a random wait time for each participant in the network for which a particular node must sleep equal to the wait time. The node with the shortest wait time will win the new block. The high-throughput and low latency make it more suitable for the IoT network.

3.3.2.8   Practical Byzantine fault tolerance

Practical Byzantine fault tolerance (pBFT) algorithm was designed to overcome the Byzantine Generals Problem of catastrophic system failure. The algorithm is designed to work in an asynchronous system. It optimizes the Byzantine Fault Tolerance.

3.3.2.9   Tangle

Tangle is a direct acyclic graph–based consensus algorithm used in IOTA, which is an open source distributed ledger and cryptocurrency designed for IoT. In the Tangle, to have a valid transaction, the peer must validate the previous two transactions. This will strengthen the verification process as more transactions are added to the tangle. However, if a number of transactions are less, then a single peer can produce one-third of the overall transactions, it can consider validations that are labeled true. It is identified as a promising algorithm for the IoT environment.

Table 3.2   Summarized comparison of IoT suitable consensus algorithms [16]

Algorithm	Accessibility	Scalability	Latency	Throughput	Suitability
PoET	Private (P)	High	Low	High	High
pBFT	Private (P)	Low	Low	High	High
Tangle	Public (PL)	High	Low	High	High
PoS	Private (P/PL)	High	Medium	Low	Medium
DPoS	Public (P/PL)	High	Medium	High	Medium
RapidChain	Public (PL)	High	Medium	High	Medium
OmniLedger	Public (PL)	High	Medium	High	Medium
Raft	Private (P)	High	Low	High	Medium
Ripple	Public (PL)	High	Medium	High	Medium
PoI	Public (PL)	High	Medium	High	Medium
Stellar	Public (PL)	High	Medium	High	Medium
dPBFT	Private (PL)	High	Medium	High	Medium

3.3.2.10   Proof-of-stake

The PoS algorithm chooses the validators of the transaction based on the quantity of blockchain tokens available with the peer. All the peers that have a stake participate in the voting procedure to select the miner. This mechanism overcomes the problem of energy consumption because of solving a complex problem as in PoW. This algorithm favors the peers that have more tokens. As a result, other peers will have less chance to become miners.

A summary of some of the algorithms, which are suitable for the IoT environment, is listed in Table 3.2.

3.3.3   Key features of blockchain technology for IoT security and privacy

Blockchain is not just a backbone network for Bitcoins, but it offers a lot more. The following key features of blockchain are responsible for gaining much popularity in the current technological transformation. Moreover, these features are responsible for converging blockchain with IoT, as shown in Figure 3.7.

- Decentralized and distributed: Blockchain uses P2P architecture as its base network architecture. This makes it decentral in nature. It does not have any governing authority or a central device managing the network. Instead, a group of devices controls the network making it distributed as well [17]. Proposed a novel solution to provide a lightweight and decentralized secure access control model to an IoT system. The authors developed blockchain managers that provide access control and ensure secure transmission between the devices. It also provides secure communication for fog and cloud computing. Due to its decentralization approach, the blockchain system overcomes the risk of a central system failure. This feature helps to eliminate the risk of network failure and the downfall of the entire system in case of a single node crash.

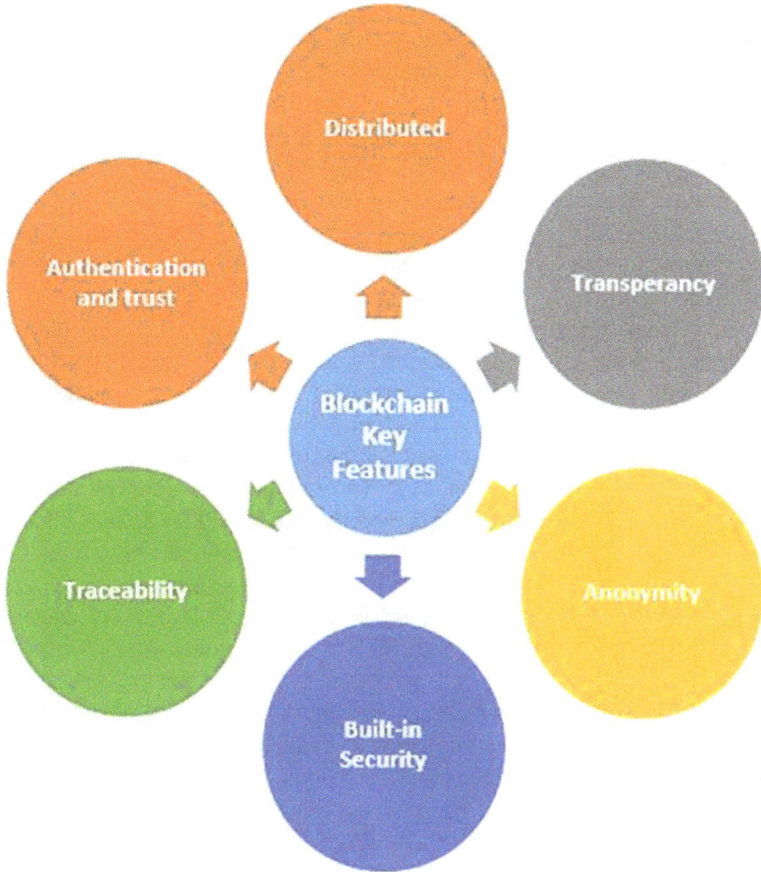

Figure 3.7   Key features of blockchain technology

- Transparency: As blockchain is distributed and decentralized system, all transactions are transparent to all nodes in its system. All the nodes keep the same
  copy of the information. This feature of blockchain improves the accountability of the transactions that happened in the network. In financial and business
  applications, this improves the trust management between the parties without
  a third MitM.
- Anonymity: Blockchain's anonymous identity is another feature that preserves
  the identity of the node that initiates the transaction. Each transaction is attached
  with the encrypted addresses of the user. The user can generate many addresses
  that preserve the identity of the user in the blockchain network.
- Security: BCT emphasizes on security more when compared to traditional
  record-keeping systems. Every new transaction is verified and validated. The
  verified transaction is then created in a new block. Each new block created is
  attached to the earlier block with the help of a hash value. The sequence of the

valid block creates the "chain." Every block provisions the hash value of the preceding block and the Merkle tree. This, along with the fact that information is stored across a network of computers instead of on a single server, makes content immutable and incorruptible. Thus, blockchain is safe from falsified information and attacks [18]. Blockchain-based smart home framework with fundamental security goals including confidentiality, integrity, and availability. The authors were able to reduce the overhead and efficiency for resource constraint IoT devices.

- Traceability: The transactions that are recorded in the digital ledger are traceable. An audit trail verifies the source of the transaction. This helps to verify the authenticity of the transaction. The traceability feature of the blockchain plays an important role in identifying business frauds. Pharmaceutical companies apply this feature to trace the drug supply chain from manufacturers to distributors. In the art industry, it helps to provide indisputable proof of ownership. In Ref [19], the authors designed and developed a software framework used to trace the food supply chain system in Industry 4.0 domain. The system allows IoT devices to communicate directly with the Ethereum blockchain. The framework also monitors the temperature of the food (fish) during the storage and delivery process.

- Authentication and trust management: Hammi *et al.* [20] proposed a decentralized system called bubbles of trust. This system provides secure virtual zones where all the devices within the zones can communicate securely. The proposed system validates the identity and authenticity of the device. Furthermore, it preserves the data integrity and ensures the availability of data. The blockchain is used to create a safe bubble to have secure data transmission. The system is implemented using the C++ language and Ethereum blockchain.

Many other types of research have been conducted by successful researchers on integrating blockchain with IoT to enhance overall security and privacy of IoT system. In summary, we can conclude that blockchain has succeeded in providing all security aspects, such as authentication, immutable, access control, and traceability under one platform. The overall impact of blockchain is depicted in Figure 3.8.

3.3.4   *Integration of blockchain in IoT-layered architecture*

The decentralized feature of the blockchain network enables IoT devices to exchange data with other IoT devices without a centralized system, thus reducing the vulnerability of the IoT network [21]. The incorporation of blockchain in IoT networks can be done in various ways. One of the simplest ways is to add an additional layer to IoT-layered architecture [22], as shown in Figure 3.9.

The brief functionality of the perception, network, and application layer is discussed in Section 2.2. In this section, we would emphasize only on IoT-blockless bandwidth and chain layer specifically for providing secure transmission of data in the IoT networks. The IoT-blockchain layer will have some of the modules that implement BCT. These modules include smart contract, identity administration,

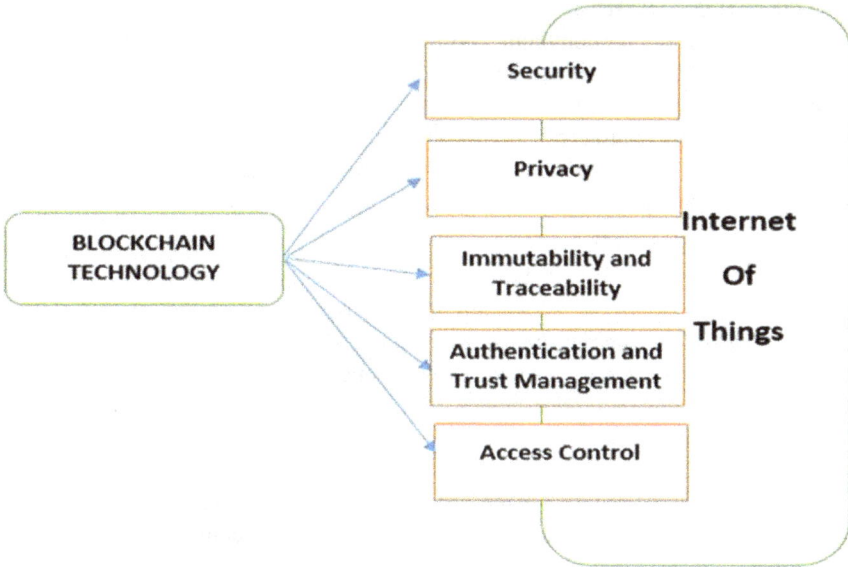

Figure 3.8   Impact of blockchain on IoT security and privacy

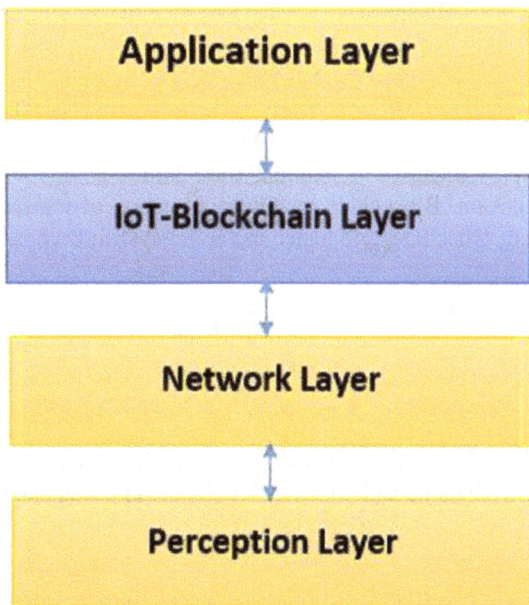

Figure 3.9   Blockchain-integrated IoT-layered architecture

distributed ledger, consensus manager, and P2P protocol. P2P protocols are necessary to permit distributed communication between numerous IoT devices. Consensus management preserves the trust among communicating devices in the IoT system. Identity administration is applied to control the addition and deletion of nodes in the IoT system. Smart contracts of the BCT enable automatic decision-making established on predetermined conditions. These modules play an important role in providing security and privacy details in IoT network.

3.3.5   Benefits of blockchain in IoT

The distributed ledger of blockchain is tamper-proof and builds trust between unknown parties. No single party has control over blockchain network. The combination of blockchain and IoT provides access control over the data generated by IoT devices. Blockchain encryption makes it virtually impossible for anyone to tamper with existing data records. It also prevents a malicious intruders from gaining access to the content of IoT data.

Protecting the information in the entire IoT ecosystem is a primary challenge for IoT networks. Security loopholes make IoT devices easy victim of various security and privacy attacks mentioned above.

The integration of blockchain and IoT encourages for a new paradigm that inherently reduces major vulnerabilities. It ensures security and transparency between all the parties and enables secure machine-to-machine transactions.

The major benefits of integrating blockchain in IoT are summarized below:

*   Enhanced security: By applying blockchain to the IoT system, the entire system gets the advantages of blockchain. The main features of blockchain, namely, encrypted data, traceability, tamper-proof, and distributive architecture prevent the entire system from major security and privacy attacks. This enhances the overall security system of the IoT in general.
*   Cost-effective: Validation of genuine transactions is another added advantage of blockchain. Blockchain enables automation of transaction validation and processing procedure, and hence the entire system becomes proactive at a reduced cost.
*   Reduced processing time: The supply chain system engages multiple parties together. The traditional approach of data exchange between suppliers, manufacturers, distributors, and consumers takes too long time that increases the delivery time of the products. With the blockchain's distributed ledger, untrusted stakeholders can exchange data directly with one another and automate the delivery process without the intervention of third parties. This will reduce the delivery time and improves supply chain management.

3.5.1.1   Comparing IoT and blockchain technology

The major security and privacy issues are discussed in Section 2. In the later section, we have seen how blockchain could be one of the potential technologies to solve major security and privacy issues of IoT. However, blockchain cannot become a solution to all the challenges of the IoT system [23]. In this section, we present the difference between these two technologies. Both technologies possess a similar type of feature on some criteria and differences on other criteria.

- System structure: IoT is built based on a centralized architecture. All the IoT devices are connected and controlled by a central server. This made IoT more vulnerable to security attacks [24]. On the other hand, a blockchain network is based on decentralized architecture. The decentralized nature of blockchain defends against security attacks. This feature of blockchain helped IoT to solve some of the major security challenges.
- Privacy: IoT network possesses low privacy to the data they carry in the network. It is a challenging task to maintain the privacy of the node and the data it transfers within the network. On the other hand, blockchain maintains high privacy of the data and the sender details with the help of cryptographic hash functions.
- Bandwidth and resources: IoT devices are resource constraint devices. Since IoT devices send a small amount of data, IoT consumes limited bandwidth. Whereas blockchain consumes high bandwidth for the data, it transfers. The size of the block determines the bandwidth requirement.
- Scalability: IoT network consists of a large number of devices. The addition and deletion of devices are not a hurdle. Therefore, IoT network scales down and up easily. Whereas blockchain scales poorly. Blockchain uses a P2P network. This makes blockchain less scalable with a large network.
- Latency: IoT demands low latency. The efficiency of the IoT network increases with the lower latency. The mining process in the blockchain is more time-consuming. The latency in a blockchain network mainly depends on the type of consensus algorithm used. For this reason, choosing of consensus algorithm plays an important role when integrating blockchain in IoT. Developing IoT suitable consensus algorithm to reduce the latency is one of the major research areas.
- Security: As discussed earlier, IoT is more susceptible to security attacks. Security is a major concern in IoT networks. On the other hand, blockchain possesses better security features. The cryptographic hash functions and decentralized nature made blockchain stronger toward security threats. These features made many researchers think about integrating blockchain with IoT. Many researchers have shown positive results in their findings.

Table 3.3 summarizes the comparison between IoT and BCT discussed above.

3.3.6 Blockchain application in solving IoT security and privacy issues

The security aspects of blockchain will make IoT devices more dependable and safe. Using blockchain can change the way the data are being communicated within IoT.

Table 3.3   Comparison of IoT and blockchain

IoT technology	Blockchain technology
Deployed as centralized architecture	Deployed as distributed/decentralized architecture
Consumes less bandwidth and devices are resource constrained	Consumes high bandwidth and resources for mining
Lack of privacy	Privacy of the peers is maintained
IoT network scales up and down easily	Scales poorly with large networks
Requires less latency for better efficiency	Mining process is time consuming that increases the latency
Security is a major issue	Possess better security

The features of blockchain make it possible in creating new business models, such as smart healthcare, supply chain management, smart energy platforms, agriculture, smart homes, and cities, as depicted in Figure 3.10. Some of the applications of BCT in IoT are discussed in the below section.

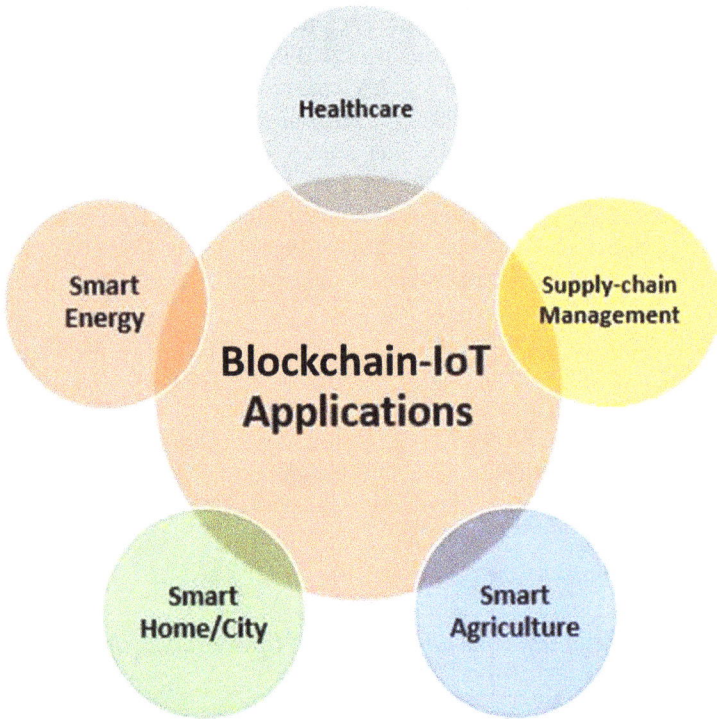

Figure 3.10   Applications of blockchain-based IoT

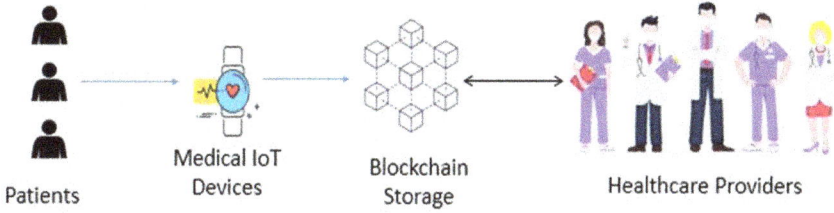

Figure 3.11   Blockchain-integrated IoT for patient data management

- Healthcare: Healthcare industry witnessed the advantage of blockchain in recent years. Blockchain can be used to communicate critical patient data securely in an effective manner. By applying blockchain, the system will reduce the possibility of mismatch in patient's data and reduce errors in patient care [25]. Only authentic relevant parties like doctors and hospital staff involved can access the patient's health records. Blockchain-based IoT has been implemented in some wearable devices to facilitate the tracking and management of diseases. They also allow for monitoring drugs for the period of their examination. Drug's efficiency and side effects are monitored by the effective use of the said technology [26]. An illustration of blockchain-IoT for patient data management is shown in Figure 3.11.
- Supply chain management: The application of IoT in supply chain management made things easy. Blockchain-applied IoT in the supply chain is an add-on advantage for IoT devices [27]. The communication technologies used like RFIDs and ZigBee lack in security and privacy. The researchers applied blockchain in supply chain management like maritime shipping [28], food supply chain [29], and pharmaceutical supply [30]. The traceability feature of blockchain enables the product transit tracing and provides the inbuilt security feature, which makes data immutable. The authors in Ref. [31] found that the combination of these two technologies could control and monitor the physical environment for spoilable food and medicines. Hence, blockchain and IoT together solve major issues in shipments and maintain a good relationship between stakeholders. A simple framework for food supply chain management between manufacturing unit, warehouse, the local distributor, and the customer is shown in Figure 3.12. Stakeholders use IoT devices for their tracking and distributing the product from source to destination. All the stakeholders store the information in the distributed ledger securely. Every stakeholder can only view their required data from the blockchain.

3.3.7   Commercial implementations of blockchain in supply chain management

BCT is applied in several commercial active projects in the area of smart supply chain management. Some of the well-known projects are listed below:

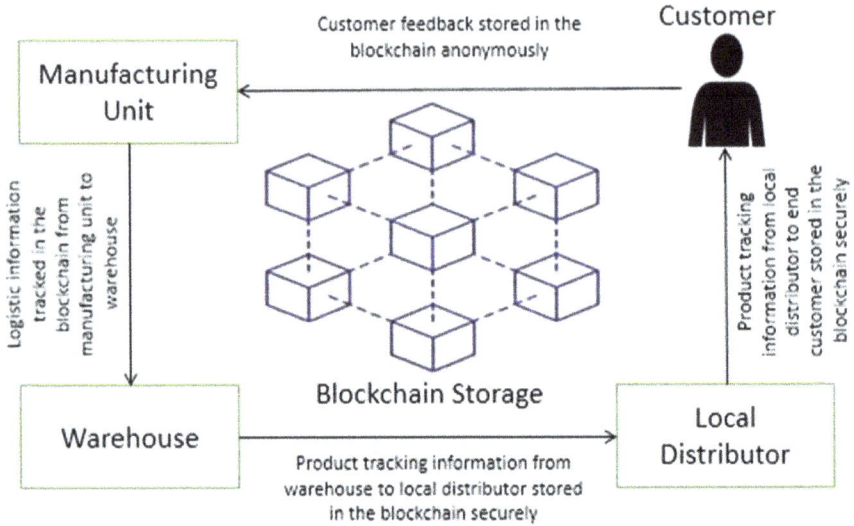

Figure 3.12   Blockchain-IoT framework for supply chain management

1. IBM Blockchain-TradeLens: TradeLens is built on a powerful permission matrix and blockchain, guaranteeing every customer to cargo has access only to their data and a secure inspection trail of all businesses. It emphasizes precisely logistics. Transparency and traceability are the most serious aspects of logistics. Blockchain is used to tackle trust challenges, deliver a common vision of the truth and deliver an immutable audit trail. TradeLens uses the IBM blockchain Platform. IBM blockchain is deployed on Hyperledger Fabric. It is an open-source permissioned blockchain [32]. IBM blockchain can rationalize trade exchanges, dealings, and transaction details with secure, worldwide commercial systems, and networks.
2. OriginTrail: OriginTrail [33] is a protocol solution that facilitates blockchain-reinforced data distribution in supply chains. It empowers transparency and tracing. Mainly protects the business from deceitful behavior and dynamic efficiencies for all stakeholders. OriginTrail is applied in the food business that leases its customers to know the location of their food products.

•   Energy sector: The combination of blockchain and IoT has great potential in improving the effectiveness of managing the energy sector. The distributed power supply of the modern era requires efficient technology to enhance the security and storage systems. The distributiveness of the blockchain has the potential to leverage this requirement. The industrial data can be exchanged securely between the devices that control the entire power generation unit. The consensus algorithm can build direct communication between customers and service providers without the involvement of a third party. In a distributed smart

Small Scale Energy Producers

Using blockchain technology energy sector can buy and sell the energy

Smart Home

Transfer of excess energy through blockchain without third party involvement

Blockchain Network

Primary Energy Producers

Smart Vehicle Producers

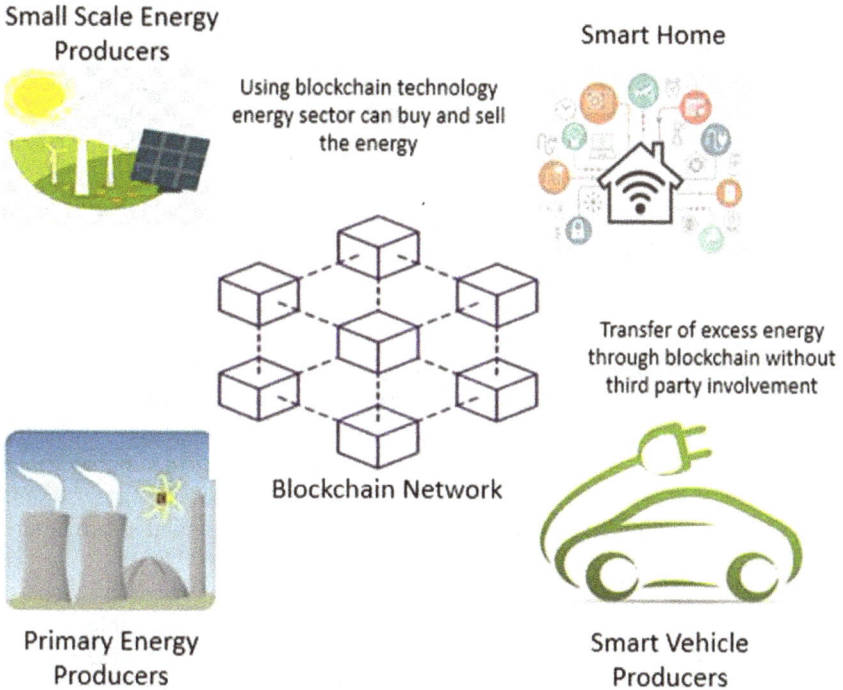

Figure 3.13   Blockchain-IoT smart energy grid management framework

grid, neighbors can trade energy themselves without the involvement of author-ities. This enhanced the trust between the customer and centralized grids, result-ing in reduced costs [34]. Diagnosis and equipment's maintenance within a smart grid becomes an easy task [35]. A simple framework for energy grid man-agement using a blockchain-enabled IoT system is presented in Figure 3.13.

• Smart homes and cities: Blockchain-integrated IoT will be an essential part of the present and future smart homes and cities. This is because the integration of the two technologies is highly versatile, as well as easy to implement [36]. One such application of these two technologies will have in smart cities is in the public transportation system. A smart and effective public transportation system is essential for smart cities. E-commerce is another area where these two technologies play an important role. With these two technologies, e-commerce will be able to encourage local business by delivering local products to homes and workplaces. Such a system will provide importance for local products and branding. Authors in Ref. [37] presented an efficient lightweight-integrated blockchain (ELIB) model and deployed it in a smart home setting. The ELIB model connects highly equipped devices to public blockchain over an overlay network. Three optimization schemes are designed in the ELIB model includ-ing a lightweight consensus mechanism, certificateless cryptography, and distributed throughput management scheme. The ELIB is successful to reduce the processing time by 50% and consume less energy of 0.07 mJ.

- Agriculture: The application of the latest technology in the agricultural sector remains behind all other sectors. It is even true with the combination of IoT and blockchain. A trustworthy and mediator-less system is very much essential to ensure maximum profit for the hardworking farmers. Blockchain and IoT have the potential to make smart farmers by including these technologies in their business. Applying blockchain and IoT in the supply chain of food products will put both farmers and consumers on a profit side. The use of blockchain with IoT would automate the entire process and build trust among all the parties. One such experiment is conducted in Ref. [38], where the author integrated blockchain and IoT in pre- and post-harvesting phases of agriculture. Blockchain acts as a backbone of the system, IoT devices collect the real data from the fields and the smart contract to regulate the interaction between the parties. The authors could ensure data availability with security and prevent manipulation along with trust amongst manufacturers and customers. Figure 3.14 shows a simple scenario of applying blockchain-integrated IoT system applied in agriculture.

Many researchers integrated blockchain and IoT to minimize security and privacy challenges in various areas. Some of them are summarized in Table 3.4.

Application areas that have benefitted from the integration of these two technologies are not restricted to the area discussed here. The chapter discussed only major areas of application.

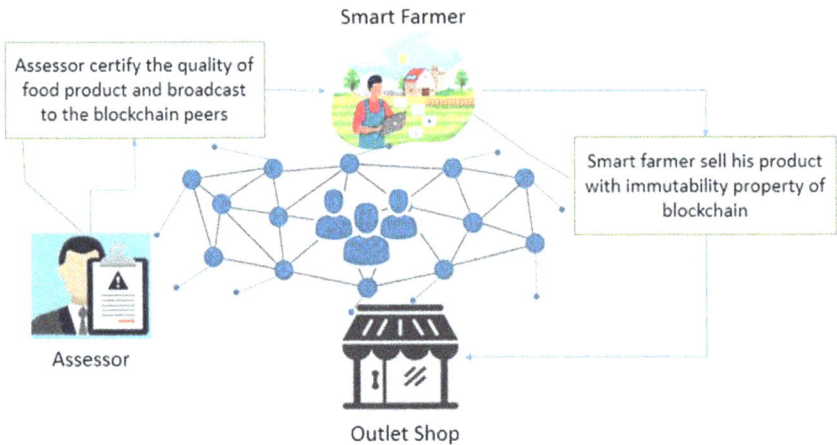

Figure 3.14   *Blockchain-IoT-enabled smart agriculture*

Table 3.4   *Recent studies on integration of IoT with blockchain to solve security and privacy issues*

Citation	Application area	Description of the work
Badr *et al.* [39]	Healthcare	Proposed a protocol named Pseudonym Based Encryption with Different Authorities (PBE-DA) to preserve privacy of patient in an e-health platform
Lin *et al.* [40]	Agriculture	Proposed food traceability system based on blockchain and IoT to ensure trustiness, self-organizable, open and ecological in agriculture system by resolving food security issues
Patil *et al.* [41]	Agriculture	Proposed lightweight blockchain-based architecture to overcome security and privacy issues in smart greenhouse farm monitoring and automation system. And developed a security framework that integrates the blockchain and IoT nodes to ensure a secure data transmission in Smart Greenhouse farming
Cadiz *et al.* [42]	Energy grid	Analyzed suitable solutions to develop a decentralized electrical power grid based on IoT sensors and blockchain. Proposed a grid model combining IoT, blockchain and smart contracts with renewable energy production
Basha *et al.* [43]	Supply chain	This paper reviews challenges in a traditional supply chain. Presented steps involved in implementation of blockchain integrated with IoT to provide secure data transmission and monitoring system in supply chain
Srivastava *et al.* [44]	Healthcare	Discussed about integration of BCT in the security of IoT based on remote patient monitoring systems. Presented advantages and practical challenges of blockchain-based security methods in remote patient monitoring using IoT system. Cryptographic technologies for deployment in IoT are evaluated
Lee *et al.* [45]	Smart home application	Proposed blockchain-based smart home gateway design for avoiding data counterfeit
Lombardi *et al.* [46]	Smart energy	Proposed a blockchain-IoT-integrated infrastructure to build reliable and cost-effective smart grid

(Continues)

Table 3.4   Continued

Citation	Application area	Description of the work
Ouaddah *et al.* [47]	Generic application	Proposed distributed pseudonymous and privacy preserving authorization managing framework called FairAccess to accomplish access control in IoT network
Pranto *et al.* [38]	Smart Agriculture	Discovered the different features of using blockchain and smart contracts with the IoT devices in pre-harvesting and post-harvesting sections of agriculture
Grecuccio *et al.* [19]	Food chain traceability	Designed and developed a software framework that allows IoT devices to interact directly with an Ethereum-based blockchain

3.3.8   Newer application areas of blockchain

Apart from healthcare, supply chain, energy grid, and agriculture, blockchain has also been introduced in other areas. Researchers also found the potential application of blockchain in the drone industry, education system, real estate, voting, and so on.

- Drone industry: Smart drone, which is also called unmanned aerial vehicles (UAVs) manufacturing, is one of the latest area where blockchain has extended its potential. UAVs are now being used for civilian purposes, other than military applications. Drone network security is an emerging exploration area. Only a few research works exist presently in the area of applying blockchain for UAV networks. In Ref. [48], researchers proposed blockchain and IoT in autonomous drone operations management. The results obtained showed that there is a reduction in signaling and operation time, improved maintenance operations successful rate, and building trust and security when handling drones in an autonomous system.
- Educational industry: Educational institutions are deploying blockchain to prevent fraudulent certifications and to ease out students and alumni record keeping. As a result of digitization, the education sector needs a robust framework to shift from a traditional paper-based system to a centralized/distributed database. Usage of blockchain would meet the desired requirement. MIT (Massachusetts Institute of Technology) of the United States is already applying for BCT. They built a platform, called Blockcerts Wallet [49], which allows the university to create, issue, view, and verify blockchain-based certificates.
- Real estate: BCT has the potential to complete digitize the real estate industry bringing transparency and trust among the parties. Blockchain along with IoT devices can improve the process of surveying the plot area. More research is required to efficiently apply these two technologies for upholding the real estate business.

- Voting: To allow decentralization of the e-voting method and its facilities, blockchain would play an important role. The technology would improve the reliability and effectiveness of the electronic voting system. South Korea piloted its part of testing associated with blockchain-based e-voting system. It was first applied in March 2018 by Gyeonggi-do Province [50]. Implementation of blockchain-integrated IoT system will ensure that voter's data are protected from altering and maintain the integrity of the voting system [50].

3.4 Open issues and research areas of blockchain in IoT

Every technology has inbuilt issues and challenges. These challenges become a problem for new research. The integration of blockchain and IoT does not solve every problem. Some of the open issues in integrating these two technologies and possible new research areas are as follows:

- Scalability and storage: Technical scalability is one of the hurdles and difficult to achieve with respect to blockchain especially in the public blockchain. As IoT network comprises a large number of devices, the introduction of new devices will put blockchain network to decreased efficiency. As per study, Bitcoin processes maximum of seven transactions per second [51]. The size of the block depends on the number of miners [52]. An increase in the number of miners will increase the size of the block. Thus, an increase in the block size will reduce the transaction speed.
- Security and privacy: The BCT is not fully secure. The vulnerability in blockchain emerges based upon the robustness of the consensus algorithm implementation. The security of blockchain also depends on the usage of software and hardware in the implementation [53]. A solution to find vulnerabilities in consensus algorithm is a new research area.
- Energy and cost efficiency: IoT devices are resource constraint devices. They have limited energy storage in them. In this regard, some of the computationally very expensive and energy-consuming consensus algorithms like PoW (Proof of Work) are not suitable for IoT system. Developing an energy proficient consensus procedure is a need of the hour. Researchers were successful in developing energy-efficient consensus algorithms such as PoS, DPoS, and Proof of Trust in recent times [54]. However, not all the transaction data are stored in the blockchain. IoT devices collect Giga Bytes of data in real time [55]. Thus, designing IoT system-oriented energy effective consensus mechanisms still remain as challenge.
- Resource constraints: Many devices in IoT network lack resources like storage and processing power. Hence, there is a need for a lightweight consensus algorithm designed especially for resource constraint devices with no deterioration in its efficiency.

- Regulations: Blockchain is a decentralized system and no single entity can control the system. No industrial standards and judiciary regulations were framed. A new blockchain framework has been formulated, but still there is a need for standard data block formats.

3.5   Conclusion

Blockchain is considered one of the trusted technologies. It provides a robust solution to major security and privacy issues faced by the IoT ecosystem. The chapter highlights major layer-wise security and privacy attacks in traditional IoT systems. Fundamentals of blockchain along with its key features associated with the Internet of Things cyber security aspects are argued to understand the basics of combining these two technologies. The features offered by the technology suit the current requirement to implement a robust IoT network. The application of blockchain-integrated IoT is vast. We have summarized the application in the various areas including healthcare, agriculture, the energy sector, and supply chain management. The healthcare industry uses it to track the drug supply and manage patient data securely. The supply chain logistics apply blockchain-integrated IoT to enable more transparent and accurate end-to-end tracking. BCT also plays an important role in improving modern agricultural process and managing smart energy grid. As blockchain is considered the new technology of the decade, it has not matured enough to tackle all the problems. Issues related with combining blockchain with IoT are discussed briefly. Every new technology evolves itself with proper research. Lastly, the latest research trends are included in the chapter to help learners to engage in continuous study of this technology.

References

[1] Knud Lasse Lueth IoT Analytics' latest reports State of the IoT Q4 2020 & 2021 outlook [online]. Available from https://iot-analytics.com/state-of-the-iot-2020-12-billion-iot-connections-surpassing-non-iot-for-the-first-time/ [Accessed 25 Jul 2021].

[2] Posey B., Rosencrance L., Shea S. Industrial Internet of Things [online]. Available from https://internetofthingsagenda.techtarget.com/definition/Industrial-Internet-of-Things-IIoT [Accessed 12 May 2019].

[3] Lin I.-C., Liao T.-C. 'A survey of blockchain security issues and challenges'. *International Journal of Network Security*. 2017, vol. 19(5), pp. 653–9.

[4] Nakamoto S. Bitcoin: A peer-to-peer electronic cash system [online]. 2008. Available from https://bitcoin.org/bitcoin.pdf [Accessed 10 May 2020].

[5] Wheelus C., Zhu X. 'IoT network security: threats, risks, and a data-driven defense framework'. *IoT*. 2020, vol. 1(2), pp. 259–85.

[6]   Burhan M., Rehman R.A., Khan B., Kim B.-S. 'IoT elements, layered archi-
       tectures and security issues: A comprehensive survey'. *Sensors*. 2018, vol.
       18(9), p. 2796.
[7]   Ali K., Askar S. 'Security issues and vulnerability of IoT devices'.
       International Journal of Science and Business. 2021, vol. 5(3), pp. 101–15.
[8]   Butun I., ÖsterbergP., Song H. 'Security of the Internet of things:
       Vulnerabilities, attacks, and countermeasures'. *IEEE Communications
       Surveys & Tutorials*. 2019, vol. 22(1), pp. 616–44.
[9]   Mishra D., Gunasekaran A., Childe S.J., Papadopoulos T., Dubey R., Wamba S.
       'Vision, applications and future challenges of Internet of things: A bibliomet-
       ric study of the recent literature'. *Industrial Management & Data Systems*.
       2016, vol. 116, pp. 1331–55.
[10]   Kozlov D., Veijalainen J., Ali Y. 'Security and privacy threats in IoT architec-
       tures'. *BODYNETS*; 2012. pp. 256–62.
[11]   Honar Pajooh H., Rashid M., Alam F., Demidenko S. 'Multi-layer blockchain-
       based security architecture for Internet of Things'. *Sensors*. 2021, vol. 21(3),
       p. 772.
[12]   Waheed N., He X., Ikram M., Usman M., Hashmi S.S., Usman M. 'Security
       and privacy in IoT using machine learning and blockchain: Threats and coun-
       termeasures'. *ACM Computing Surveys*. 2020, vol. 53(6), pp. 1–37.
[13]   Al-Rubaie M., Chang J.M. 'Privacy-preserving machine learning: Threats
       and solutions'. *IEEE Security & Privacy*. 2019, vol. 17(2), pp. 49–58.
[14]   Wu J., Tran N. 'Application of blockchain technology in sustainable energy
       systems: An overview'. *Sustainability*. 2018, vol. 10(9), p. 3067.
[15]   Merkle R.C. 'A digital signature based on a conventional encryption function'.
       Conference on the Theory and Application of Cryptographic Techniques;
       1987. pp. 369–78.
[16]   Salimitari M., Chatterjee M. 'A survey on consensus protocols in blockchain
       for IoT networks'. *arXiv:1809.05613*. 2018.
[17]   Algarni S., Eassa F., Almarhabi K., *et al.* 'Blockchain-based secured access
       control in an IoT system'. *Applied Sciences*. 2021, vol. 11(4), p. 1772.
[18]   Dorri A., Kanhere S.S., Jurdak R., Gauravaram P. 'Blockchain for IoT se-
       curity and privacy: The case study of a smart home'. *IEEE International
       Conference on Pervasive Computing and Communications Workshops
       (PerCom Workshops)*; 2017. pp. 618–23.
[19]   Grecuccio J., Giusto E., Fiori F., Rebaudengo M. 'Combining blockchain
       and IoT: Food-chain traceability and beyond'. *Energies*. 2020, vol. 13(15), p.
       3820.
[20]   Hammi M.T., Hammi B., Bellot P., Serrhrouchni A. 'Bubbles of trust: A de-
       centralized blockchain-based authentication system for IoT'. *Computers &
       Security*. 2018, vol. 78(10), pp. 126–42.
[21]   Jesus E.F., Chicarino V.R.L., de Albuquerque C.V.N., Rocha A.A. 'A sur-
       vey of how to use blockchain to secure Internet of things and the Stalker
       attack'. *Security and Communication Networks*. 2018, vol. 2018(7), pp.
       1–27.

[22] Atlam H.F., Azad M.A., Alzahrani A.G., Wills G. 'A review of blockchain in Internet of things and AI'. *Big Data and Cognitive Computing*. 2020, vol. 4(4), p. 28.

[23] Tawalbeh L.A., Muheidat F., Tawalbeh M., Quwaider M. 'IoT privacy and security: Challenges and solutions'. *Applied Sciences*. 2020, vol. 10(12), p. 4102.

[24] Kumar N.M., Mallick P.K. 'Blockchain technology for security issues and challenges in IoT'. *Procedia Computer Science*. 2018, vol. 132(7), pp. 1815–23.

[25] Esposito C., De Santis A., Tortora G., Chang H., Choo K.-K.R. 'Blockchain: A panacea for healthcare cloud-based data security and privacy?' *IEEE Cloud Computing*. 2018, vol. 5(1), pp. 31–7.

[26] Salahuddin M.A., Al-Fuqaha A., Guizani M., Shuaib K., Sallabi F. 'Softwarization of Internet of things infrastructure for secure and smart healthcare'. *arXiv:1805.11011*. 2018.

[27] Venkatesh V.G., Kang K., Wang B., Zhong R.Y., Zhang A. 'System architecture for blockchain based transparency of supply chain social sustainability'. *Robotics and Computer-Integrated Manufacturing*. 2020, vol. 63(4),p. 101896.

[28] Howson P. 'Building trust and equity in marine conservation and fisheries supply chain management with blockchain'. *Marine Policy*. 2020, vol. 115(1), p. 103873.

[29] Kayikci Y., Subramanian N., Dora M., Bhatia M.S. 'Food supply chain in the era of Industry 4.0: Blockchain technology implementation opportunities and impediments from the perspective of people, process, performance, and technology'. *Production Planning & Control*. 2020, vol. 1, pp. 1–21.

[30] Hastig G.M., Sodhi M.S. 'Blockchain for supply chain traceability: Business requirements and critical success factors'. *Production and Operations Management*. 2020, vol. 29(4), pp. 935–54.

[31] Chang Y., Iakovou E., Shi W. 'Blockchain in global supply chains and cross border trade: A critical synthesis of the state-of-the-art, challenges and opportunities'. *International Journal of Production Research*. 2020, vol. 58(7), pp. 2082–99.

[32] Solution Architecture [online]. Available from https://docs.tradelens.com/learn/solution_architecture/ [Accessed 25 Mar 2021].

[33] Rakic B., Levak T., Drev Z., Savic S., Veljkovic A. First purpose built protocol for supply chains based on blockchain. Tech Rep. OriginTrail, Ljubljana, Slovenia; 2017. p. 1.

[34] Hossein Motlagh N., Mohammadrezaei M., Hunt J., Zakeri B. 'Internet of things (IoT) and the energy sector'. *Energies*. 2020, vol. 13(2), p. 494.

[35] Alladi T., Chamola V., Rodrigues J.J.P.C., Kozlov S.A. 'Blockchain in smart grids: A review on different use cases'. *Sensors*. 2019, vol. 19(22), p. 4862.

[36] Khrais L.T. 'IoT and blockchain in the development of smart cities'. *International Journal of Advanced Computer Science and Applications*. 2020, vol. 11(2), pp. 153–9.

[37] Mohanty S.N., Ramya K.C., Rani S.S., *et al.* 'An efficient lightweight integrated blockchain (ELIB) model for IoT security and privacy'. *Future Generation Computer Systems.* 2020, vol. 102(10), pp. 1027–37.

[38] Pranto T.H., Noman A.A., Mahmud A., Haque A.B. 'Blockchain and smart contract for IoT enabled smart agriculture'. *PeerJ Computer Science.* 2021, vol. 7(15–16), p. e407.

[39] Badr S., Gomaa I., Abd-Elrahman E. 'Multi-tier Blockchain framework for IoT-EHRs systems'. *Procedia Computer Science.* 2018, vol. 141, pp. 159–66.

[40] Lin J., Shen Z., Zhang A., Chai Y. 'Blockchain and IoT based food traceability system'. *ICCSE'18: Proceedings of the 3rd International Conference on Crowd Science and Engineering*; 2018. pp. 1–6.

[41] Patil A.S., Tama B.A., Park Y., Rhee K.H. 'A framework for blockchain based secure smart green house farming'. *Advances in Computer Science and Ubiquitous Computing.* Singapore: Springer; 2017. pp. 1162–7.

[42] Cadiz J.V., Mariscal N.A., Ceniza-Canillo A.M. 'An empirical analysis of using blockchain technology in E-voting systems'. *1st International Conference in Information and Computing Research (iCORE)*; 2021. pp. 78–83.

[43] Basha M.M., Nikitha A.P., Gupta C.N., Subramanya K.N. 'Blockchain and IoT for enhancing supply chain security–A review'. *International Journal of Innovative Science and Research Technology.* 2020, vol. 5(9), pp. 545–51.

[44] Srivastava G., Crichigno J., Dhar S. 'A light and secure healthcare blockchain for IoT medical devices'. *IEEE Canadian Conference of Electrical and Computer Engineering (CCECE)*; IEEE; 2019. pp. 1–5.

[45] Lee Y., Rathore S., Park J.H., Park J.H. 'A blockchain-based smart home gateway architecture for preventing data forgery'. *Human-centric Computing and Information Sciences.* 2020, vol. 10(1), pp. 1–4.

[46] Lombardi F., Aniello L., De Angelis S., Margheri A., Sassone V. 'A blockchain-based infrastructure for reliable and cost-effective IoT-aided smart grids'. 2018, pp. 42–6.

[47] Ouaddah A., Elkalam A.A., Ouahman A.A. 'Towards a novel privacy-preserving access control model based on blockchain technology in IoT'. *Advances in information and communication technologies.* 2017, pp. 523–33.

[48] Dawaliby S., Aberkane A., Bradai A. 'Blockchain-based IoT platform for autonomous drone operations management'. *Proceedings of the 2nd ACM MobiCom Workshop on Drone Assisted Wireless Communications for 5G and Beyond*; 2020. pp. 31–6.

[49] MIT, USA. *About Blockcerts [online].* Available from https://www.blockcerts.org/about.html [Accessed 21 Nov 2021].

[50] Othman A.A., Muhammed E.A., Mujahid H.K., Muhammed H.A., Mosleh M.A. 'Online voting system based on IoT and ethereum blockchain'. *International Conference of Technology, Science and Administration (ICTSA)*; 2021. pp. 1–6.

[51] Bitcoin and Ethereum vs Visa and PayPal Transactions Per Second [online]. Available from http://www.altcointoday.com/bitcoin-ethereum-vs-visa-paypal-transactions-per-second/ [Accessed 21 Jun 2019].

[52] Wu H.-T., Tsai C.-W. 'Toward blockchains for health-care systems: Applying the bilinear pairing technology to ensure privacy protection and accuracy in data sharing'. *IEEE Consumer Electronics Magazine*. 2018, vol. 7(4), pp. 65–71.

[53] ElMamy S.B., Mrabet H., Gharbi H., Jemai A., Trentesaux D. 'A survey on the usage of blockchain technology for cyber-threats in the context of industry 4.0'. *Sustainability*. 2020, vol. 12(21), p. 9179.

[54] Alladi T., Chamola V., Parizi R.M., Choo K.-K.R. 'Blockchain applications for industry 4.0 and industrial IoT: A review'. *IEEE Access*. 2019, vol. 7, pp. 176935–51.

[55] Reyna A., Martín C., Chen J., Soler E., Díaz M. 'On blockchain and its integration with IoT. Challenges and opportunities'. *Future Generation Computer Systems*. 2018, vol. 88(3), pp. 173–90.

Chapter 4

Secured energy-efficient routing protocol with game-based fuzzy Q-learning approach for smart grid systems

*P Tamil Selvi[1], Tadele Degefa Geleto[2],
Suresh Gnana Dhas C[2], Karthikeyan Kaliyaperumal[3],
C Sivapragash[4], and N Kousik[5]*

In recent years, enormous growth has taken place in the field of electric power systems. Smart grids (SGs) were introduced to address the issues in traditional power grids that are authentic, safe, and secure. The SG collects sensor nodes from various networks that use various communication technologies for safe communications. With wireless sensor network, the SG permits bidirectional energy flow between the service distributors and users. Direct communications take place between the sensor nodes, whereas they are in the communication range. The sensor nodes beyond the communication limit in the networks need the support of the routing protocols. The protocols that are used for the routing process reduce the path overhead in the network topology. However, it causes larger power consumption. Therefore, routing and power control become significant issues in a grid structure. Hence in this research, we proposed an improved hybrid zone-based routing protocol (IZCG) for neighborhood area network (NAN) in SG communications. The proposed scheme consists of different techniques like managing network topology, controlling dynamic congestion, isolating selfish nodes, and utilizing the multiple channels in the network based on the game theory-based routing protocol. Moreover, a game-based fuzzy quality learning approach is implemented to prevent DoS attacks, which would increase the accuracy rate in the prevention of attacks and decrease the rate of false alarm that identifies the various denial of service attacks. Finally, the performance evaluation

[1]Michael Job College of Arts and Science for Women Coimbatore, India
[2]Ambo University, Hachal Hundesa Campus, Oromia, Ethiopia
[3]School of Informatics & Electrical Engineering, Ambo University, Ethiopia
[4]Swarnandhra College of Engineering and Technology, Narsapur, Andhra Pradesh, India
[5]School of Computing Science and Engineering, Galgotias University, Uttarpradesh, India

of the proposed IZCG demonstrates that it achieves better performance than other existing protocols for the routing process.

4.1 Introduction

A smart grid (SG) is a power grid that consists of several active energy sources with smart meters, smart devices, and renewable energy sources in which the data are being collected in real time. The key features of SG networks are integration, reliability, scalability and security.

In Figure 4.1, the architecture of SG consists of power system, power flow, and information flow. The power system delivers the electricity from producers to consumers. There are four subsystems in power flow. They are generations, transmission, distribution, and consumer. The power flows through the power system and the information flow through the networks in terms of block chain technology, which forms the grid to decentralize its operations [1].

The communication in SG consists of home area network (HAN) layer, neighborhood area network (NAN) layer, and wide area network (WAN) layer. The top layer is HAN. It controls and manages the consumer's on-demand power needs for smart and electronic devices, appliances at home, electric vehicles, and renewable energy sources [2]. It consists of a service unit and metering unit (MU). The details

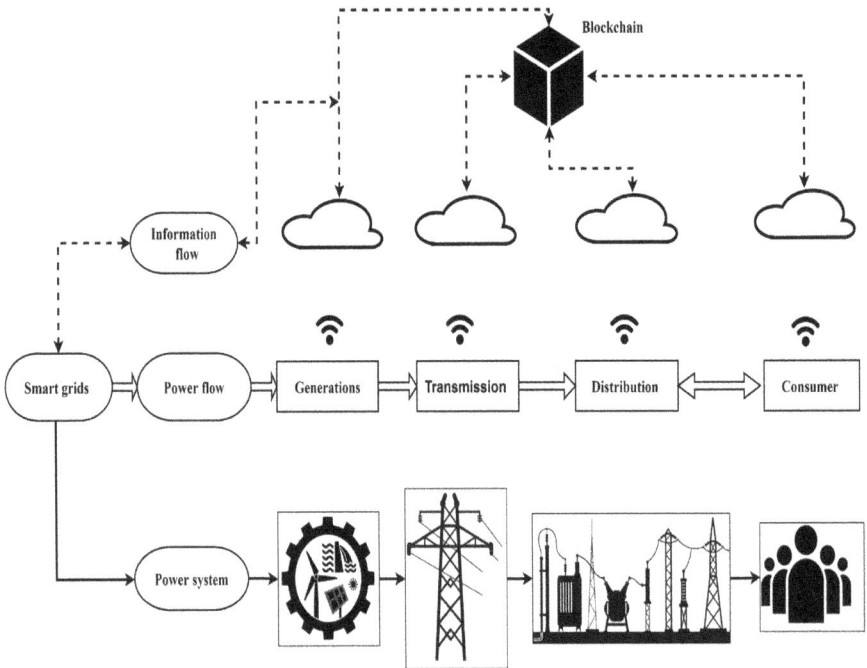

Figure 4.1   Architecture of smart grid

of energy expenditure and energy consumption were generated based on the usage details of in-home meter. It covers an area of 1–100 m.

The second layer of an SG is neighborhood area networks (NAN). It is called field area networks. It is a collection of multiple HANs, centralized control unit (CCU), smart meter recording unit, and NAN ids. The CCU supports communication between energy dealers and various HAN's. The data collectors collect the information about its services and meter reading from multiple HANs and transmit it into NANs to WAN. It covers an area of 100 m to 10 km.

The final layer of SG is the WAN layer, the backbone for SG communications and connecting the grid and the system's utilities. It consists of distribution systems, WAN ids, and controllers. It accelerates the communication among the various systems like power transmitter, bulk generator, and renewable energy resource and power controller. Additionally, the SG had video surveillance systems to safeguard the assets and fire alarms for safety purposes. It covers an area of 10–100 km [3, 4].

The remaining of the chapter is organized as follows. Section 4.2 describes routing protocols for SG communications. Section 4.3 gives a detailed introduction and implementation of a novel routing algorithm. Section 4.4 deals with the development of efficient dynamic congestion detection and control (EDCDC)-based routing protocol. Section 4.5 describes how the record- and trust-based detection techniques (RTBD) detect the selfish nodes in the networks. Section 4.6 explains the development of cognitive radio multichannel message authentication code (MAC) protocol for second channel utilization (SCU). Section 4.7 describes the game-based fuzzy Q-learning (G-FQL) approach. Section 4.8 narrates about overall performance analysis of the proposed approaches with existing protocol in terms of different metrics. The conclusion and further enhancement are drawn in Section 4.9.

4.2 Routing protocols in SGs

Routing is a process of exchanging information between one sensor node into another sensor node by using the most efficient path [5]. The SG includes smart meters with MU, various home appliances and others. Utility and MU routing can be done to facilitate efficient data exchange through various routing protocols. The routing categorized depends upon the structure of the network. The flat routing is used to allow every node with identical tasks. In hierarchical routing, all the nodes were allotted different roles that describe the system's structure. In location-based routing, the efficient routes are found depending upon the sensor node's locality in the networks [6].

It is further categorized into proactive, reactive, and hybrid routing protocols. The proactive-based routing protocol is used to periodically maintain the paths and destinations that share the routing information. The main issue is restructuring the network topology due to node failures. It also not suited for large size of networks, which require node information for maintenance of each node in the routing table and cost is also high due to high channel usage overhead.

The reactive protocol floods route request messages throughout the networks and find its route path. The maximal time of latency and clogging of the network are the major issues in reactive protocols [7, 8]. In hybrid-based protocols, the merits of proactive and reactive protocols are integrated. At first, proactive paths perform the routing, and the reactive flooding technique serves the demands from excess activated nodes. Examples of hybrid-based protocols are zone-based routing protocol and zone-based hierarchical link-state routing protocol [9, 10]. But still it has disadvantages like low scalability, large area memory and high overlapping in zones. This leads to idea for finding the most suitable routing protocols for SG.

The SG networks are usually composed of multiple nodes, it is very essential to find the routing protocol, which ensures better performance and guarantee of quality of service (QoS). The fundamental research issues deal with the routing protocols in SG communications, which consists of series of activities like discovery of optimum routes and maintenance of routes between the nodes in grid. It also selects the optimum path from the available multiple paths based on energy efficiency of the node in SG. The changes in network structure and less QoS in terms of bandwidth, packet loss, etc. were not controlled due to the movability of the node.

The main intention of this work is as follows: (a) the network structures are maintained by estimating the node die-out rate based on the energy dropout of sensor nodes, (b) to achieve the reliable communication in SG by detecting the dynamic congestion (DC) and controlling the routing protocol, (c) to enhance the fairness among the nodes by preventing the zone nodes from the selfish sensor nodes, (d) the SG's performance was raised by utilizing the second channel for reducing the delay, bandwidth congestion and increasing the throughput of the network, (e) to develop a game theory approach-based routing mechanism for discovering the reliable nodes and paths which improves the throughput, packet delivery ratio (PDR), etc., in SG.

The remaining of chapters are organized as follows: Section 4.3 summarizes the review of related works in this field. Section 4.4 gives a detailed introduction and implementation of a novel routing algorithm. Section 4.5 deals with the development of efficient dynamic congestion detection (EDCDC) and control-based routing protocol. Section 4.6 describes how the RTBD detect selfish nodes in the networks. Section 4.7 explains the development of cognitive radio multichannel MAC protocol for SCU. Section 4.8 describes the G-FQL approach. Section 4.9 narrates about overall performance analysis of proposed approaches with existing protocol in terms of different metrics.

4.3   Review of related works

A novel zone-based energy-efficient routing protocol was introduced by J. R. Srivastava and Sudarshan, which utilizes both movable and stationary nodes. It does not require any additional mechanisms such as detection of route paths and security in path [5]. The computation overhead was minimized by the dynamic forwarding concepts. By utilizing this protocol, the energy of the nodes was measured. But, here the packets are forwarded by zone heads only which leads to high computation time.

A new technique called as passive clustering strategy was proposed by Ramalingam Vinodha and Sundaraj, which boost the efficiency of the network in SG communications. The chicken swarm optimization mechanism is used to obtain the optimum paths in the cluster which also reduces the efficiency of the sensor nodes and increases the life span of sensor nodes in SG communications [11].

The combined optimized technique called as DHOCSA method, which is the combination of deer hunting optimization technique and crow search algorithm, was introduced by Priya and Josephkutty Jacob. It controls the power and resources distribution in SG [12]. S.J. Kampli, D. Ramesh and K.B. Shivakumar proposed the protocol for grid WSN, which uses Markov model. It uses Reed Solomon encoding techniques to control the link failure and errors in the networks. It also maximizes the reliability of the nodes by dynamic chain formation techniques, which leads to the improvement of QoS in WSN.

Secure zone routing protocol was presented by B. R. Devi *et al.*, which integrates the concepts of proactive and reactive mechanisms. It integrates the mechanisms of cryptography and digital signatures. It improves the integrity, confidentiality and end-to-end verification at IP layer [8]. But still, the path acquisition latency was high due to generation of digital signatures and authentication process.

An improved zone-based routing protocol was proposed by R.S. Malwe *et al.*, which is based on selective border cast and also depends upon the location of the nodes in the networks. Instead of normal border cast technique, selective border cast technique was used to find out the optimum path, whereas path request was transmitted only by peripheral nodes. To reduce the storage complexity of the routing table, the table entries within the zone were neglected. It also reduces the control overhead of the nodes in the networks. Here, end-to-end delay and energy consumption were not considered.

To improve the QoS in video traffic, zone-based routing protocol was introduced by S.A. Selvi and A. Vijayaraj. Instead of OLSR protocol in Video Load Balancing Solution (ViLBaS) principle, the zone routing protocol was utilized based on hybrid concepts. If both sender and receiver were located within the zone, the proactive technique is used to transmit the packets else reactive technique is used to forward the packets and QoS was measured [13]. However, the security issues were not considered while transmitting the packets in the networks.

S.S. Rajput and M.C. Trivedi proposed security enhanced zone routing protocol, which utilizes the verification mechanism for securing the packet routing in Zone based Routing Protocol (ZRP) and MAC mechanism to maintain the privacy of the messages in the networks. It also prevents the commonly occurred attacks in the networks. The overhead due to dissemination and sharing the keys was minimized by key pre-distribution mechanism in it [14]. However, the energy consumption was maximized due to the security issues.

To optimize the clustering and cluster head (CH) selection, an optimized zone-based energy efficient routing protocol based on genetic fuzzy system was described by J.R. Srivastava and T.S.B. Sudarshan. The fuzzy inference system selects the CH based on factors like energy, distance, density, and mobility. By using the genetic algorithm, the optimal CH was selected, which maximizes the life time of the nodes in the networks [5]. But still, the latency and network-wide transmissions were not minimized.

N. Sharma and G. Patidar introduced an enhanced congestion control techniques by using modified hybrid-TCP (H-TCP) in Mobile Ad-hoc Netwoks (MANET), which avoids the congestion and packet loss in wireless networks. It is the variant of H-TCP and TCP Reno which estimates retransmission time accurately [15]. However, it does not consider the mobility or interferences in the networks.

P. Keerthana and S. Chandra Mohan proposed a new joint scheduling and congestion control algorithm for multihop wireless networks with dynamic route flows. It also integrates window-based flow control of new rate based on distributed scheduling algorithm to achieve the quality of the system [16]. It randomly selects the set of routes based on the traffic load and achieves better throughput, end-to-end delay, and PDR.

S.V. Sangoli and J. Thyagarajan described an efficient congestion control scheme using cross-layered mechanisms, which optimizes the transport and network layers by collaborating with physical layer to procure information such as signal power and battery level. It is alternatives to AODV/DSR protocols. The mathematical model on window-based exponential averaging of received signal at the transport layer resolves the congestion and makes an appropriate decision to address the node for disconnection problem [17]. It also achieves end-to-end guarantee at the transport layer and controls the flooding overhead.

The game theory approach-based detection of selfish node in MANET was proposed by D. Das *et al.* It uses least total cost factor mechanism that transmits the data packets from source to destination through the least cost path (LCP). While transmitting the packets, the path will be broken due to the presence of selfish nodes in the networks [18]. If so, the next best path will be selected for data packet transmission. So, it guarantees the data transfer with low cost and less idle time. Thus, it effectively detects the selfish nodes. However, the communication cost was high.

X. Cao *et al.* [19] proposed a two-stage selfish misbehavior detector for 802.11-based ad hoc networks. It checks whether the measured performance matches the model-based expectations. In first stage, it utilizes neighbor-broadcasted information to establish the relationship between channel availability and transmission rate. It also checks whether the relationship matched with theoretical model or not. In second stage, it checks whether the model-based throughput meets the observations or not [19]. However, still proposed approach cannot achieve better performance.

An evaluation of large-scale IEEE 802.11b wireless network was presented by A.P. Jardosh *et al.*, which investigates about the factors like congestion level in the networks, throughput, busy time of the channels for transmission of packets, Ready-to-send/clear-to-send (RTS–CTS) mechanism, frame transmission, and acceptance in delay. Initially, how channel was utilized for identifying different states of congestion in the wireless medium was explained. Moreover, utilization of channel utilization was based on the behavior of MAC layer [20]. However, it requires further improvement.

Game theory-based routing algorithm for multihop wireless networks was described by Z. Ren *et al.* In this approach, the fundamental concepts of game theory and its routing algorithms were investigated for multihop wireless networks [21]. Moreover, their characteristics were studied and compared with various routing algorithms. However, overhead and complexity of the networks model were high.

An obstacle avoidance routing protocol based on game theoretical approach was introduced by X. Guan *et al.* for solving the obstacle problems in wireless sensor networks. By using this algorithm, concave regions were formed, which improves the success rate of packet transmission and reduces the delays during packet transmissions. Moreover, the idle energy, out-degree for forwarding angle were determined for the forwarding probability and payoff function of forwarding candidates. Furthermore, a game theoretical approach was introduced based on forwarding probability of forwarding candidates, which is proved in Nash equilibrium [22]. However, in this approach, all nodes were assumed as stationary. Therefore, the effect of node's mobility was required to consider for verifying the effectiveness of this approach.

A secured routing mechanism based on game theoretical approach was proposed by D. Das *et al.* It is based on the packet forward rate and route density factor. It also detects the selfish nodes and avoids the malicious nodes in the networks during packet transmission. If any nodes out of the radio range in selected path during the transmission of data packets, an alternate path will be formed from the position without stopping the data transmission. It guarantees secured uninterrupted data transmission from source to destination [18]. However, this routing protocol does not used for removing the malicious nodes from the networks.

A novel routing scheme known as GTR was introduced by Y. Qin *et al.*, which is based on the game theoretical approach in opportunistic networks. In this approach, the message transmission was mapped with model of multiplayer bargain. This approach saves the network resources by constructing the proper utility function and achieves the better performance in dense opportunistic networks [23]. However, the average delay of this approach was high and PDR was less.

4.3.1   Framework of secured energy-efficient routing protocol for SG communication

The main aim is to obtain the most suitable protocol for routing in SG communications, which addressed the limitations in existing protocols. The framework of proposed protocol is illustrated in Figure 4.2. Therefore, an enhanced energy-efficient zone routing protocol (IZCG) is proposed for SG's communications, which controls the frequent changes in structure of the networks based on node die-out rate calculations. The estimation of die-out rate is used for estimating how long the node can remain active and continue its service. The reliable nodes and paths are discovered based on the game theoretic approach mechanism.

In second phase, an efficient dynamic congestion detection and control (EDCDC) mechanism proposed for detecting and controlling the DC level in the routing paths in the networks. It is based on estimation of average queue length, which detects the current congestion level and its changes. It sends a warning message to the neighbors based on the congestion level in the networks. When it receives a warning message, the sensor nodes in the grid find the alternative congestion-free path to the destination.

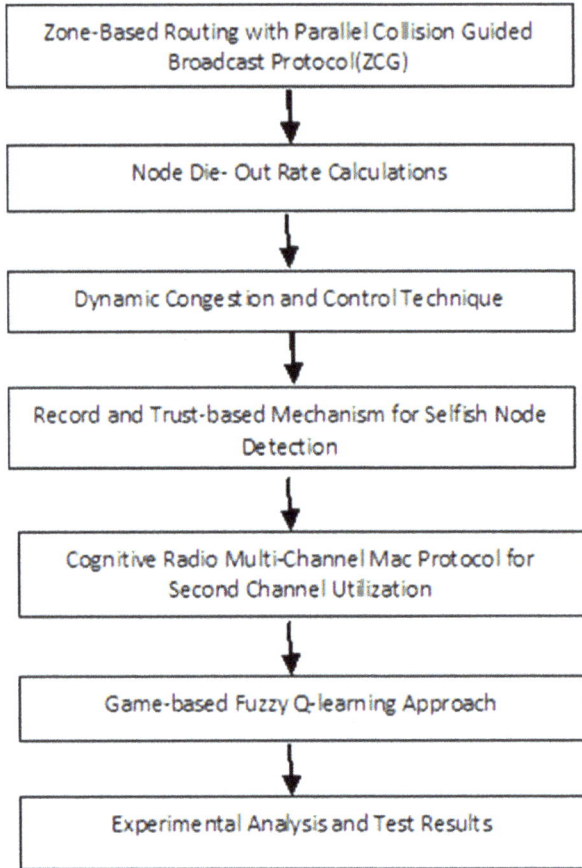

Figure 4.2   Framework of proposed secured energy efficient routing protocol for smart grid communications

In third phase, selfish node detection (SND) mechanism is proposed based on the RTBD approaches. It is used to detect the selfish nodes and protect the other sensor nodes in the networks. When the selfish nodes were detected, then the other nodes in the networks avoid the packet transmission between them and improve the packet transmission rate without the intervention of selfish nodes in the networks.

In last phase, cognitive radio multichannel MAC protocol is proposed for utilizing the SCU, which improves the fairness of the grid. The game theory approach-based Fuzzy Q-learning approach is used that maximizes the security levels and minimizes routing cost efficiently. Finally, the results of stimulation process demonstrates that the proposed IZCG protocol compared with ZCG protocol and achieves the better performance in terms of throughput, ratio of packet delivery, drop rate of packet, delay, and control overhead.

4.4 An improved energy-efficient zone-based routing protocol (IZCG) for SG communications

An energy-efficient zone-based routing protocol with parallel collision guided broadcasting mechanism (ZCG) is used [24]. The network topology was spitted into different zones. The routing protocols handled the packet transmissions between the zones. The packets were delivered to the destination with minimal path overhead. The significant issues in ZRP are reliability, requirements of large memory area and extensive overlapping in zones [12]. Therefore, an improved energy-efficient zone-based routing protocol (IZCG) is introduced for SG communications.

Initially, the networks are constructed and split into many zones as per requirements. The nodes with the attributes like maximum power of the battery, connectivity degree, and less mobility are assumed as zone header (ZH). Initially, a message with Hello is received by an idle node, and it sets a fixed value to the timer. The node with a maximum degree of connectivity becomes Zone Construction Organiser (ZCO), which broadcasts the calls to sensor nodes. Then, it calculates direct neighborhood active links, which should not be the member of the neighbor zone and regulates its countdown timer speed. The zone construction end time was determined based on the sensor nodes' time to live [25].

When node receives the call, it becomes ZCO and cancels its countdown timer. It changes the state to participant called zone construction participant (ZCP), and its timer reaches to end time of zone construction. The nearby nodes also receive the call for zone construction. Every member of the zones changes its status within the allocated time, and the number of hops between the two sensor nodes was calculated based on the number of iterations.

The weighted clustering algorithm calculates the fitness factor of the sensor nodes. ZCP sorted the received fitness factors. The nodes that had high fitness factor remain at the top of a sorted list. The topmost node is treated as a CH. The ZCP identifies the CH. Then ZCP constructs the zone with one hop length and becomes a member who puts its internet protocols address in Hello headers.

After the transmission of the node, the energy of the node reduces. The energy meter is fixed to calculate the energy of the node in it. Based on the energy level, the CH is selected. The transmission of packets takes place between it, and the power of each active node is decreased. The node with the maximum energy level is elected as CH for the next iteration onward. The transmission takes place between it [13].

Consider there are T clusters such as $C = \{C_1, \ldots, C_T\}$ and $C_T = \{N_{i1}, \ldots, N_{im}\}$, which refer the set of clusters with m nodes where for each $N_{ij} \in C_i$. The fitness factor fitness function of the node (Fij) based on the congestion level for each node N_{ij} is calculated during the zone construction time based on corresponding neighbors N_{im} in the cluster C_i. The distance d is calculated based on the following equation:

$$d = \frac{c \left(t_{remote} - t_{local}\right)}{2}, \tag{4.1}$$

In (4.1), c refers the speed of light $c \approx 3 \times 10^8$ m/s, t_{remote} is the time gap exists between the transmission of sending a Hello packet and the receiving the acknowledgement, and t_{local} is the time gap between the receiving one Hello packet and transmitting an acknowledgement. The sum of distances to all its neighbors N_{il} is computed as,

$$k_{ij} = \sum_{i=1, i\neq j}^{m} d_{avg}\left(N_{ij}, N_{il}\right),$$
(4.2)

where,

$$d_{avg}\left(N_{ij}, N_{il}\right) = \frac{\sum_{\Delta=1}^{n} d_{\Delta}\left(N_{ij}, N_{il}\right)}{n},$$
(4.3)

In (4.3), d_{Δ} refers the distance d mentioned in (4.1). The fitness function is used to differentiate the reliable ZH which are at the top of the routing table will be stable for the longest time duration. In addition, the link stability is also estimated based on link expiration time, which denotes the stable link connections between two sensor nodes in the grid [26].

The node movement is measured as follows:

$$\lambda_{ij}^{l}(N_{ij}, N_{il}^{t1}, N_{il}^{tn}) = \frac{d_{\Delta}(N_{ij}, N_{il}^{t1})^{\frac{1}{n}}}{d_{\Delta}(N_{ij}, N_{il}^{tn})} - 1$$
(4.4)

In (4.4), N_{il}^{t1} is N_{ij} node position at time t_1 and $\frac{1}{n}$ denotes total number of Hello message received in specific time. For every node N_{ij}, there is a series of m values of λ_{ij}^{l} such that $\lambda_{ij} = \{\lambda_{ij}^{1}, \ldots, \lambda_{ij}^{m-1}\}$. Then, the degree difference of the node N_{ij} mobility relates to all neighbors sensor node in the cluster Ct where, $\bar{\lambda} = \frac{\sum_{l=1}^{m} \lambda_{ij}^{c}}{m}$. The mobility of a node N_{ij} can be found by standard variation α_{ij} of the entire set of sensor node with mobility values λ_{ij}^{l} as follows:

$$\alpha_{ij} = \sqrt{\frac{1}{m-1} \sum_{l=1}^{m} \left(\lambda_{ij} - \bar{\lambda}\right)^{2}}.$$
(4.5)

ZCP uses the weighted mean formula for calculating fitness function by using local data, where β_{ij} finds out power of battery consumed.

$$F_{ij} = \left(\omega_1 \cdot \alpha_{ij}\right) + \left(\omega_2 \cdot k_{ij}\right) + \left(\omega_3 \cdot \beta_{ij}\right),$$
(4.6)

here ω_1, ω_2, and ω_3 are weights such that $\sum_{i=1}^{3} \omega_i = 1$ and α_{ij} is the mobility of sensor nodes and direction of movement. In this approach, high weights are given to α_{ij} and k_{ij} then to β_{ij} for maximizing the connectivity between ZH and other sensor nodes in the grid [8]. Therefore, it maximizes the lifetime of the zones and ZH communicates an essential updates to the member nodes in the networks.

When zone construction time ends, the ZCP sorts the received fitness function. The nodes that had high fitness factor remain at the top of a sorted list. The topmost node is treated as a CH. The ZCP identifies the CH. Then ZCP constructs the zone with one hop length and becomes a member who puts its internet protocols address in Hello headers.

The energy of the node reduces while transmitting the nodes. So the energy meter is fixed to calculate the energy of the node in it. The CH is selected depends upon the energy level. The transmission of packets takes place between it, and the power of each active node is decreased. The node with maximum energy level is elected as CH for the next iteration onward. The transmission takes place between it [26].

4.4.1   Estimation of node die-out rate

During the ZH selection process, the network group of live sensor nodes detected each zone. ZH is selected from the sensor nodes with maximum energy and less mobility through fitness function in the entire iteration. Each sensor node maintains the energy meter, which calculates the residuum energy of the sensor node after each transmission of packets. The power or energy level of the sensor node would be declined after every transmission of packets.

Randomly, ZH is elected in the initial state due to an equivalent power level is assumed. ZH transmits an aggregated packet for each live node in the zone in the first iteration and calculates residuum power level [27]. For the next iteration onward, the node with the most significant energy level is elected as ZH. The power level declined after every transmission was measured. The energy drop-out rate for each sensor node maintains the network structure based on the node die-out ratio level.

4.5   An energy-efficient DC and control technique

The die-out rate of each node is computed, which reduces the path overhead. However, the reliable communication is affected by DC. The major aim of congestion control is to utilize the network resources and maintains the load level below the node's capability. Therefore, the level of congestion in the network is detected dynamically by using a new technique was proposed which discovers the non-congested path in the networks. It consists of four processes. They estimate DC, construction of congestion-free set (CFS), congestion-free path discovery, and congestion-free alternate path discovery.

The congestion level at each node in network is detected by mean value of queue size. Every time, the possession of nodes in the link-layer queue is checked by the DC technique, an algorithm for managing queue [28]. It utilizes the three parameters such as $(Min)_{th}$, $(Max)_{th}$, and w_q for standardizing its performance. Here, $(Min)_{th}$ and $(Max)_{th}$ are the queue thresholds for a present position in the queue. From the queue's current length, the queue's mean size is computed, denoted by w_q.

Therefore, the DC performance depends on the values of the thresholds. If the values of the threshold are low, then utilization of the link is also reduced. If the values of thresholds are very high, then high traffic occurs in the network. Here, $(Min)_{th}$ and $(Max)_{th}$ values are selected based on the following conditions:

$$Min_{th} = 35\% \, Queue\_size, \tag{4.7}$$

$$Max_{th} = 2 * Min_{th}. \tag{4.8}$$

Then, the average queue length is calculated as follows:

$$Avg_{Que_{new}} = \left(1 - w_q\right) * Avg_{Que_{old}} + Inst_{Que} * w_q. \tag{4.9}$$

If the value of mean queue distance is less than the threshold value Min_{th} and the present queue length is also less than half of the total size of the queue, then the sensor node is in a safe zone. If the value of mean queue distance is larger than the value of the threshold Min_{th} and less than the value of the threshold Max_{th}, then it is in a state of congestion. Therefore it initiates other substitute path discovery processes. At that same time, if the current length of the queue is higher than the value of the threshold Max_{th} due to heavy traffic, the status of alternative discovery of path also becomes a false state. Hence, the value of Max_{th} changes until an alternative path becomes true by using the following equation.

The network congestion is regulated by the weight parameter $(w)_q$, which denotes constant time for low-pass filter. The recurrent network congestion, which occurs over long time duration, is identified by the mean queue length. If w_q value is too high, then the instantaneous queue length is followed by the average length of the queue, which minimizes the congestion level. Therefore, the w_q denotes the traffic flow in the queue. The w_q values are assigned dynamically by using the following equation where N is the number of flows that is in active state and P denotes the packets transmitted per second.

$$w_{q\_new} = w_{q\_old} * N * P, \tag{4.10}$$

$$Queue\_utilization = \left(Max_{th} + Min_{th}\right)/2, \tag{4.11}$$

If the distance of the mean queue is greater than the value of the threshold Max_{th}, then it becomes a congested zone. By selecting the non-congested one-hop neighbors' sensor node in the network, CFS is constructed. The CFS(S) refers to the source host S, a one-hop neighbor of non-congested sensor node S. It should satisfy the condition that each sensor node within the two-hop neighborhood distance of senor node S has links toward CFS(S). It should not be in a zone of congestion state. CFS is constructed during the initial process, where each mobile host estimates the congestion status based on the DC algorithm. Then, the congestion status of each host is broadcasted to its one-hop neighbors of sensor nodes [17].

Each mobile host identifies non-congested neighborhood sensor nodes within a one-hop distance, and its information is added to the non-congested one-hop list. After that, each mobile host also identifies non-congested neighborhood sensor nodes within a two-hop distance and its information is inserted in the non-congested two-hop list. The information of each mobile host is updated in the routing table and its format is as follows:

$$< src_{addr}, \; dst_{addr}, \; hop_{cnt}, \; CFS_{Node}, \; CFS_{SET}, \; congest_{status} >, \tag{4.12}$$

here src_{addr} refers address of the source, dst_{addr} refers address of the destination, hop_{cnt} denotes the number of hop, CFS_{Node} is the address of the non-congested node in the grid, CFS_{SET} is the set of non-congested neighbors and $congest_{status}$ is the congestion status of the sensor nodes. Based on the routing table updates, each mobile host shares details of a two-hop non-congested neighborhood to non-congested neighbor sensor nodes within a one-hop distance.

The source host generates RREQ and transmits the data packet to the destination through the CFS nodes. To accomplish this, the source finds out its non-congested node in the two-hop list and transmits the packet through the path specified in the table. If two-hop non-congested is unavailable in the list, then it broadcasts RREQ to the CFS. The node that first receives the RREQ would respond to the destination, and the new entry is added to the routing table. Thus, the new path becomes the primary path between source and destination [17, 21].

The new CFS node is identified by the ancestor CFS node from its list of neighbors in the routing table. The new CFS is constructed for a destination, and it is exchanged with its neighbor sensor node in the network. While receiving the new CFS, it compares the details in the routing table. Suppose the detail does not match the entry in the routing table. In that case, the new entry is created in the routing table, which is utilized to discover the new path to the destination in the network [29].

4.5.1   Algorithm for DC estimation

1. Initialise $Avg\_Que_{new} = 0$; $Avg\_Que_{old} = 0$; $Inst\_Que = 0$;
2. Set threshold values $Min_{th} = 35\%$ of the size of the queue; $Max_{th} = 2 *$ value of Min_{th};
3. Set utilisation of $queue[i] = \{.9, .925, .95\}$;
4. Initialize $warn\_line = Queue\_size/2$; $w_q = 0.002$;
5. For every arriving packet in queue
6. $Inst\_Queue + 1$
7. If($Que \neq 0$)
8. Compute average queue length using (4.3)
9. If($Avg\_Que_{new} < Min_{th}$ && $Inst\_Que < warn\_line$)then
10. Set $Queue_{Status} = Not - congested$(Safe Zone);
11. Else
12. Set $Queue_{Status} = $ Most Likely to be congested (Non $-$ safe);
13. Initiate an alternative path
14. If ($Inst_Queue > $ value of Max_{th} && $alter\_path = False$)
15. Set value of $Max_{th} = Queue\_utilization[i + 1] * buffer\_size$;
16. Else
17. Set $Queue_Status = $ Congested(nonsafe zone);
18. Set $Avg\_Que_{old} = Avg$
19. $w_{q\_new} = w_{q\_old} * P * N$
20. Stop
21. Stop
22. For each transmitting the packet in instant queue
23. $Inst_{Que}-$

4.5.2   Algorithm for path discovery process during packet transmission

1. For all mobile hosts construct CFS
2. For every node pair of $(Src, Dst)_i, i = 1$ to $(N - 1)$
3. Hop_Count = 0; $Path_i$ = NULL;
4. If(Dst_i is present in the two $-$ hop list of Src_i in the routing tabl
5. Generate the path for pair$(Src, Dst)_i$
6. Set $Path_i$ = True; Hop_Count = 2;
7. Else
8. CFS = Src_i;
9. Call$_{Procedure}$PATH with (input: CFS, Dst_i; output: $Path_i$)
10. End
11. Procedure PATH with (input: CFS, Dst_i; output: $Path_i$)
12. If(value of Dst is in CFS)
13. Generate the path for$(Src, Dst)_i$
14. Set $Path_i$ = True; Hop_Count $+ +$;
15. Return
16. Else
17. If$\big(($Unavailable CFS_SET in $Path_i)$AND$($Unavailable two $-$ hop CFS'_{SET}s in the list to $Dst_i)\big)$
18. Hop_Count $+ 1$;
19. Include $the\ value\ of$ CFS_SET to value of $Path_i$;
20. For every neighbour node$Nei_b$of node CFS_SET do
21. PATH $(Nei_b, Dst_i, Path_i)$
22. End
23. End

4.5.3   Algorithm for predecessor CFS node receive CSP

1. For all valid entries, Input P = (congestion_status, src_address, dst_address)
2. Initialise the values
3. Form new CFS from current CFS node to dst_node;
4. Call New Path Discovery Process;
5. Update new CFS by inserting all two_hop neighbours CFS node details;
6. Assign Path = True;
7. End

Therefore, an average queue length is measured at the node level for detecting the congestion level dynamically and transmitting the warning message to its neighbors in order to choose the congestion-free path from source to destination node. Thus, it improves the reliable communication through the networks in grid.

4.6 Record- and trust-based mechanism for SND in SG communications

Generally, the node misbehavior is referred to as the fraction from the original routing and transmission. In an SG, the packets are relayed from source to destination through the other sensor nodes. The nodes with selfish characteristics would not participate in the routing process, leading to delays or discarding the packet. This misbehavior of nodes may affect the network's efficiency, reliability, and fairness [30, 31]. Therefore, as explained below, this research introduces RTBD to detect the sensor nodes with selfish behavior for SG communication.

4.6.1 Selfish node detection

All sensor nodes achieve the routing function in zone wherein few nodes act like a selfish node, which consists of minimal resources [32]. The manners of selfish sensor nodes are given below:

- RREQ messages not forwarded by selfish nodes: The selfish nodes would not forward the RREQ message and discard such data packets for avoiding the route members from the other sensor nodes in the network.
- Data messages not forwarded by selfish nodes: This type of selfish node would transmit the data messages. However, the data messages do not relay by them and may be discarded by them.
- RREP messages not forwarded by selfish node: The selfish node discards all received Route Reply messages (RREP).
- Route Request (RREQ) messages forwarded with delay by selfish node: The selfish nodes forward RREQ after the upper limit's timeout.

Such kind selfish sensor nodes are detected by using this mechanism which is presented in the following section.

4.6.2 RTBD mechanism

In this model, the trust table holds the list of all selfish behavior sensor nodes. The trust table with two fields such as node id n_{id} and trust value t_{val}. The state of trust for all sensor nodes is updated in the trust table. In contrast, a sensor node receives the new trust certificate, which is determined based on the authentication from each neighbor in a specific zone [33].

The updated trust value is done as follows function:

$$(1 - T_{new}) = a (1 - T_{old}) + b (1 - T_c) - F, \tag{4.13}$$

In (4.13), a and b are the weights of the old and new trust value, F denotes the trust replenishment factor over time and the value of b based on the three factors such as $a_1, a_2,$ and a_3. The parameters of b and a_1 are computed as follows:

$$b = a_1 \times a_2 \times a_3, \tag{4.14}$$

$$a_1 = \frac{\sum_{maj} W_i T_i}{W_n},$$ (4.15)

In (4.15), W_i denotes the weight and T_i refers to the trust values of the majority group of the neighbors' node and W_n denotes the size of the network. The weight a_2 refers the value of new trust rate and a_3 is computed as follows:

$$a_3 = \begin{cases} 1 \text{ if } k = 1 \\ 0 \text{ if } k > 1. \end{cases}$$ (4.16)

Based on this algorithm, the number of forwarded packets to the selfish sensor nodes is declined, which mitigates the misbehavior routing in the network. For avoiding the dropping of a data packet, the nodes are verified. If a node is detected as a selfish node, then it is reported in the block list. This process is continued for every node in all zones. Due to this, a set of a selfish nodes is acquired. The block list refers to the list of packets that are available or not.

The above processes continue to transfer each packet. At each iteration, one by one, the packets are transmitted until all the packets are transmitted. Each trust node in the network receives the trust reports for the transmission of each data packet. For further authentication and protection, the records are signed. The trust value is calculated as follows:

$$T_c^{ij} = \frac{t_s + p/2}{t + p}, \quad t_s, t \geq 0, p > 0,$$ (4.17)

In (4.17), T_c^{ij} is the trust calculation of the node i to j, t_s is the time success, t refers to the time transactions, and p denotes the positive real number. This process identifies the selfish nodes in the network, which is shown in Table 4.1.

Where Q_{rs} refers the success rate of query request and computed based on the effective receivable of RREQ from source, Q_{rf} denoted as the failure rate of query request and computed based not successfully received RREQ. Also, Q_{ps} is the success rate query receivable from source and Q_{pf} is the failure to reply to the query. Moreover, Q_{ds} refers the rate of data success depends on data transmitted successfully and Q_{df} is the failure rate of data that fails to reach the destination node.

$$Q_{req} = \frac{Q_{rs} - Q_{rf}}{Q_{rs} + Q_{rf}},$$ (4.18)

Table 4.1   Parameter types for trust calculations

Type of parameter	RREQ	RREP	Data
Parameters for the success rate	Q_{rs}	Q_{ps}	Q_{ds}
Parameters for failure rate	Q_{rf}	Q_{pf}	Q_{df}

$$Q_{res} = \frac{Q_{ps} - Q_{pf}}{Q_{ps} + Q_{pf}}, \qquad (4.19)$$

$$Q_{data} = \frac{Q_{ds} - Q_{df}}{Q_{ds} + Q_{df}}, \qquad (4.20)$$

In (4.18)–(4.20), Q_{req}, Q_{res}, and Q_{data} denote a rate of node request, response, and transmission, respectively, which are the intermediate values calculated depends upon utilization of sensor nodes in networks. Then, the trust level value TV is estimated as,

$$V = T(RREQ) \times Q_{req} + T(RREP) \times Q_{res} + T(DATA) \times Q_{data}, \qquad (4.21)$$

In (4.21), $T(RREQ)$, $T(RREP)$, and $T(DATA)$ denote the time taken for request and response of route and also for transmission of data. Thus, the RTBD mechanism detects and prevents selfish nodes from the network and routing paths.

4.6.3   Algorithm: SND–EDCDC–IZCG mechanism

//Path Discovery
1. Source S transmits an RREQ to Destination D.
2. Destination D receives the RREQ from source S and transmits RREP to source S.
//Trust Estimation
3. Collect the neighbour node information such as energy, packet count and queue size.
4. Generate a report and validate the report rules.
5. Calculate the trust value by using (5.5).
6. Retrieve the current trust value(C_TV).
7. $If(C\_TV > 0.7)$ then
8. {
9. $If(Node = Selfish\ node)$ then
10. Add selfish node to block list (B_L)
11. $Else$
12. Transmit the data to the destination node.
13. }
14. End

This approach overcomes the issue of misbehavior of selfish nodes in SG communications. The selfish node's misbehavior is detected by proposing SND approach which uses RTBD algorithm. The proposed RTBD algorithm detects selfish nodes in the network based on estimation of trustworthiness of each node in the network. Based on the estimated trust values, the selfish nodes are identified and neglected from the network efficiently for improving the fairness of the networks. Thus, the proposed protocol is an effective mechanism that enhances the SG's performance in the networks.

4.7   Cognitive radio multichannel MAC protocol for second channel utilization

In this section, the proposed SCU–SND–EDCDC–IZCG protocol is discussed in brief. Multiple channels are integrated to network for improving the performance of network, which has proper infrastructure. In an SG, this is achieved by utilizing a common channel in the network to communicate all the other nodes locating inside the communication range or not [34].

It is done by distributed coordination function where node reserves a data transmission channel by sending the messages like ready-to-send (RTS) and CTS to other nodes [35]. The packet (RTS) is transmitted to the destination whenever it intends for transmitting the packets to the other nodes. Then, a CTS packet is replied to the receiving node. The time taken by the channel to be occupied is estimated by both RTS and CTS packets. Hence, each node maintains the network allocation network, a time interval of postponed transmissions. This process is called carrier virtual detection that facilitates the reservation of the area around the source and destination.

4.7.1   *Cognitive radio multichannel MAC protocol*

Every cognitive radio (CR) channel maintains two channels for data communications. The first channel is used for data transmission, and the second channel is used for channel communication. The spectral image of primary (SIP) users are used to identify the network channel for data transmission or not. The vector determines about the requirement of new scans during data transmission. The SIP value indicates the uncertainty of requirements of the new scan scheduling. The secondary users' channel load vector (SCL) selects the communication channel and $SIP\lceil n \rceil$ vector estimates the consumption of spectrum for n^{th} channel which holds the following values:

- If PU is idle over channel c, $SIP\lceil c \rceil = 0$.
- If PU is active over channel c, $SIP\lceil c \rceil = 1$.
- When there is uncertainty about the presence of PU, $SIP\lceil c \rceil = 2$.

A node enters into the network and scans on Ad-hoc Traffic Indication Message window (ATIM). Its result stores in SIP vector, which determines the requirement of new scan during data transmission. The time slot is split into two windows, namely, the ATIM and data window. The node performs a rapid scan, and control information is exchanged in the ATIM window. The data window achieves the exchange of data and accurate sensing. The transmission of the scan result packet initializes the mini-slot protocol. The mini-slot identifies the value of SIP value. The node with the stored packets denotes transmitting the ATIM frames.

A node can insert a preferred channel in the ATIM window for transmission. Each node listens to the ATIM window and updates its SCL vector. The ATIM frames remain active until the end of exchange of data. The nodes that neither transmit nor receive ATIM frames remain inactive till next interval starts. The nodes with uncertainty SIP value perform an accurate scan during the data window, which is achieved in parallel with the other channel. Then for this channel, the value of SIP is updated. This protocol has additional features: the nodes are facilitated for standing idle after the completion of exchange within the data window, i.e., while the transmission queue is null.

4.8   G-FQL approach

The G-FQL approach integrates game theory with the fuzzy Q-learning approach for SG communication to enhance performance. The game theory-based routing mechanisms are followed, which reduces the path overhead and routing cost. The nodes in the network are split into three types such as source node v_S, intermediate node v_K and destination node v_D. Then, all the nodes are assumed as honest and willing to pay a fixed set of path prices [36].

Also, consider that each node can compute its requirements and maximum price. If $P_{willing-to-pay}$ is the maximum price and v_S is willing to pay for transmitting a packet, then the utility function of v_S is defined as follows:

$$U_S = P_{willing-to-pay} - C_S\left(v_D\right), \tag{4.22}$$

In (4.22), $C_S\left(v_D\right)$ is the cost for transmitting a packet from the node v_S to v_D. If the routing path does not exist, then U_S should be equal to zero. After that, consider v_K with the information known only to itself, which sets prices accordingly as C_{v_K}. This node may utilize distance to neighboring nodes, available energy, bandwidth, and other factors. This node also has a utility function as follows:

$$U_S = P_{v_K} - l\left(v_K\right), \tag{4.23}$$

In (4.2), $l\left(v_K\right)$ refers the neighboring node and node v_S is required for paying the cost of v_K transmitting a single packet, where the cost is denoted as P_{v_K}, if v_K is not in the routing path, then $P_{v_K} = 0$.

4.8.1   Pricing mechanism

The cost $c\left(P\right)$ for path P from a source to destination calculated for the proposed approach is $c\left(P\right) = \sum_{v \in P, v \notin \{v_S,\ v_D\}} l\left(v_K\right)$. The selected path should be LCP. For any intermediate node v_K on the LCP, $c\left(P^{-v_K}\right)$ is used for the cost of a path that bypasses the node included on the LCP [18, 23]. According to these assumptions, the network topology of the link is bi-connected. Therefore, the path P^{-v_K} should exist, and the node v_K requires for receiving the payment $pay\left(v_K\right)$ which is given by,

$$y\left(v_K\right) = c\left(P^{-v_K}\right) - c\left(LCP\right) + l\left(v_K\right), \tag{4.24}$$

By using the (4.24), the package cost $C_S\left(v_D\right)$ is determined.

4.8.2   Routing process

When v_S requires to find the destination node v_D, it would generate the route request (RREQ) packet and utilize the broadcast transmission. This packet consists of the address of the source and destination node and also unique information of $P_{willing-to-pay}$. If it can find an immediate path to a destination node and the total cost of $C_S(v_D)$ is less than $P_{willing-to-pay}$, then the source node can initiate transmitting packets and pay the cost for each intermediate node v_K.

The received RREQ packet format is as follows:

$$EQ\left[v_S,\ v_D, \ldots, P_{willing-to-pay}, l\left(v_{K1}\right), \ldots, v_{K1-1}, l\left(v_{K1-1}\right)\right], \tag{4.25}$$

In (4.25), $v_{K1}, \ldots, v_{K1-1}$ is the path from source to next node v_{K1-1} and $l\left(v_{K1}\right), \ldots, l\left(v_{K1-1}\right)$ is the payment to the intermediate nodes. When each node v_K receives this packet, it would include its private message v_K in the packet in the corresponding field and $l\left(v_{K1}\right)$ would continue to broadcast. This action would continue until the packet reaches its destination node.

4.8.3   Routing cost

In LCP, the selected path aims to the minimum the accumulated cost of paths computation. The cost of each path is given by,

$$C_{\text{LCP}}(v_K \mid v_K \neq v_S) = \sum_{(v_K \to m) \in (v_S \underset{\text{paths}}{\longrightarrow} v_D)} C_{rc}(v_K) \times (E_t + E_r), \tag{4.26}$$

In (4.26), C_{rc} is the routing cost, E_t is the required energy for transmitting a packet from source, E_r is the required energy by the destination that receives each packet and $v_S \underset{\text{paths}}{\longrightarrow} v_D$ is the selected shortest and a reliable path that accumulates cost from node v_S to node v_D.

The fuzzy Q-learning approach is implemented to identify future attacks regardless of regular or irregular basis. It is a three-player game with sink nodes, base station, and attacker [18]. Through this game theory-based fuzzy Q-learning approach, the sink node monitors attacks. The attacker assaults the sensor node, and an alarm can be transformed from sink node to base station after completing the first stage [37].

A detection fitness test is used based on the attack patterns and severity when an abnormal alarm is received from the sink node [22, 38]. The abnormal alarm indicates that the affected sink nodes need to guard themselves against the offending attack. If the sink node remains in irregularity, it requires revising the strategy of detection. It continues until the attack condition is resolved and turns to a back defense strategy. Thus, the fuzzy Q-learning approach is integrated with the game theory, which maximizes the security levels and degrades the energy efficiency of nodes in the network.

4.9   Simulation results

In the proposed IZCG protocol, Network Simulator 2 simulation tool is used and compared with zone-based routing protocol with parallel collision guided-broadcasting

Table 4.2   Simulation parameters and energy consumption model for proposed protocol

Simulator	NS-2.35
Network size	1 000 × 800 m
Mode of configuration in topology	Randomized
Sensor nodes count	1 000
MAC layer's data rate	2.5 Mbps
Sensor node's transmission limit	45–75 m
Packet loss rate	0.25%
Common node – primary energy level	0.3 w
Video node – primary energy level	5 w
Consumption of energy level for sending m_{tx}	0.760 w/s
Consumption of energy level for receiving m_{recv}	0.495 w/s
Consumption of energy level for overhearing $m_{overhearing}$	0.295 w/s
Consumption of energy level for idle m_{idle}	0.025 w/s

mechanism (ZCG) in terms of throughput, a ratio of packet delivery, rate of packet dropout, delay, and control overhead. The parameters for simulation are depicted in Table 4.2.

4.9.1   Throughput

It is a proportion of packets transmitted and the time taken. It gives information about whether the destination nodes correctly receive the data packets or not. It denotes in terms of kilobits per second (Kbps).

The comparison of throughput is depicted in Table 4.3.

Figure 4.3 compares the proposed IZCG with the other protocol, such as ZCG for throughput. In the graph, nodes were taken in the *x*-axis and *y*-axis, represented by throughput values in terms of Kbps. From the analysis, it is cleared that when

Table 4.3   Comparison of throughput (Kbps) for IZCG with ZCG protocol

Number of nodes	ZCG (Kbps)	IZCG (Kbps)
100	32.1	72.3
200	51.3	92.2
300	82	105.8
400	97.2	122
500	109.1	132.29
600	131	162
700	140.8	177
800	155.6	182.2
900	173	201.8
1 000	192.25	217.52

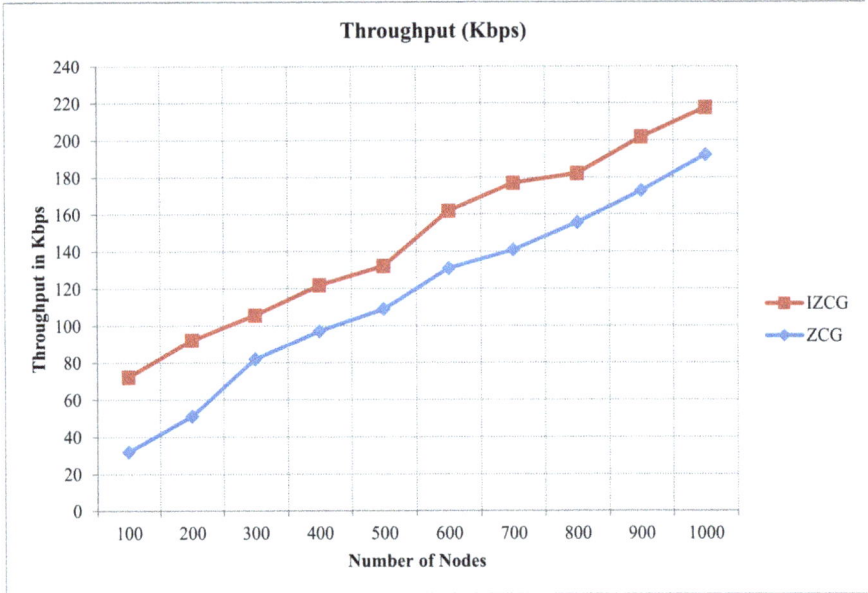

Figure 4.3   Comparison of throughput (Kbps) for IZCG with ZCG protocol

the nodes are 1 000, the throughput of IZCG protocol is 25.27 Kbps higher than the ZCG protocol.

4.9.2   Packet delivery ratio

It is proportion of packets received by destination and the packets transmitted by source. The Packet Delivery Ratio (PDR) is compared with other protocols are shown in Table 4.4.

Table 4.4   Packet delivery ratio (%) for IZCG compared with ZCG protocol

Number of nodes	ZCG (%)	IZCG (%)
100	80	83.78
200	81.8	85
300	82.9	87.5
400	83.7	89.32
500	84.9	90.3
600	85.8	91.62
700	86.6	93.1
800	87.8	94
900	89	95.8
1 000	91.2	98.5

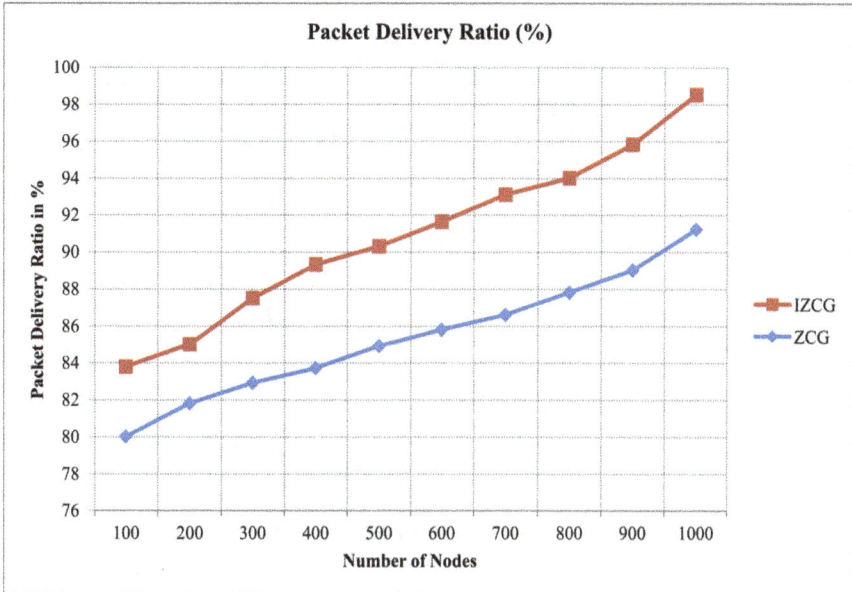

Figure 4.4   Values of packet delivery ratio (%) for IZCG compared with ZCG protocol

Figure 4.4 exhibits the values of PDR comparison of the proposed IZCG with the other protocol, such as ZCG. In the graph, the *x*-axis denotes the total number of nodes taken for simulation. The *y*-axis refers to the ratio of packet delivery in terms of %. This analysis shows that if the nodes are 1 000, then the PDR of the IZCG protocol is 7.3% higher than the ZCG protocol.

4.9.3   Packet drop ratio

It is the proportion of dropped packets at destination and packets generated at source. The values of packet drop ratio compared with other protocols are displayed in Table 4.5.

Figure 4.5 narrates the percentage of packet drops out rate comparison of the proposed IZCG with the other protocol, such as ZCG. In Figure 4.5, the *x*-axis denotes the nodes taken for simulation and the percentage packet drop rate taken in the *y*-axis. The analysis clearly shows that if the total nodes are 1 000, then the ratio packet dropout of the IZCG protocol is 5.3% lower than the ZCG protocol.

4.9.4   Delay

Delay is the time gap between the packet departures from the source to the arrival of packets to the destination node. It consists of delays that take place in the queue, delays of switching and propagation, and others. A comparison of latency values with other protocols is shown in Table 4.6.

Table 4.5   Packet drop ratio (%) for IZCG compared with ZCG protocol

Number of nodes	ZCG (%)	IZCG (%)
100	5.8	2.2
200	6.5	2.8
300	7.4	3.1
400	8.1	3.8
500	8.9	4.2
600	9.7	4.7
700	9.9	5.4
800	10.3	5.2
900	11.1	5.9
1 000	11.9	6.6

Figure 4.6 compares the proposed IZCG with the other protocol, such as ZCG, in terms of delay. In Figure 4.6, the number of nodes used for simulation is shown on the x-axis, and the axis of y represents the delay in milliseconds. The analysis clearly states that if the nodes are 1 000, then a delay of IZCG protocol is 10.93 ms lesser than the ZCG protocol.

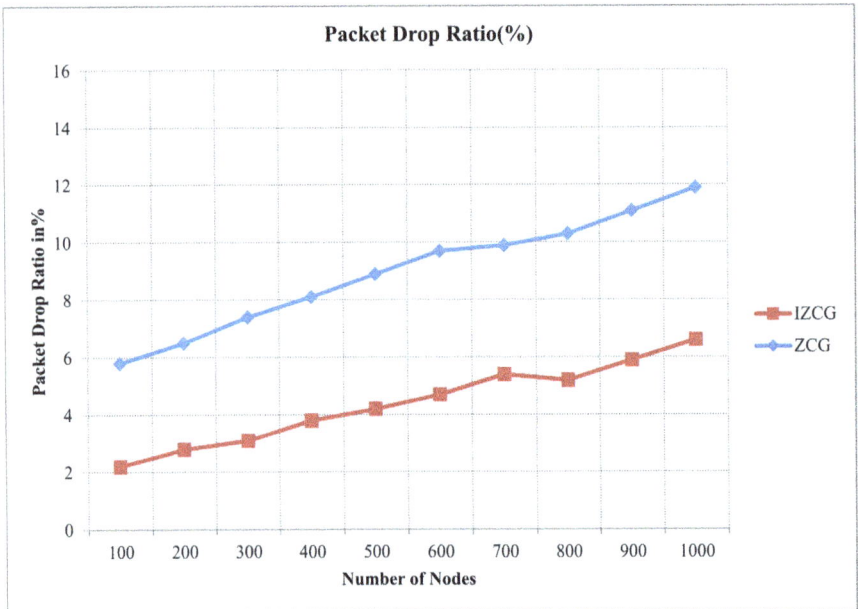

Figure 4.5   Percentage of packet drop ratio (%) for IZCG compared with ZCG protocol

Table 4.6   Comparison of delay (ms) for IZCG with ZCG protocol

Number of nodes	ZCG (%)	IZCG (%)
100	55.8	47.29
200	57.2	49.3
300	59.9	50.5
400	60.21	53.85
500	63.5	56.1
600	65.1	59.2
700	69.5	63.1
800	73.2	66.23
900	79.65	68.91
1 000	81.23	70.3

4.9.5   Control overhead

It is the ratio of the total number of packets transmitted and the total number of packets transmitted successfully. It is in the form of kilobits/second (Kbps). In order to receive 1 KBits of data packets in the sink node, it needs the control packets. Such control packets are present in it. Table 4.7 compares the control overhead of each protocol with other protocols.

Figure 4.7 displays the values of control overhead of the proposed IZCG compared with the other protocol such as ZCG. In Figure 4.7, the nodes taken for simulation are represented on the axis of *x*, and the values control overhead in Kbps are shown on the axis of *y*. The analysis demonstrates that if the nodes reached 1 000, then the control overhead of the IZCG protocol is 10 Kbps lesser than the ZCG protocol.

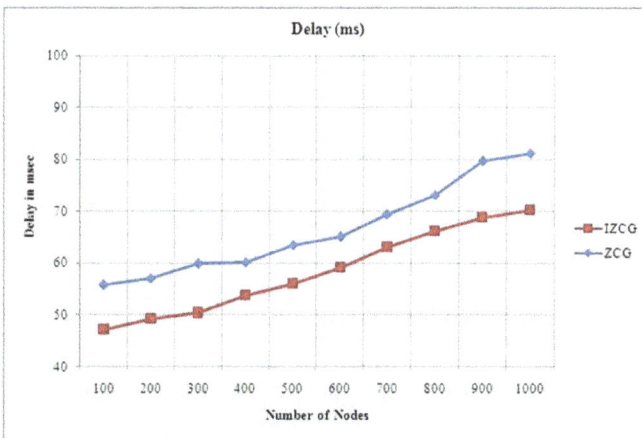

Figure 4.6   Comparison of delay (ms) for IZCG with ZCG protocol

Table 4.7   Values of control overhead (Kbps) for IZCG compared with ZCG
protocol

Number of nodes	ZCG (Kbps)	IZCG (Kbps)
100	10	4
200	16	8
300	22	13
400	27	15
500	33	21
600	38	26
700	44	32
800	50	37
900	56	44
1 000	59	49

4.10   Conclusion

In this research, an improved hybrid zone-based routing protocol (IZCG) for sensor networks is used in neighbor area network (NAN) for SG communications. In this research, zone-based routing protocol with parallel collision guided broadcasting mechanism (ZCG) is upgraded with different techniques such as managing network

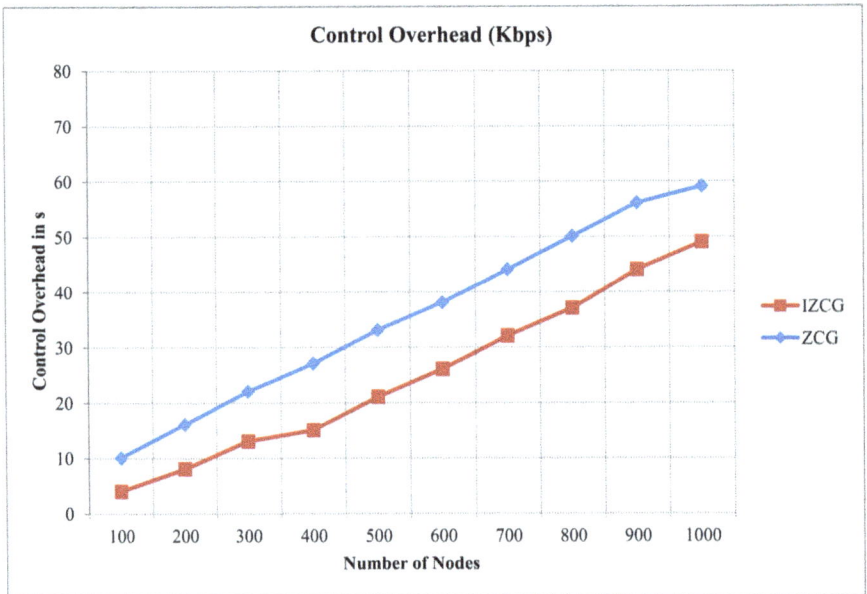

Figure 4.7   Values of control overhead (Kbps) for IZCG compared with ZCG
protocol

topology, controlling DC, isolating selfish nodes, and utilizing the multiple channels in the network based on the game theory-based routing protocol. Moreover, the fuzzy quality learning approach in game theory is implemented to prevent DoS attacks, increasing the accuracy rate in preventing attacks and decreasing the false alarm rate that identifies the various denial of service attacks. Finally, the performance evaluation of the proposed hybrid ZCG demonstrates that it achieves better performance than other existing protocols for routing processes in an SG. This work can be extended by integrating with cloud and fog computing in EMOR. The novel CR sensor networks need to be applied to resolve the predictive channel assignments and opportunistic spectrum access problems. Moreover, a game theory-based routing fuzzy quality learning approach is implemented to prevent DoS attacks, which maximizes levels of security and minimizes energy usage in every node in the network.

References

[1]   Alladi T., Chamola V., Rodrigues J.J.P.C., Kozlov S.A. 'Blockchain in smart grids: A review on different use cases'. *Sensors*. 2019, vol. 19(22), pp. 4862–87.

[2]   Deepa K., Khurana M.S. 'Optimisation of routing in smart grids using intelligent techniques'. *Proceedings of 3rd International Conference on Internet of Things and Connected Technologies (ICIoTCT), 2018 held at Malaviya National Institute of Technology*; Jaipur; 2018. pp. 562–8.

[3]   Patil S., Chaudhari S. 'DoS attack prevention technique in wireless sensor networks'. *Procedia Computer Science*. 2016, vol. 79(5), pp. 715–21.

[4]   Anees J., Zhang H.-C., Baig S., Lougou B.G. 'Energy-efficient multi-disjoint path opportunistic node connection routing protocol in wireless sensor networks for smart grids'. *Sensors*. 2019, vol. 19(17), p. 3789.

[5]   Srivastava J.R., Sudarshan T.S.B. 'ZEEP: Zone-based energy-efficient routing protocol for mobile sensor networks'. *2013 International Conference on Advances in Computing Communications and Informatics (ICACCI)*; IEEE, Mysore, India; 2013. pp. 990–6.

[6]   Ahmed U., Hussain F.B. 'Energy-efficient routing protocol for zone-based mobile sensor networks'. *7th International Wireless Communications and Mobile Computing Conference*; Istanbul, Turkey; 2011. pp. 1081–6.

[7]   Tamil Selvi P., Suresh GhanaDhas C. 'A novel algorithm for enhancement of energy efficient zone based routing protocol for MANET'. *Mobile Networks and Applications*. 2019, vol. 24(2), pp. 307–17.

[8]   Devi B.A.S.R., Murthy J.V.R., Narasimha G. 'Secure zone-based routing protocol for mobile Adhoc networks'. *International Mutli-Conference on Automation, Computing, Communication, Control and Compressed Sensing (iMac4s)*; IEEE, Palai, Kerala, India; 2013. pp. 839–46.

[9]   Nasser N., Al-Yatama A., Saleh K. 'Zone-based routing protocol with mobility consideration for wireless sensor networks'. *Telecommunication Systems*. 2013, vol. 52(4), pp. 2541–60.

[10]   Malwe S.R., Rohilla S., Biswas G.P. 'Location and selective-bordercast based enhancement of zone routing protocol'. *3rd International Conference on Recent Advances in Information Technology (RAIT)*; Dhanbad, India; 2016. pp. 83–8.

[11]   Vinodha R., Durairaj S. 'Energy-efficient routing protocol and optimized passive clustering in WSN for smart grid applications'. *International Journal of Communication System*. 2021, vol. 3, pp. 150–68.

[12]   Tamil Selvi P., Ghana Dhas S. 'Implementation of energy efficient load balanced adaptive zone routing protocol for mobile ad hoc networks'. *International Journal of Linguistics & Computational Application (IJLCA)*. 2017, vol. 4(2), pp. 90–5.

[13]   Tamil Selvi P., Ghana Dhas S. 'Performance improvement of energy-aware hybrid routing protocol for mobile ad hoc networks'. *Australian Journal of Basic & Applied Science*. 2015, vol. 9(10), pp. 99–106.

[14]   Rajput S.S., Trivedi M.C. 'Securing zone routing protocol in MANET using authentication technique'. *International Conference on Computational Intelligence and Communication Networks (CICN)*; Bhopal, Madhya Pradesh, India; 2014. pp. 880–97.

[15]   Sharma N., Patidar G. 'Improved congestion control mechanism using modified hybrid-TCP in mobile ad-hoc networks'. *International Conference on Computational Intelligence & Communication Technology (CICT)*; Ghaziabad, India; 2016. pp. 315–25.

[16]   Keerthana P., Chandra Mohan S. 'Adaptive approach based joint-scheduling & congestion control in wireless networks'. *International Conference on Innovations in Information, Embedded and Communication Systems (ICIIECS)*; Coimbatore, India; 2015. pp. 115–35.

[17]   Sangolli S.V., Thyagarajan J. 'An efficient congestion control scheme using cross-layered approach and comparison of TCP variants for mobile ad-hoc networks (MANETs)'. *First International Conference on Networks & Soft Computing (ICNSC2014)*; Guntur, Andhra Pradesh, India; 2014. pp. 30–4.

[18]   Das D., Majumder K., Dasgupta A. 'A game-theory based secure routing mechanism in mobile ad hoc network'. *International Conference on Computing, Communication and Automation (ICCCA)*; IEEE, Greater Noida, India; 2016. pp. 437–42.

[19]   Cao X., Liu L., Cheng Y., Cai L.X. 'A two-step selfish misbehavior detector for IEEE 802.11-based ad hoc networks'. *Global Communications Conference (GLOBECOM)*; San Diego, CA, USA; 2015. pp. 226–46.

[20]   Jardosh A.P., Ramachandran K.N., Almeroth K.C. 'Understanding congestion in IEEE 802.11b wireless networks'. *5th ACM SIGCOMM Conference on Internet Measurement*; Philadelphia, PA; 2005. pp. 25–45.

[21]   Ren F., Zhang J., Wu Y., He T., Chen C., Lin C. 'Attribute-aware data aggregation using potential-based dynamic routing in wireless sensor networks'. *IEEE Transactions on Parallel and Distributed Systems*. 2013, vol. 24(5), pp. 881–92.

[22] Guan X., Wu H., Bi S. 'A game theory-based obstacle avoidance routing protocol for wireless sensor networks'. *Sensors*. 2011, vol. 11(10), pp. 9327–43.

[23] Qin Y., Li L., Liu H. 'GTR: A novel routing scheme based on game theory in opportunistic networks'. *IEEE/CIC International Conference on Communications in China (ICCC)*; IEEE, China; 2014. pp. 775–9.

[24] Basurra S.S., De Vos M., Padget J., Ji Y., Lewis T., Armour S. 'Energy efficient zone based routing protocol for MANETs'. *Ad Hoc Networks*. 2015, vol. 25(1), pp. 16–37.

[25] Tamil Selvi P., Ghana Dhas S. 'A review of energy-aware routing protocols for mobile ad hoc networks'. *International Journal of Applied Engineering Research*. 2015, vol. 10(52), pp. 370–7.

[26] Tamil Selvi P., GhanaDhas S. 'Performance improvement of energy-aware hybrid routing protocol for mobile ad hoc networks'. *Australian Journal of Basic & Applied Science*. 2015, vol. 9(10), pp. 99–106.

[27] Ng J., Yu M. 'A new model for the efficient channel utilisation of wireless networks and applications'. *IEEE Wireless Communications and Networking Conference Workshops (WCNCW)*; New Orleans, LA, USA; 2015. pp. 176–81.

[28] Senthilkumaran T., Sankaranarayanan V. 'Dynamic congestion detection and control routing in ad hoc networks'. *Journal of King Saud University - Computer and Information Sciences*. 2013, vol. 25(1), pp. 25–34.

[29] Chang H.-P., Kan H.-W., Ho M.-H. 'Adaptive TCP congestion control and routing schemes using cross-layer information for mobile ad hoc networks'. *Computer Communications*. 2012, vol. 35(4), pp. 454–74.

[30] Hussain M.A., Nadeem A., Khan O., Iqbal S., Salam A. 'Evaluating network layer selfish behaviour and a method to detect and mitigate its effect in MANETs'. *15th International Multitopic Conference (INMIC)*; Islamabad; 2012. pp. 283–9.

[31] Samar P., Pearlman M.R., Haas Z.J. 'Independent zone routing: An adaptive hybrid routing framework for ad hoc wireless networks'. *IEEE/ACM Transactions on Networking*. 2004, vol. 12(4), pp. 595–608.

[32] Subramaniyan S., Johnson W., Subramaniyan K. 'A distributed framework for detecting selfish nodes in MANET using Record- and Trust-Based detection (RTBD) technique'. *EURASIP Journal on Wireless Communications and Networking*. 2014, vol. 2014(1), p. 205.

[33] Nobahary S., Garakani H.G., Khademzadeh A., Rahmani A.M. 'Selfish node detection based on hierarchical game theory in IoT'. *EURASIP Journal on Wireless Communications and Networking*. 2019, vol. 2019(1), pp. 255–74.

[34] Pedraza L.F., Paez I.P. 'An evaluation of MAC protocols running on a MANET networks'. *3rd International Conference on Ambient Systems, Networks and Technologies, Procedia Computer Science*; Niagara Falls, Ontario, Canada; 2012. pp. 86–93.

[35] Tlouyamma J., Velempini M. 'Channel selection algorithm optimized for improved performance in cognitive radio networks'. *Wireless Personal Communications*. 2021, vol. 119(4), pp. 3161–78.

[36] Feng R., Li T., Wu Y., Yu N. 'Reliable routing in wireless sensor networks based on coalitional game theory'. *IET Communications*. 2016, vol. 10(9), pp. 1027–34.

[37] Lazrag H., Saadane R., Aboutajdine D. 'A game-theoretic approach for optimal and secure routing in WSN' in Abraham A., Haqiq A., Ella Hassanien A., Snasel V., Alimi A. (eds.). *Advances in Intelligent Systems and Computing*. 565. Cham: Springer; 2018.

[38] vent T.L., Preux P., Penne E.L., Badosa J. 'Energy management for microgrids: Are inforcement learning approach'. *Innovative Smart Grid Technologies*. 2019, vol. 1, pp. 1–5.

Chapter 5

Blockchain-based secured IoT-enabled smart grid system

Sivasankar P[1], Rathy G A[1], John Deva Prasanna D S[2], Karthikeyan Perumal[3], Gunasekaran K[1], and Sivapragash C[4]

The electric grid is a transmission infrastructure that connects the power plant to houses and industries. Industry inclusion of automation, communication, and IT systems in the electrical grid will make it a smart grid (SG) to monitor power transmission from generation to consumption, control the power flow, and load shedding to match power generation in real life-time as per the requirement. The major areas of SG are power generation, energy storage, power transmission and distribution, and energy consumption. In the SG, it is essential to monitor the power generation, maintain the power quality in the grid, locate the faults in transmission and distribution, and prevent smart meter tampering. The challenges in running SGs are maintaining high reliability and power quality, robust cyber attacks, and minimize operation and maintenance costs. Among the various challenges, cyber attacks are the major threats in establishing an efficient SG system. Disclosure, deception, and description are some of the popular cyber attacks in the SG. Nowadays, blockchain technology is becoming popular in securing data in these kinds of cyber-physical systems (CPSs). Blockchain is a secured technology to maintain confidential and authentic data in a distributed environment. The important components of blockchain technology are shared ledger, smart contract, and consensus mechanism. In this technology, data blocks are chained together in a sequence ordered by the time of block generation.

The data blocks data stored in a smart ledger that maintains in a distributed fashion across several servers. Every data is updated in the block after thorough verification through a consensus mechanism. In this work, a blockchain-based security model is developed for IoT-enabled SG system. The smart meters update the

[1]NITTTR, Chennai, Tamil Nadu, India
[2]Hindustan Institute of Technology and Science, Chennai, Tamil Nadu, India
[3]IntelliSoft Technologies Inc, Irving, TX, USA
[4]Swarnandhra College of Engineering and Technology, Narsapur, Andhra Pradesh, India

meter reading along with the digital signature to the nearby server periodically. The distributed server verifies the authenticity of the received meter reading using a consensus mechanism. The meter reading is verified against tampering using the smart contract codes running in the server. The verified meter reading is then updated to the distributed smart ledger. In this way, the proposed model prevents unauthorized access to confidential data and the transmission of false metering data by an adversary. Hence, wrong demand estimation and erroneous billing are avoided. We develop a foolproof billing and consumption tracking mechanism by using Advance Metering Infrastructure and blockchain-inspired security model. The goal of this chapter is to develop a distributed infrastructure that maintains power consumption data that are obtained from Advanced Metering Infrastructure (AMI).

5.1   Introduction

An SG is a secured model for transmitting electric power through modern automated controls and communication systems and other types of information technology. It incorporates cutting-edge techniques and technology from power production, the transmission of power, and distribution to utensils. This concept incorporates energy infrastructure, gadgets, information, and markets into a collaborative and coordinated process that benefits the environment that improves the effectiveness and efficiency of energy generation, distribution, and consumption. It is an electrical network that allows for the two-way flow of power and data. It uses digital communication technology to monitor, react, and respond to changes in consumption and various concerns. SGs have the ability to self-heal and allow power customers to become active participants. The model aims to provide consumers, distributors, and grid operators with optimized information and load control to minimize system overheads and costs while boosting energy efficiency. Based on the design, the primary difference between the SG is that the traditional power grid is based on supply and demand, whereas the SG is based on supply and demand, whereas the conventional grid is based on supply follows demand. The ultimate purpose of a SG is to save energy, reduce cost, and increase reliability. It provides everyone with plentiful, inexpensive, clean, efficient, and dependable electric power at all times and from any location.

Despite the benefits of SGs, improving access to dispersed and scalable energy resources, ensuring energy security, and incorporating other ways to increase energy efficiency and dependability are difficult. By integrating the intelligent grid context with Internet technology, the energy Internet, also referred to as the Internet of energy or Smart Grid 2.0, is being created to advance it and solve its current limitations [1–9]. The energy Internet contrasts the intelligent grid, an Internet-based solution for energy problems. An SG is a collection of intelligent digitalization designed to address energy-related challenges. On the other hand, the energy Internet accomplishes this through Internet of Things (IoT), enhanced power system components, and other energy networks. Energy can be connected anywhere with these emerging and innovative approaches.

In conclusion, the principles were established to ensure that all participants and components communicate effectively.

- to be close to one another
- to make their independent judgments
- to communicate energy and relevant metadata in a variety of methods
- to have seamless access to many sorts of distributed energy resources on a wide scale
- to adapt with both centralized and distributed energy sources
- to achieve energy supply and demand balance through energy sharing
- to provide flexible energy generation/sale and purchasing/consumption

Due to the accelerating pace of connectivity, one of the biggest challenges is integrating and coordinating distributed energy producers, consumers, electric vehicles, intelligent devices, and CPSs within the traditional centralized grid system. A centralized approach to continuous network expansion would require complex and costly information and communication systems. Consequently, an intelligent grid moves toward decentralization to integrate all its components dynamically. According to its vision, decentralization is another fundamental requirement for the electronics industry's growth in SGs. A decentralized SG system may potentially offer security, privacy, and trust nightmare if its pieces and connections are complicated and many. New and novel technologies may be necessary to address these concerns [10–16].

Blockchain facilitates the development and maintenance of a decentralized accounting system that is tamper-resistant, traceable, and highly trustworthy as another disruptive technical evolution post-Internet. Blockchain has the potential to improve the security of grid data and aid in the development of a dependable, effective, and trusted distributed SG system.

This chapter focuses on blockchain-based secured IoT-enabled SG systems. The following subsections will detail the paradigm shift of conventional power grid system to the SG system and its importance and challenges, AMI, need for security measures in AMI, and blockchain technology and its components to improve the security measures in the SG system through strengthening the AMI. Furthermore, the subsequent sections of the chapter will present the various existing works that support the need for the proposed work. The proposed blockchain-based secured IoT-enabled SG system, necessary analysis, and evaluation of the proposed work and finally summarized the proposed work in the conclusion section.

5.1.1 Paradigm shift of conventional power grid system to the SG system

The concept of the microgrid is introduced to improve the performance of the electric grid system. A microgrid is an electrical system with various loads and energy resources that may run parallel with larger grids or smaller independent power systems. It has enhanced dependability through distributed generation (DG), increased efficiency through decreased transmission length, and made alternative energy sources easier to integrate. Three key characteristics define the microgrid. They are given as follows: (i) microgrid is local, (ii) microgrid is independent, and (iii) microgrid is intelligent.

Microgrids are composed of many components such as regulated loads, uncontrolled loads, DG units, and storage devices that function in combination with controlled power electronic devices (active and reactive power flow controllers, frequency and voltage regulators, and protective devices). It is a self-sufficient energy system that serves discrete power utilization units such as a college campus and business centers. A further microgrid is categorized as AC microgrid and DC microgrid.

The most prevalent type of microgrid is an AC microgrid. An AC bus will connect AC microgrids to the distribution network, which the Power Control Centers's (PCC) circuit breaker controls the microgrid's interconnection and disconnection from the distribution network. Since it is connected to the power grid through an AC bus, it does not require an inverter to supply electricity to AC loads. The inverter connects DG and energy storage systems (ES) to the AC bus. It has the disadvantage of being difficult to manage and operate.

As per the work of the authors in ref. [17] DC microgrid will be connected through a DC bus, which feeds DC loads linked to it. DC loads are often low-power electronic equipment such as laptops, mobile phones, wireless phones, and other electronic devices. Sources with DC output are directly connected to the DC bus in a DC microgrid structure, whereas sources with AC output are interfaced to the DC bus through an AC/DC converter. As there are more DC-generating sources than AC-generating sources, fewer converter units are required. The DC microgrid's total efficiency increases and the harmonics problem caused by the power electronic converter is mitigated due to the DC nature of the power. In both the cases of AC and DC microgrid systems, automation and control are big challenges. Intelligent electronic devices (IEDs) have been introduced to overcome these issues.

The IED is a microprocessor-based integrated controller of power system equipment. Sensors and power devices provide data to IEDs. If IEDs detect voltage, current, or frequency abnormalities, they can trip circuit breakers or raise/lower tap positions to uphold the appropriate voltage level. Protective relays, tap changer controllers, circuit breakers, capacitor bank switches, recloser controllers, and voltage regulators are some examples of IEDs. The introduction of IED in conventional Grid initiates the need for a SG system.

5.1.2   SG system

The inclusion of automation, communication, and IT systems in the electrical grid will make it an SG to monitor power flows from generation to consumption. Figure 5.1 shows the SG power from generation to consumption and the power flow.

As part of the SG infrastructure, smart meters are essential. Automated meter reading (AMR) and advanced metering infrastructure (AMI) have been improved to create a smart meter, an enhanced version of a conventional power meter. Intelligent meters are pretty sophisticated and feature-rich with their sophisticated and detailed

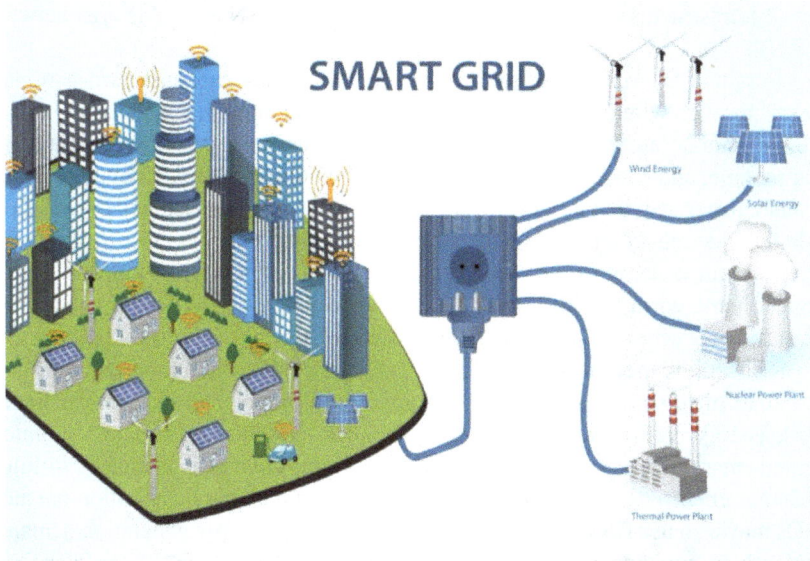

Figure 5.1   The SG power from generation to consumption and the power flow

Information and Communication Technology (ICT) interfaces. Smart meters can also provide additional information such as power factor and total harmonic distortion (THD), predict power consumption regularly, and provide metering data. Smart meters are also called intelligent sockets since they can distribute electricity from the grid to the house. Since AMI provides measurements of electricity consumed, power demand rates, and power quality of an entire grid, it is used to describe the intelligent meter-based infrastructure and SG applications. Grids and smart meters typically communicate two-way, and data are stored and transmitted to monitoring centers regularly, as well as a remote disconnect switch, a HAN interface, the ability to record and store blackouts, voltages, and currents, and a secure data network. [18–21].

Today, over several million smart meters are used worldwide [22]. Approximately 400 million smart meters are expected to be installed by 2020. AMR functions such as detailed consumption storage and previously mentioned features are provided on almost all smart meters installed. Meter data management systems (MDMS) and outage management system (OMS) perform several tasks required by smart meters and AMI because of the close relationship. As a result of smart meters' advanced remote monitoring and control functions, they allow for two-way communication, immediate data collection, and remote bill-paying capabilities. Power quality issues are being addressed via smart metering, such as automatic voltage restoring, frequency and voltage control, active and reactive power control, decentralized generation in microgrids, and cyber-secure communication systems. AMR, Distribution Management System (DMS), and time-of-use (TOU) pricing are the three key subsections of the metering section. Meanwhile, the communication system comprises control infrastructure and network interfaces such as HAN and wide area network (WAN).

Users and Distribution System Operator (DSO) may exchange two-way data streams through wireline and/or wireless communication modes. A smart meter can also incorporate auxiliary modules such as a power supply, controller, data logging module, and encoding/decoding module, in addition to the two main sections. A smart meter requires a data logging module that stores consumer information, including identification, energy logging, timestamps, and outage histories. The metering part consists of an analog interface that communicates with the grid to connect home wiring to the distribution network and is outfitted with voltage and current transducers to install a metering interface. The paying system is integrated into a scheduling mechanism, generating TOU price data with a timestamp [18].

Smart meters provide a variety of add-on applications and services, including DSM, energy theft protection, and CPS security, in addition to remote monitoring and smart metering. The DSM and DR programs have been enhanced to fulfill consumer energy demand while maintaining the generation-consumption balance. DSOs may also use DSM and DR algorithms to manage energy generation demands efficiently by avoiding bulk generating at inconvenient periods. DSM's primary goal is to regulate peak clipping, load shedding, valley filling, and peak shifting techniques, one of its most essential goals. The balance is set based on load sufficiency and demand rates to run these load shifting systems, with peak clipping allowing for lower energy consumption during peak hours and valley filling allowing for higher energy consumption during off-peak hours.

Peak shifting, which combines peak clipping and valley filling, makes it easier to reduce peak loads while still meeting base-load energy demands. Smart meters should be used to meet the criteria of the DSM and DR programs, which call for quick and exact metering. The DR programs are based on price-based or incentive-based methods, with TOU, critical peak pricing, critical peak rebate, and real-time pricing being the most prevalent price-based programs [9, 23].

SG is a concept for revolutionizing the electric power grid through advanced automatic control and communications systems and other information technology components that employ cutting-edge technologies at every value chain stage, right from power generation, transmission, and distribution. Using Digital Technology, a SG transports power from providers to consumers. Figure 5.2 shows the SG conceptual diagram.

The primary difference between the SG and conventional grid is that SGs are designed on demand follows supply, whereas conventional grids are designed based on supply follows demand. The purpose of a SG is to increase energy efficiency, minimize cost overheads, and increase reliability. One of the important components of an SG system is automatic meter reading that can be greatly achieved through AMI.

5.1.3   Advanced metering infrastructure

The AMI enables interaction between utilities and customers in real-time, which involves smart meters, communications networks, and data management systems.

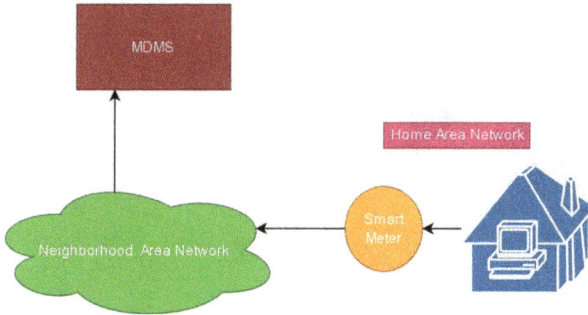

Figure 5.2   SG conceptual diagram

Smart meters can provide other important functions such as remote connection management, corruption detection, outage monitoring, voltage monitoring, and bidirectional measurement of electricity use, in addition to remote meter reading, to facilitate the adoption of DG and dynamic pricing. AMI includes an interface for a display device at the end-premises that shows the current level of energy consumption, the cost of energy used, and the tariff presently in effect. It also provides an interface for load control devices that remotely turn on and off appliances and equipment (demand response with Home Energy Management System (HEMS)). A supply capacity control (circuit breaker) will disconnect the supply if the demand exceeds a preset value at the end user's premises (demand limiting by utility). Figure 5.2 illustrates the simple AMI systems with their necessary components.

The components of AMI are:
HANs
Smart meters
Wide area communications systems
MDMS

- **Home area networks**

HANs and home energy management systems are two concepts that can use interchangeably to describe all the competence and activity in home energy management systems, and this section describes both concepts. According to [24], HANs are like local area networks but within the home, and they can extend intelligent technology and communication structures [25]. Instead of servers, printers, copiers, and computers, HANs link devices that can send and receive signals from metering systems and/or home energy management system applications. There are trade-offs involved in installing wired or wireless equipment, such as power consumption, signaling distance, interference sensitivity, and security.

An HAN is not an energy management application but allows energy management applications to watch, monitor, and control devices on the home network. An in-home display shows electricity rates through limited data input and display capabilities. The home energy management is a two-way communication system,

Figure 5.3   Smart meter

meaning users can only monitor, not take real-time actions, and communicate with the HAN like an in-home display. For HANs and in-home displays to derive the most benefit from these SG components, energy management applications—home energy management solutions—are still needed [26, 27].

Utility billing and demand response programs are best operated through a web-based portal for a home energy management system. With this program, intelligent appliances can be easily enrolled and controlled. In a home energy management solution, users would recall the optimized settings for sustainable energy savings, receive suggestions for improvement, and compare their energy management to other buildings in their neighborhood or peer group [24].

- **Smart meter**

A smart meter is an ecofriendly device that measures electrical energy in KWh units (Kilowatt hours). It is a gadget that has the advantage of reducing the end user's power cost. They are part of the advanced meter infrastructure division and automatically report meter readings to the energy supplier. The image of a smart meter is shown in Figure 5.3.

The features of smart meters include non-volatile data storage, remote connection management, tamper detection, and two-way communication. They send gathered data to the central meter through remote reporting. This central meter oversees the smart meter's operation. Smart metering provides for better monitoring and control of the power grid from an operational standpoint.

- **Wide area communications systems**

A WAN connects many Neighborhood Area Networks (NAN) to collect data from them, which it then sends to a private utility network, which serves as a central controller. Long-distance communications between various DAPs of power generating systems, distributed energy resources, transmission and distribution systems, and management systems are made feasible by the WAN. The network scheme's bidirectional communication infrastructures enable several utility applications, such as AMI, DR, DA, power quality monitoring, and demand-side management. The WAN coverage area is quite large, covering hundreds of square miles, and the data throughput is roughly 10–100 Mbps. Furthermore, huge bandwidths are necessary for these networks' operation and control. The WAN was designed to connect utility systems and SG applications. As a result, it encompasses two distinct network topologies known as core networks and backhaul communication networks. The backhaul network is linked to the network's DAPs (NANs), while the core network is linked to the utility's metropolitan network and substations. Data acquisition, status monitoring, fault detection, control, and power grid management are all possible with these network methods [28–31].

Because the quantity of sent data and transmission settings have a significant impact on the performance of SG communication networks, each segment of SG communication networks must use separate communication technology. The most often used communication technologies in WANs are public networks, such as wired broadband technology and cellular networks [32]. programmable logic controller (PLC), fiber optics, Ethernet, DSL, IEEE 802.15.4, IEEE 802.11/Wi-Fi, Cellular (3G, LTE, LTE-A 5G), IEEE 802.16/WiMAX, and so on are some of the possible communication technologies for WAN applications. It's also worth noting that numerous criteria may be sorted out to determine which technology is ideal for exciting applications in any SG communication network, including installation and maintenance costs, necessary data rate, coverage, power consumption, network deployment characteristics, and network scalability.

In [33], the authors studied the performance of a ZigBee-based communication network in an SG system. Another strategy for SG communications based on Bluetooth, Wi-Fi, and WiMAX is investigated in [34–37]. Furthermore, Spano *et al.* [38] discuss the use of the IoT idea in SG communications, while Garcia-Hernandez [39] discusses advancements in sophisticated metering [20, 40, 41]. SG systems can be monitored remotely using PLC-based applications. As a result, security poses a significant challenge for public networks used in the WAN architecture of SG systems.

PLC technologies in SG WANs benefit from covering communication networks cheaply by utilizing existing power lines primarily designed for energy transmission. PLC technology may also give complete control based on utilities' broad coverage solely maintained. PLC systems strive toward minimal latency by providing direct connection between control systems and subsystems. Because numerous types of electrical equipment are linked to power lines and act as noise generators, various noise types adversely influence the operation of PLC systems.

Furthermore, the variable impedance of power lines and the electromagnetic interference problem cause significant distortion and signal attenuation. These are regarded as significant concerns for PLC-based SG systems. Another option for SG WANs is to leverage IP-based technology. IP-based communication systems can give a broad coverage advantage for connecting the entire SG system's components. Differentiated services and multi-protocol label switching methods can provide excellent Quality of service (QoS) and secure connections in these systems. Furthermore, IP-based systems use enhanced security mechanisms such as Internet protocol security (IPSec). On the other hand, latency is a significant issue for IP-based communication systems since slave nodes cannot send data in a master/slave setup.

It is the primary shortcoming of IP-based communication systems in SG applications. The real benefit of wireless networks in SG WANs is that they give extensive coverage at a lower cost than the previously discussed communication methods. Additionally, wireless communication technologies, particularly cellular communication technologies, can enable high-speed data rates for SG applications. Furthermore, some wireless technologies can allow mesh network schemes to increase network resilience. A connection to the wireless network is required before data can be transmitted, which might be a problem in the case of emergencies or outages. Long-term latency can pose significant challenges for SG systems' real-time services. These are the primary challenges with wireless technology in SG WANs.

- **Meter data management systems**

A database is meter data management systems (MDMS). It contains analytical tools that communicate with other information systems such as client information scheme, billing software, load shedding management system, enterprise resource planning, management of power quality, demand forecasting systems, mobile workforce management, and transformer demand management.

- **Benefits of AMI**

(i) Accurate meter reading and customer service benefits
The use of sophisticated metering may considerably cut meter reading costs while simultaneously improving the accuracy and timeliness of meter reading and billing.
(ii) Asset management benefits
They assist in the management of energy network assets. Data about consumer demand and usage trends are detailed and precise. Proper equipment size, predictive maintenance of equipment, detection of theft, and the ability to utilize precise data on client demand to postpone investment in networking infrastructure.
(iii) Additional tariff options (particularly time-varying pricing)

Adjustable billing periods, real-time information on actual power usage and its cost, along with web-enabled billing and interface services.

(iv) Outage management

It has the capacity to identify outages by giving exact information about their location, allowing maintenance workers to schedule their work more efficiently. Another significant advantage is the ability to respond to minor disruptions more quickly, which helps to improve customer service.

(v) Billing improvements

Increased accuracy in billing prevents consumers from underpaying (risk of debt) and overpaying (potential cash flow issues). There is little need for a manual meter reading. As a result, users are no longer required to allow meter readers access to their homes; customers also have numerous payment alternatives.

5.1.4 Challenges and issues in SG system

Applications and challenges of the SG system are presented in the book authored by [42] Mariana Hentea (2021), as summarized in Table 5.1.

One of the important issues is secured meter reading data communication within the AMI of SG. Nowadays, blockchain technology is becoming popular in securing data in these kinds of CPSs.

5.1.5 Blockchain-based security

Blockchain is an immutable ledger that records transaction details in a secured fashion. Transactions pertaining to tangible and intangible assets can be recorded in a blockchain. Once entered in the blockchain, transactions cannot be changed in the future; instead, new records can be added with updated transaction values. The ledger is maintained in a distributed manner. Hence, all the system participants can have a copy of the ledger, but updating the ledger can only be done by designated system participants. Every transaction is verified for its genuineness through a technique called consensus mechanism. Blockchain has gained focus as it can be used as a security model for emerging from Bitcoin-based transactions and various applications. In this chapter, a blockchain-based security system is used for avoiding and detecting the tapering of data read from smart meters in electrical grids. Blockchain largely depends on cryptography, ledger immutability, distributed maintenance of ledger, smart contract, and consensus mechanism.

- **Shared ledger**

A shared ledger is a modern version of book keeping in which the data are immutable. The data or transaction, once entered in the ledger, cannot be changed afterward. Any changes to the existing data have to add to the ledger with a time-stamp as a new transaction. The ledger can be updated only by the designated members of the system, and every transaction is verified for legitimacy by the verifier member and the verified data/transactions. The ledger is maintained in a distributed manner and every member of the system. It is allowed to have an individual copy of

Table 5.1   Applications and challenges of SG system

Level of network	Properties	Applications	Issues
Power plant	Advanced circuit breakers and switches/Regulating Inverter Distribution System Automation SCADA Loading Monitor Phase Angle Regulating Transformer	Metering (event log, data log, customer's authorization, and device identification) Disaster digital evidence (breakdown of equipment, sensor of equipment health tracking Bitcoin payments for energy use. Management of distributed equipment Ecore software (full nodes)	Personal Information Privacy Security of data storage and collection costs of data storage and processing (partially) throughput. Deficiency of real-time analytic tools Accessible Information Collection Advanced information processing. Hardware for event logging for mission-critical assets (similar to flight data recorders)
Transmission System	Flexible AC transmission system Device Fault Current Controller (FCC) Micro Controller PMU Technology SCADA	Metering (event log, data log, customer's authorization, and device identification) Healthcare sensor Device Highly decentralized application (Device Management) Light Vertices Getting a pattern collection for old ENF recordings. Attacks on GPS spoofing Analog and digital signal processing for audio and video recordings (partially)	Metadata collection during severe disasters Management of the SG in a communications network outage or damage. Hardware for event logging for critical assets (similar to flight data recorders)
Home	AMI/Smart Meters Smart Appliances and Equipment Customer EMS/Display/Portal	Metering (event log, data log, customer's authorization, device identification, and Healthcare sensor Device) Highly decentralized application (Device Management, invoice payment) Light Vertices	
Network	RTU Distribution Management System	IEC 61158	The private key provides the user with access. Only authenticated devices have access. The information saved on all

the ledger. The transactions recorded in the ledger have restricted access, and any user can only access the permitted transactions for viewing purposes.

- **Consensus mechanism**

All transactions or data are verified thoroughly for the authenticity of the source, tampered data, and validity. It is done by the designated members of the blockchain network. There are several consensus mechanisms in blockchain, and a few are discussed here.

(i) Proof-of-work

In this technique, the member of the blockchain network should solve complex mathematical puzzles to create and add a block in the blockchain. These members are called miners, and they mine using CPUs or GPUs to solve the mathematical puzzles. The puzzles get complicated as the speed of block addition is increased. It is done to control the speed of the addition of blocks. This technique necessitates the members to have a high processing capacity.

(ii) Proof-of-stake

In this technique, the verifiers lock the Bitcoins they own for a certain amount of time to become a validator. In other words, validators stake the coins they own to become the producer of the next block. The validators who stake the Bitcoins for a longer time will have a higher probability of creating the next block. Validators usually get rewarded for the block they have created. Proof-of-stake technique is more energy-efficient than proof of work mechanism.

(iii) Delegated proof-of stake

It is a modified version of proof-of-stake where various delegates stake their coins in favor of a particular validator. If two delegates want to become a validator, and if one has 20 coins and another has coins, the delegate with the maximum number of coins at stake will be selected as the next validator.

(iv) Proof-of-uthority

In proof-of-authority, the delegates put their reputation at stake. Moreover, it became a validator. In this technique, only the authorized delegates can become validators and add new blocks to the blockchain. Compared to other techniques, proof-of-authority has the smallest number of validators.

Apart from the above techniques, there are more techniques to proof-of-identity, proof-of-elapsed time, and proof-of-capacity that can be used in suitable venues.

- **Smart contract**

The smart contract is a set of codes that is executed automatically on every ledger update. Business agreements are usually implemented in these contracts, and the transactions are validated. For example, the smart contract can create a warning message if updated values exceed the threshold value.

In this chapter, the necessary feature of blockchain is extracted and used to alleviate various issues related to Smart Metering Systems such as disclosure, disruption, deception, usurpation, and interrupting the measurement. Hence, the integration of blockchain technology with the AMI of SG will improvise the secured communication among the components of SG, which in turn improvise the efficient billing system from the tampering of electric power consumption in the SG system. Hence, the chapter proposes a blockchain-based secured IoT-enabled SG system. The following section details the literature survey.

5.2   Literature survey

The authors present SG CPS is based on a centralized supervisory control and data acquisition (SCADA) system hierarchically networked with several elements, including master terminal units, remote terminal units, phasor measurement units, and other sensors. The SCADA system is frequently used to monitor and operate electricity grids. IoT smart devices, sensors, and phasor measurement units (PMU) often gather power device status information and exchange it with Master Terminal Units (MTU) through remote terminal units (RTU), with MTUs serving as regulate units and primary storage. The SG system uses the measurements between CPS components, multiple grid operators, suppliers, and consumers to provide intelligent control, wide-area monitoring, and governance to better manage their grid's safety, stability, reliability, and monitoring power theft and loss. Malicious attackers or insiders can conduct cyber-attacks in various ways. Some examples are changing data in central controllers, initiating availability attacks, and providing incorrect data via sensors and PMUs. As a result, the attacker can seize control channels and issue malicious orders. The blockchain provides additional monitoring, measurement, and control capabilities in the decentralized SG system [43].

A SG surveillance method based on blockchain and intelligent protocols is proposed to provide energy usage secure data transmission. Smart meters, consensus nodes, and utility corporations are the three nodes that make up this blockchain network. Smart meters frequently send energy data to the blockchain network in digital form. On the other hand, consensus nodes maintain energy usage statistics, record individual utility provider subscription details, validate the latest modules, and transmit them to the primary channel. However, before generating the latest modules, these nodes make interim phases for individual customers, including meters IDs and other user data [44].

Prosumers and consumers trade peer-to-peer without relying on a centralized authority thanks to blockchain inclusion in SGs. The use of blockchain technology in SGs can be beneficial [45]. A SG with blockchain integration provides a safe trade infrastructure with restricted participation from prosumers and consumers for effective transactions.

The authors proposed a blockchain-based security mechanism for MANET. In this work, the MANET is divided into a grid, and every grid would have a miner node to verify the transactions. The algorithm builds a trust value for nodes based on their correctness in delivering the packets to the destination. Based on the trust value, the maliciousness of the nodes is identified. This algorithm is more suitable for a static environment as a dynamic environment would attract additional control overhead when the nodes move from one grid to another. Moreover, some nodes are dedicated as miner nodes, which would perform the verification process alone that is not practical in MANET with less computational capability [46].

The authors proposed a technique to secure a MANET using trust chain optimization. This technique uses a hierarchical model based on stochastic Petri nets. Through this technique, the nodes develop a trust value based on social trust and Quality of Service. The trust values are developed based on direct and indirect interactions with the node, and the longer the indirect interaction, the longer the trust chain is. Though the algorithm efficiently attains secured transactions in MANET, this algorithm would be more suitable for static systems, and though the concept of stochastic Petri nets is used, backbone formation using Connected Dominating Set (CDS) is not implemented [47].

The authors devised a technique for detecting gray and black hole nodes in the Cloud MANET. The transactions in the MANET are verified using the digital signatures of the nodes attached to the acknowledgments, and the elected verifier nodes perform this verification process. Moreover, the legitimacy of the transactions is estimated and updated in the shared ledger. This technique effectively minimizes the overhead and computational complexity of blockchain technology. This technique solves the gray/blackhole attacks in conventional MANET, and work is not extended for Cloud MANETs with CDS [48].

The authors proposed a reinforcement learning-based trust value development in which the nodes learn the trust value by interacting with other nodes. This technique maintains a packet delivery ratio of the neighboring nodes in a table format. Every node maintains information only about its neighbors, and hence any gray hole node that lies beyond the one-hop cannot be identified by the node. This technique detects the gray hole behavior only after flooding the MANET with data packets, creating a broadcast storm and not being adopted for CDS-based communication [49].

The authors introduced establishing the route between nodes through smart contracts without using encryption or public-key cryptography. The nodes would reactively make a route request using a smart contract. The smart contract maintains a timer and ratifies whether the route reply is legitimate based on the time taken for the reply. This algorithm suffers from a setback as message flooding has to be done for route requests, increasing the chances for a broadcast storm. The messages' authenticity could not be verified as the algorithm does not adopt any authentication technique or public-key cryptography [50].

The authors proposed a blockchain-based algorithm for establishing a route in Vehicular Ad hoc Network (VANET). The algorithm creates a connected dominating set for miner nodes in the blockchain. The algorithm considers various network performance parameters like connectivity degree and average link quality along with social behavior of the nodes to rank them and select as mining nodes in the CDS. This algorithm demands higher processing capacity and dynamic computation of CDS is unavoidable due to the frequently changing topography of the VANET. Adopting this algorithm to Cloud MANET is hard as the algorithm depends on RSUs processing capacity. Moreover, the algorithm focuses less on particular network attacks like gray and black holes [51].

The authors proposed a virtual backbone construction using connected dominating sets in MANET. The reinforcement learning algorithm is used to learn about the neighboring nodes, and the Q value is estimated using feedback-based learning. It gives a fair assumption of nodes and helps in efficiently establishing a backbone. Though this solves routing issues in MANET, detecting gray hole attacks cannot be detected using an algorithm, and hence the performance of the algorithm degrades in the presence of spurious nodes [52].

The authors proposed a context-sensitive downloading scheme that dynamically decides about downloading content from cloud services based on node resource availability. The algorithm stands as proof for forming MANET of smart devices and connecting the same to the cloud to avail cloud computing services in real-time [53].

The authors [54] suggested a secured routing scheme for linked dominant cast effective communication. Every transaction is acknowledged with a digital signature from all CDS nodes sending the data, and the technique verifies the acknowledgments sent by the various nodes to detect the black hole and gray hole attacks. The CDS would be reconstructed when the packet delivery ratio is lesser than the threshold value [54].

The authors present a comprehensive survey on various security models of blockchains and their pros and cons. The work explores unspent the transaction output mode and verifies the transaction's legitimacy using the sender's unspent amount as proof of transaction. In contrast, the account-based online transaction model uses the amount being sent by the sender as proof of transaction by verifying it against the sender's account statement. The work also ascertains the consistency availability and partitioning tolerance features in the blockchain, which can be used for devising various security models for various applications [55].

The authors comprehend the adoption of blockchain-based security systems in IoT networks [56]. As IoT comprises various connected cyber-physical devices, this work proposes a way to implement the end-to-end traceable and verifiable transaction in IoT systems. The author explores the pros and cons of the public blockchain, private blockchain, and consortium blockchain. The shared ledger and the consensus mechanisms are cast to the resource-rich IoT gateways for verification.

From the various literature, it is inferred that the SG system should be strengthened by improvising the security measures in AMI. Similarly, blockchain technology is a more opt distributed security mechanism suitable for SG systems that are distributed in nature. Hence, the chapter proposes a blockchain-based secured IoT-enabled smart SG. The following section details the proposed blockchain-based secured SG system.

5.3 Blockchain-based secured IoT-enabled SG system

In this work, a blockchain-based security system is proposed to avoid various security threats, including tampering of data. In this infrastructure, every device consuming power supply will be connected to the smart meter and periodically update the power consumption to the smart meter. The updates will have digital signatures attached to them, and hence the smart meter can verify the authenticity of the data received. All transactions have the data's hash value attached, making any data

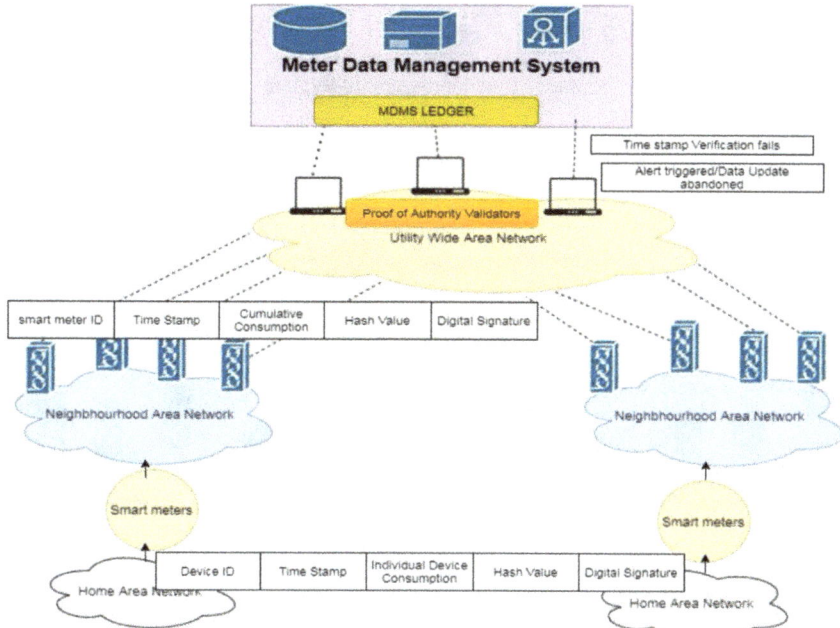

Figure 5.4   Normal scenario of blockchain-based secure meter reading communication in AMI

modification during transit easy to detect. Figure 5.4 illustrates the normal scenario of blockchain-based secure meter reading communication in AMI.

The smart meter then forwards cumulative verified data from the HAN to the utility wide area network with its digital signature and hash value added to the meter reading. It ensures security between the smart meter and the utility head end. The computational facilities at utility head end verify the sent data and updates in the MDMS, replaced by the blockchain ledger concept called MDMS Ledger. Figure 5.5 illustrates the detection of tampered meter reading using blockchain security system in AMI of neighborhood area network.

Figure 5.4 illustrates the behavior of the server on receiving tampered or delayed data. After verifying the hash value and the digital signature, the server verifies the authentication and integrity of the periodical update. The server also checks the period between two updates, and if the updates are delayed, the server will discard an update and alert the administrators of a possible attack in the smart meter.

5.3.1   Immutability of data in MDMS ledger

The records that will be added to the MDMS ledger will be verified for their authenticity and integrity. The verified transactions are then added to the MDMS ledger along with the timestamp. Periodical updates received from various smart meters

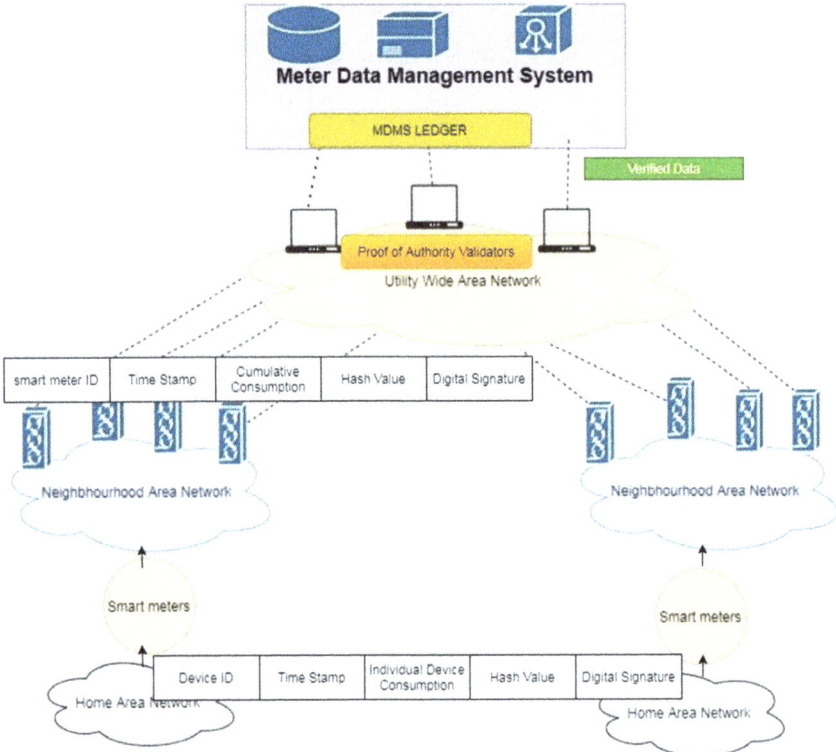

Figure 5.5   Detection of tampered meter reading using blockchain security system in AMI of neighborhood area network

are updated along with the smart meter ID and the timestamp. The data added to the MDMS ledger are immutable and hence changes to the existing data are not possible. Whenever an updated meter reading is received from a smart meter, it will be added as a new record in the blockchain. In this way, tampering of meter reading is minimized as changes to the existing record are not possible.

- **Consensus by proof-of-authority**

   The proposed security model follows a modified version of the proof-of-authority concept. The verification servers placed at the utility head end will be the authorized entity to make additions to the MDMS ledger. All the smart meters and the connected devices need to be registered in the servers connected to the edge routers. Hence, the servers hold all the details like device ID, public key, and location and maintain the information in a separate database. These servers act as the verifying authority and add the transactions to the MDMS ledger after verifying the data's digital signature and hash values. Any device/smart meter newly added to the grid needs to be registered with the verifying server and agree on a pair of private and public keys. This technique avoids the election of verifier nodes in real blockchain as it is time-consuming, and participating devices like smart meters cannot involve in the verification process. The servers maintain and preserve the data, which are needed for verification and execute the smart contract. There may be more than one validation server attached to the edge routers, and all routers maintain their copy of the MDMS ledger.

- **Smart contract**

The smart contract is a code, which is executed when the shared ledger is updated. The smart contract verifies the time gap between the previous and current updates in the proposed security model. If the smart meters fail to send updated meter readings in a specified time, smart contract sends an alert to the server cautioning about a possible adversarial activity in the site. The following listing provides the abstract code of the smart contract.

Algorithm for smart contract mechanism implementation:

```
1.   If ((time of the previous update - time of current update) > prescribed time
period)
2.   {
3.           trigger alarm
4.   }
5.   else
6.   {
7.           verify Digital_Signature
8.           verify Hash_Value
9.           If(Digital_Signature==correct and Hash_Value=correct)
10.          {
11.                 Update data into the ledger
12.          }
13.          else
14.          {
15.                 Discard data
16.                 Trigger alert
17.          }
18.  }
```

Public key infrastructure in SG

Public key infrastructure protects all devices connected to the SG. All devices would possess a private key and a public key. The public key of any device would be shared with all the other devices in the smart. Hence, any device connected to the SG would hold the public key of all the other devices. On the other hand, the private key of the

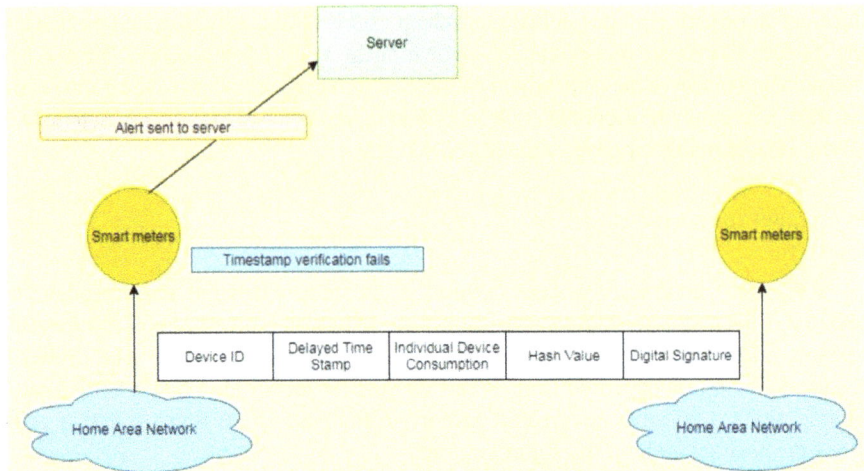

Figure 5.6   The detection of tampered meter reading and adversaries like a jammer, using blockchain security system in AMI of HAN

smart device would not be shared with any other device in the network. Through, this cracking of the private key is completely avoided.

Smart devices generate digital signatures based on their respective private keys. The digital signature can be verified for its authenticity by any other device that possesses the sending device's public key. All transactions made by the smart devices would be digitally signed before being sent to the destination. The receiving end would verify the digital signature using the public key using the respective public key. When a new device like a smart meter is added to the grid, the device would be assigned with a private–public key pair. The public key of the newly added devices is sent to all the other devices in the grid, whereas the private key would be preserved in the smart device.

5.3.2   Adversary avoidance in HAN

This technique also vies to avoid tampering with data inside the HAN. In the HAN, the smart meter has the ability to verify the hash values, digital signature, and timestamp of data updates received from the devices. The smart meter in the HAN has the capacity to verify the hash value and digital signature timestamp of the data updates received from the devices is recorded. The smart meters validate the digital signature and hash value of consumption updates received from various devices. The cumulative consumption over a specific time period is subsequently transmitted to the verification servers. The following Figure 5.6 illustrates the detection of tampered meter reading and adversaries like a jammer, using a blockchain secured system in AMI of HAN.

In the HAN, connected devices sent periodical updates on their consumption through the sensor network. The adversary can be caused by placing a jammer in the area, stopping the connected devices from sending periodical updates to the smart meter. It can be thwarted by adding a time stamp with the updates. The smart meter will calculate the time difference between two consecutive updates, and if the verification fails smart meter sends an alert to the server along with its Id cautioning about a possible attack.

- **Disclosure**

Disclosure refers to unauthorized access and disclosure of confidential data. For example, location and ownership information available in a smart meter are prone to data theft due to sensor vulnerability. The proposed security model brings in the public key cryptography concept, and hence all communications are done with digital signatures. Hence, attempts made from adversarial sources can be easily detected and discarded.

- **Deception**

The proposed security model thwarts the deception or injection of false data into the system. It is achieved as all devices add a hash value of the updates sent from the devices to the smart meter in the home network area and the updates sent from the smart meters to the verification servers. If any alterations are done to the data, the server can easily identify the hash value mismatch.

- **Disruption**

Disruption of periodical updates can be avoided using the proposed security model. Disruption is usually caused using jammer devices, thereby preventing the devices from sending periodical updates. It is identified by the smart contract code, as it verifies the time difference between the previous update and the current update of a particular device. If the current update is not received within the stipulated time, the smart contract code will automatically alert the servers prompting a disruption.

- **Usurpation**

The proposed security model can avoid usurpation or gaining unauthorized access to the devices or data by extracting passwords from smart meters. Since the AMI implements public key cryptography, every transaction is fortified with the digital signatures of the devices, and hence there is no possibility of pretending like a member device in the system. The data maintained in the servers are immutable and cannot be changed even if an attacker gains access to the server. Moreover, the data are updated only after verification of digital signature and hash value. Hence, the proposed system is protected from usurpation too.

5.4   Result and analysis

The proposed blockchain-based secured IoT-enabled SG system is implemented using MATLAB® software, and the data acquisition and data logging from the various devices can be maintained using SCADA software. The communication among the various devices with the AMI system follows Zigbee and IEEE 802.11 standards for HAN and neighborhood area networks. The experiments have been carried out by injecting the attacks through various devices by tampering with digital signature, hashing, and timestamping attributes of the data, which are to be periodically updated. This research has injected the attacks into the network through various devices linked to it once an hour. In a scenario, 144 assaults have been injected through six devices and evaluated using the suggested blockchain-based secured method. A confusion matrix was developed utilizing the true positive, false positive, true negative, and false negative values to estimate the accuracy, precision, sensitivity, and specificity parameters.

True positive (TP) means that the proposed work identifies the actual attacks as attacks during testing; True negative (TN) detects updated original data as actual data during testing. False negative (FN) and false positive (FP) are terms used when attacks are detected as updated data, and updated data are identified as attacks.

Definition of accuracy, precision, sensitivity, and specificity metrics are as follows:

Accuracy = (TP+TN)/ (TP+TN+FP+FN)
Precision = TP/(TP+FP)
Specificity = TN/(TN+FP)
Sensitivity = TP/(TP+FN)

- **Confusion matrix**

The proposed mechanism is evaluated daily using the information by randomly injecting the attacks through different devices in the HAN and neighborhood area network. During the experiment, among the total injected attacks through updated data using six devices are 144; in this, 120 updated data either have modifications in any one of the attributes such as digital signature, hashing and timestamping attributes at least. However, to validate the effectiveness of the proposed work, randomly, 24 actual data which are to be updated are also transmitted. For the particular scenario of six devices that inject the attacks, the confusion matrix is developed using the proposed algorithm for a particular day is presented, and the respective metrics have been evaluated. Table 5.2 illustrates the confusion matrix for the specific scenario of six connected devices that inject attacks, and Table 5.3 presents the estimated performance metrics for the said scenario.

Finally, the experiments are carried out by varying the number of attacks by varying the devices as 6, 12, 18, and 24, respectively, and the devices inject attacks once in every hour. However, among 24 attacks per device per day, the device will randomly send the actual original data occasionally in a day, which is consistent for

Table 5.2   Confusion matrix for the scenario of six connected devices that inject attacks

Actual tampered data	Deducted as tampered data		
		Positive	**Negative**
	Positive	TP = 117	FN = 2
	Negative	FP = 4	TN = 22

all the devices, and in the experiment, four constant actual data are injected in addition to 20 attacks. The accuracy of the proposed work is evaluated under the above stipulated constraints and presented in the graph shown in Figure 5.7 and the same quantitative analysis is presented in Table 5.4.

Figures 5.7–5.10 illustrate the accuracy, precision, sensitivity, and specificity analysis with respect to the varying numbers of devices that inject attack. Figure 5.7 presents the accuracy metric in terms of varying numbers of adversaries. When the adversaries are increasing, naturally, the accuracy and precision will be reduced. Because of the proposed technique that uses blockchain technology to prevent data tampering, though the adversaries are increasing considerably, accuracy, and sensitivity are not proportionally degraded. Similarly, the precision and specificity performances have been showing improved performance though the adversaries are increasing considerably.

5.5   Conclusion

In this work, a blockchain-based secured IoT-enabled SG system is developed to achieve more secured meter reading communication among the AMI system. The work implemented the blockchain mechanism in both HAN and neighborhood area

Table 5.3   Estimated performance metrics for the scenario of six connected devices which inject attacks

	No. of devices that inject attacks = 6	
S. No.	**Metrics**	**Value**
1	Accuracy	0.9653
2	Precision	0.9669
3	Sensitivity	0.9832
4	Specificity	0.8462

Table 5.4   Number of adversaries versus performance metrics

Sl. No.	No. of connected devices which inject attacks	Accuracy	Precision	Sensitivity	Specificity
1	6	0.9653	0.9669	0.9832	0.8462
2	12	0.9444	0.9587	0.9748	0.8
3	18	0.9421	0.9638	0.9665	0.8243
4	24	0.9323	0.9766	0.9387	0.8764

networks to prevent tampering of electric meter reading from the intruders, so that the billing system is efficiently developed without any economic loss to the government. The smart meters periodically update the meter reading and the digital signature to the nearby server. The distributed server verifies the authenticity of the received meter reading using a consensus mechanism. The meter reading is verified against tampering using the smart contract codes running in the server. The verified meter reading is then updated to the distributed smart ledger. In this way, the proposed model prevents unauthorized access to confidential data and the transmission of false metering data by an adversary. Hence, wrong demand estimation and erroneous billing are avoided.

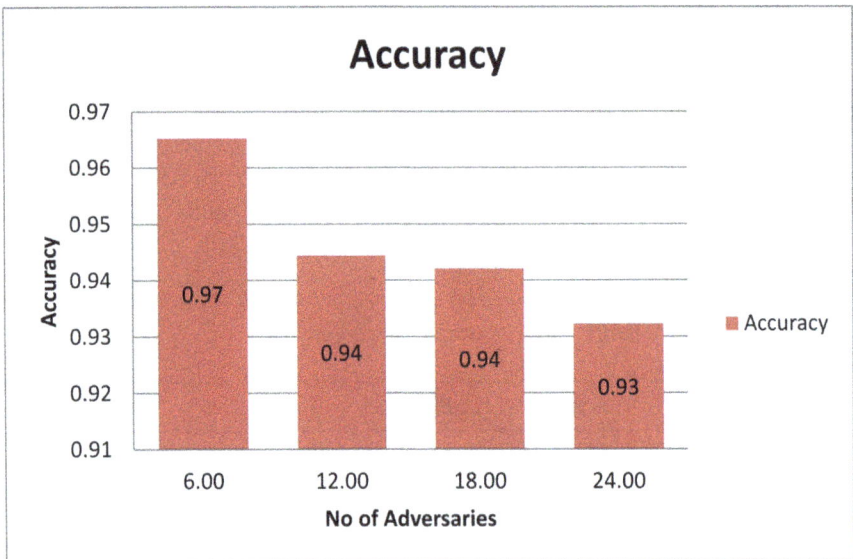

Figure 5.7   Number of adversaries versus accuracy

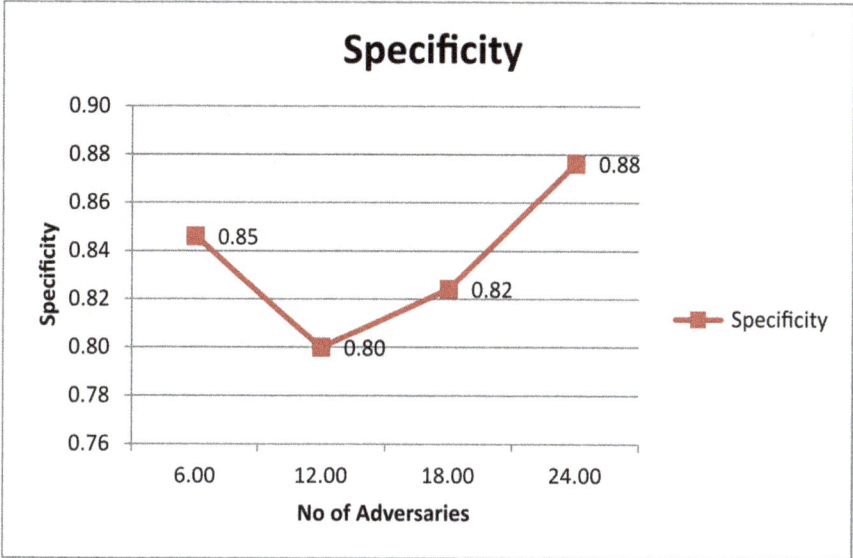

Figure 5.8   Number of adversaries versus specificity

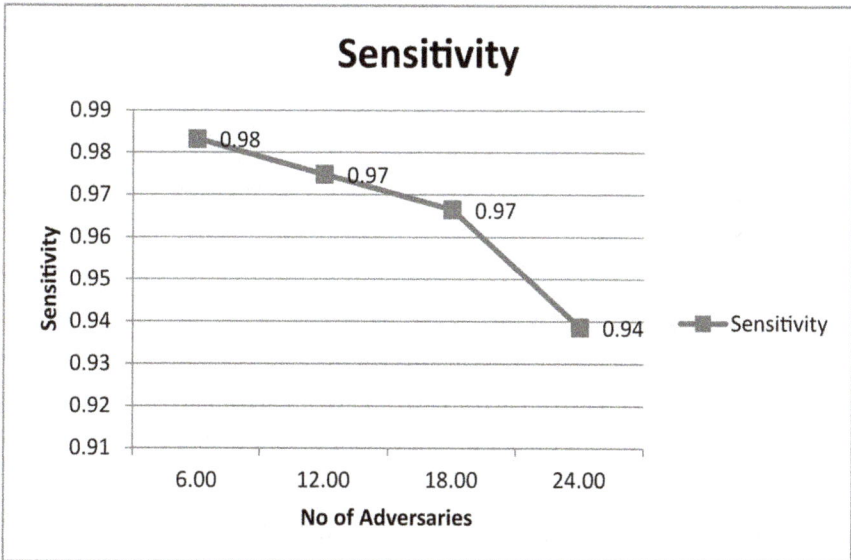

Figure 5.9   Number of adversaries versus sensitivity

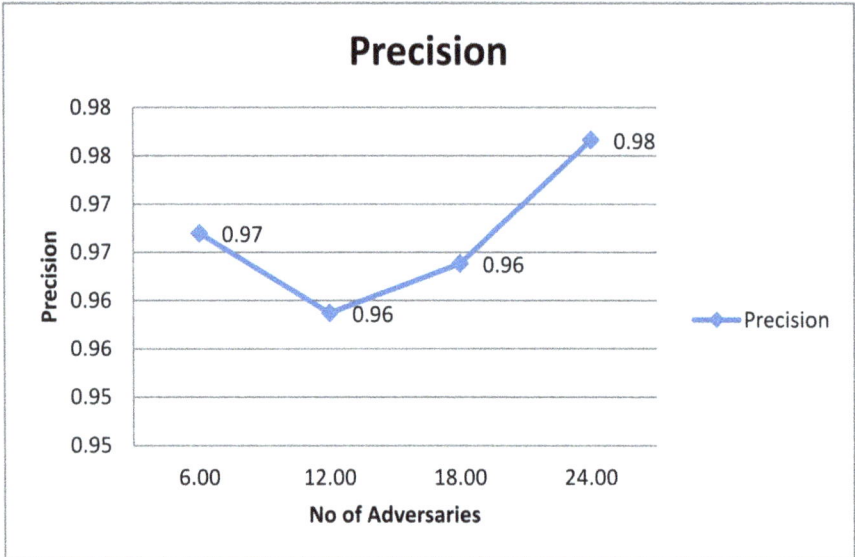

Figure 5.10   Number of adversaries versus precision

References

[1]   Mahmud K., Khan B., Ravishankar J., Ahmadi A., Siano P. 'An Internet of energy framework with distributed energy resources, prosumers and small-scale virtual power plants: An overview'. *Renewable and Sustainable Energy Reviews*. 2020, vol. 127(3), p. 109840.

[2]   Al-Turjman F., Abujubbeh M. 'IoT-enabled smart grid via SM: An overview'. *Future Generation Computer Systems*. 2019, vol. 96(6), pp. 579–90.

[3]   Saleem Y., Crespi N., Rehmani M.H., Copeland R. 'Internet of things-aided smart grid: Technologies, architectures, applications, prototypes, and future research directions'. *IEEE Access*. 2019, vol. 7, pp. 62962–3003.

[4]   Kabalci Y., Kabalci E., Padmanaban S., Holm-Nielsen J.B., Blaabjerg F. 'Internet of things applications as energy Internet in smart grids and smart environments'. *Electronics*. 2019, vol. 8(9), p. 972.

[5]   Wang K., Yu J., Yu Y., *et al.* 'A survey on energy Internet: Architecture, approach, and emerging technologies'. *IEEE Systems Journal*. 2018, vol. 12(3), pp. 2403–16.

[6]   TAN Y., LI Q., TAN Y. 'A comprehensive review of energy Internet: Basic concept, operation and planning methods, and research prospects'. *Journal of Modern Power Systems and Clean Energy*. 2018, vol. 6(3), pp. 399–411.

[7]   Zhaoyang D., Junhua Z., Fushuan W.E.N., *et al.* 'From smart grid to energy Internet: Basic concept and research framework'. *Automation of Electric Power Systems*. 2014, vol. 38(15), pp. 1–11.

[8]   Chen S., Wen H., Wu J., *et al.* 'Internet of things based smart grids supported by intelligent edge computing'. *IEEE Access*. 2019, vol. 7, pp. 74089–102.

[9]   Hussain S.M.S., Nadeem F., Aftab M.A., Ali I., Ustun T.S. 'The emerging energy Internet: Architecture, benefits, challenges, and future prospects'. *Electronics*. 2019, vol. 8(9), p. 1037.

[10] Gunduz M.Z., Das R. 'Cyber-security on smart grid: Threats and potential solutions'. *Computer Networks*. 2020, vol. 169(11), p. 107094.

[11] Islam S.N., Baig Z., Zeadally S. 'Physical layer security for the smart grid: Vulnerabilities, threats, and countermeasures'. *IEEE Transactions on Industrial Informatics*. 2019, vol. 15(12), pp. 6522–30.

[12] Kumar P., Lin Y., Bai G., Paverd A., Dong J.S., Martin A. 'Smart grid metering networks: A survey on security, privacy and open research issues'. *IEEE Communications Surveys & Tutorials*. 2019, vol. 21(3), pp. 2886–927.

[13] Ghosal A., Conti M. 'Key management systems for smart grid advanced metering infrastructure: A survey'. *IEEE Communications Surveys & Tutorials*. 2019, vol. 21(3), pp. 2831–48.

[14] De Dutta S., Prasad R. 'Security for smart grid in 5G and beyond networks'. *Wireless Personal Communications*. 2019, vol. 106(1), pp. 261–73.

[15] Rastogi S.K., Sankar A., Manglik K., Mishra S.K., Mohanty S.P. 'Toward the vision of all-electric vehicles in a decade [Energy and Security]'. *IEEE Consumer Electronics Magazine*. 2019, vol. 8(2), pp. 103–7.

[16] Shayeghi H., Shahryari E., Moradzadeh M., Siano P. 'A survey on Microgrid energy management considering flexible energy sources'. *Energies*. 2019, vol. 12(11), p. 2156.

[17] Arif M.S.B., Hasan M.A. 'Microgrid architecture, control, and operation'. *Hybrid-Renewable energy systems in Microgrids*. Elsevier; 2018. pp. 23–37.

[18] Kabalci Y. 'A survey on smart metering and smart grid communication'. *Renewable and Sustainable Energy Reviews*. 2016, vol. 57–pp. 302–318.

[19] Salman S.K. 'Introduction to the smart grid: Concepts, technologies and evolution'. *Institution of Engineering and Technology Energy Engineering*. 2017, vol. 96, pp. 1–260.

[20] Kabalci Y., Kabalci E. 'Modeling and analysis of a smart grid monitoring system for renewable energy sources'. *Solar Energy*. 2017, vol. 153(3), pp. 262–75.

[21] Borlase S. *Smart Grids: Infrastructure, Technology, and Solutions*. 1 Edition. Boca Raton, FL: CRC Press; 2013. pp. 1–608.

[22] Mah D., Hills P., Li V.O.K., Balme R. in in Mah D., Hills P., Li V.O.K. (eds.) *Smart Grid Applications and Developments. Green Energy and Technology*. London: Springer; 2014. pp. 331–7.

[23] Sun Q., Li H., Ma Z., *et al.* 'A comprehensive review of smart energy meters in intelligent energy networks'. *IEEE Internet of Things Journal*. 2016, vol. 3(4), pp. 464–79.

[24] Hertzog C. *Smart grid dictionary*. Greenspring Marketing LLC; 2009.

[25] Farhangi H. 'The path of the smart grid'. *IEEE Power and Energy Magazine*. 2010, vol. 8(1), pp. 18–28.

[26]   Inoue M., Higuma T., Ito Y., Kushiro N., Kubota H. 'Network architecture for home energy management system'. *IEEE Transactions on Consumer Electronics*. 2003, vol. 49(3), pp. 606–13.

[27]   Kushiro N., Suzuki S., Nakata M., Takahara H., Inoue M. 'Integrated residential gateway controller for home energy management system'. *IEEE Transactions on Consumer Electronics*. 2003, vol. 49(3), pp. 629–36.

[28]   Chhaya L., Sharma P., Bhagwatikar G., Kumar A. 'Wireless sensor network based smart grid communications: Cyber attacks, intrusion detection system and topology control'. *Electronics*. 2017, vol. 6(1), p. 5.

[29]   Hartmann T., Fouquet F., Klein J. 'Generating realistic smart grid communication topologies based on real-data'. *IEEE International Conference on Smart Grid Communications (SmartGridComm) (IEEE 2014)*. 2014, pp. 428–33.

[30]   Parikh P.P., Kanabar M.G., Sidhu T.S. Opportunities and challenges of wireless communication technologies for smart grid applications. *IEEE PES General Meeting (IEEE 2010)*; 2010. pp. 1–7.

[31]   Yan Y., Qian Y., Sharif H., Tipper D. 'A survey on smart grid communication infrastructures: Motivations, requirements and challenges'. *IEEE Communications Surveys & Tutorials*. 2013, vol. 15(1), pp. 5–20.

[32]   Terzija V., Valverde G., Deyu Cai, Cai D., *et al.* 'Wide-Area monitoring, protection, and control of future electric power networks'. *Proceedings of the IEEE*. 2011, vol. 99(1), pp. 80–93.

[33]   Batista N.C., Melício R., Mendes V.M.F. 'Layered smart grid architecture approach and field tests by ZigBee technology'. *Energy Conversion and Management*. 2014, vol. 88, pp. 49–59.

[34]   Amin R., Martin J., Zhou X. 'Smart grid communication using next generation heterogeneous wireless networks'. *IEEE Third International Conference on Smart Grid Communications (SmartGridComm) (IEEE)*; 2012. pp. 229–34.

[35]   Bera S., Misra S., Obaidat M.S. 'Energy-efficient smart metering for green smart grid communication'. *IEEE Global Communications Conference (IEEE, 2014)*; 2014. pp. 2466–71.

[36]   Kaebisch S., Schmitt A., Winter M., Heuer J. 'Interconnections and communications of electric vehicles and smart grids'. *2010 First IEEE International Conference on Smart Grid Communications (IEEE)*; 2010. pp. 161–6.

[37]   Wang B., Sechilariu M., Locment F. 'Intelligent DC Microgrid with smart grid communications: Control strategy consideration and design'. *IEEE Transactions on Smart Grid*. 2012, vol. 3(4), pp. 2148–56.

[38]   Spano E., Niccolini L., Pascoli S.D., Iannaccone G. 'Last-meter smart grid embedded in an Internet-of-Things platform'. *IEEE Transactions on Smart Grid*. 2015, vol. 6(1), pp. 468–76.

[39]   Garcia-Hernandez J. *Recent Progress in the Implementation of AMI Projects: Standards and Communications Technologies. International Conference on Mechatronics, Electronics and Automotive Engineering (ICMEAE) (IEEE)*; 2015. pp. 251–6.

[40] Kabalci E., Kabalci Y. 'A measurement and power line communication system design for renewable smart grids'. *Measurement Science Review*. 2013, vol. 13(5), pp. 248–52.

[41] Kabalci E., Kabalci Y. 'Multi-channel power line communication system design for hybrid renewables'. *4th International Conference on Power Engineering, Energy and Electrical Drives (IEEE)*; 2013. pp. 563–8.

[42] Hentea M. 'Building an EffectiveSecurity program for distributed energy resources and systems'. Wiley; 2021.

[43] Mollah M.B., Zhao J., Niyato D., *et al.* 'Blockchain for future smart grid: A comprehensive survey'. *IEEE Internet of Things Journal*. 2021, vol. 8(1), pp. 18–43.

[44] Gao J., Asamoah K.O., Sifah E.B., *et al.* 'Grid monitoring: Secured sovereign blockchain based monitoring on smart grid'. *IEEE Access*. 2018, vol. 6, pp. 9917–25.

[45] David B., Dowsley R., Larangeira M. 'Mars'. *Proceedings of the 1st Workshop on Cryptocurrencies and Blockchains for Distributed Systems'*. 2018, pp. 82–6.

[46] Careem M.A.A., Dutta A. 'Reputation based routing in MANET using blockchain'. *International Conference on COMmunication Systems & NETworkS (COMSNETS) (IEEE)*; 2020. pp. 1–6.

[47] Cho J.-H., Swami A., Chen I.-R. 'Modeling and analysis of trust management with trust chain optimization in mobile ad hoc networks'. *Journal of Network and Computer Applications*. 2012, vol. 35(3), pp. 1001–12.

[48] Lwin M.T., Yim J., Ko Y.B. 'Blockchain-based lightweight trust management in mobile ad-hoc networks'. *Sensors*. 2020, vol. 20(3), p. 698.

[49] Mayadunna H., De Silva S.L., Wedage I., *et al.* Improving trusted routing by identifying malicious nodes in a MANET using reinforcement learning. *2017 Seventeenth International Conference on Advances in ICT for Emerging Regions (ICTer) (IEEE)*; 2017. pp. 1–8.

[50] Ramezan G., Leung C. 'A blockchain-based contractual routing protocol for the Internet of things using smart contracts'. *Wireless Communications and Mobile Computing*. 2018, vol. 2018(1), pp. 1–14.

[51] Yahiatene Y., Rachedi A. Towards a blockchain and software-defined vehicular networks approaches to secure vehicular social network. 2018 *IEEE Conference on Standards for Communications and Networking (CSCN)' (IEEE)*; 2018. pp. 1–7.

[52] John Deva Prasanna D.S., John Aravindhar D., Sivasankar P., Perumal K. 'Reinforcement learning based virtual backbone construction in MANET using connected dominating sets'. *Journal of Critical Reviews*. 2020, vol. 7(09), pp. 146–52.

[53] Zhou B., Dastjerdi A.V., Calheiros R.N., Srirama S.N., Buyya R. 'A context sensitive offloading scheme for mobile cloud computing service'. *2015 IEEE 8th International Conference on Cloud Computing' (IEEE)*; 2015. pp. 869–76.

[54] Prasanna D.S.J.D., Aravindhar D.D.J., Sivasankar D.P. 'Block chain based grey hole detection Q learning based CDS environment in cloud - MANET'. *Webology*. 2021, vol. 18, pp. 88–106.

[55] Zhang R., Xue R., Liu L. 'Security and privacy on blockchain'. *ACM Computing Surveys*. 2020, vol. 52(3), pp. 1–34.

[56] Mathonsi T.E., Tshimangadzo M., Tshilongamulenzhe B.E.B. 'Blockchain security model for internet of things'. *ACADEMICS WORLD 158th INTERNATIONAL CONFERENCE' (Academic World)*; 2019. pp. 52–6.

Chapter 6

Deployment of IoT-based sensor data management in smart grids

Goutham B[1], Sunil Kumar B R[1], Gururaj H L[1], Ravikumar V[1], and Francesco Flammini[2]

The Internet of Things (IoT) is a widely recognized technology that connects ordinary devices to the Internet to provide convenience and a variety of functions, while the smart grid (SG) is described as a power grid that is connected to a broad network of Information and communications technology (ICT). This chapter discusses SG and its importance and requirement for the present trend. The deployment of IoT-based sensor data management in SGs is found to be a tedious process when it comes to data processing. The huge data of SGs has to be processed for increasing the adaptability of the system. There are various data-centric models for IoT-based data management. This chapter provides an insight into different data processing techniques, their advantages, and disadvantages.

6.1 Introduction

There is a steep increase in demand for electrical power. The traditional power system suffers from low quality of power supply along with unreliability. This is because of the inefficient monitoring of the system, employment of less automation techniques, and improper fault diagnostic. In the early 21st century, the improvements in the development of electronic equipment and communication technology resolve many problems in traditional grids, making the grid smarter. SG is the evolution of traditional grids with enhanced effectiveness, autonomous nature, and improved efficiency of power delivery [1]. The traditional grid was just capable of transmitting and distributing electric power. This modern grid is able to store, communicate, and make decisions. The SG is the intelligent power grid with a large installation of IoT networks. Where there will be the development of intelligent algorithms that can collect the information and process them, thus making the system self-reliant [1].

[1]Vidyavardhaka College of Engineering, Mysore, India
[2]Mälardalen University, Västerås, Sweden

IoT is a growing technology in recent times, which connects both machines and people [2]. The evolution of IoT is Machine to Machine (M2M) communication. Commercial & Industrial Security Corporation (CISCO) stated that by the year 2020, against the population of 7 billion, there will be more than 50 billion connections of the object. The connecting object can be anything such as a device or equipment with large storage and communication capabilities. Entire IoT is made of three layers: the application layer, the network layer, and the perception layer [3]. The top layer will be the application layer and in that layer, the processing of incoming information is done. This layer is mainly responsible for the integration of renewable energy, demand-side energy management, and power system monitoring. The work of the network layer is transferring data from the perception layer to the application layer. In this network layer, many constraints of devices and their network limits will be considered. The bottommost layer is the perception layer, which includes a group of many that can percept and exchange information through Internet communication networks [3].

With the deployment of the IoT, sensors are also considered as critical infrastructure that is more prone to cyberattacks because monitoring and controlling are done through standard Internet-based protocols and solutions [4]. They rely on these public communication infrastructures which leads to threats and may lead to financial losses and damage to the assets. If data are not managed properly, then it can be easily manipulated, leading to serious problematic issues [4].

6.2   Internet of things

IoT is a popularly growing technology in this digital transformation era. Due to the presence of various types of things, such as sensors, radio-frequency identification (RFID) tags, mobile phones, and actuators, IoT has become more progressive [5]. The greatest potential for IoT development is its impact on day-to-day life. IoT has a strong influence in the field of SGs' health and other. There is also the development of threats to the technology, influencing all the objects resulting in information security risks. IoT architecture is made of four main things, as shown in Figure 6.1.

The IoT development in the world has made billions of objects to be control with intelligence, communication means, sensing and actuation capabilities will connect via Internet protocol networks. Our modern Internet has experienced a fundamental transformation, moving away from hardware-driven opportunities (fibers, computers, and Ethernet connections) and toward market-driven ones (Facebook and Amazon). This has occurred as a result of fast-growing technology [5].

IoT performs its operation in four stages, as shown in Figure 6.2.

1. In the first stage, sensors collect the data.
2. In the second stage, connectivity is done by sending the data to the cloud.
3. The third stage is processing the data and making it useful.
4. The final stage is providing the interface.

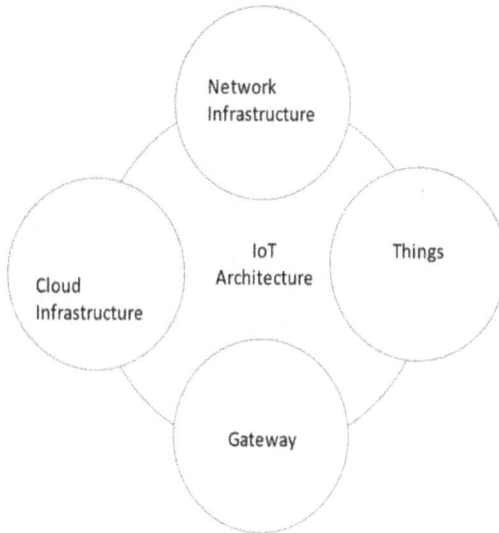

Figure 6.1   IoT architecture

The IoT necessitates open environments and a platform architecture that is interoperable. Objects being smart and cyber-physical systems, or simply "things," are the new IoT entities: everyday objects augmented with microcontrollers, optical and/or radio transceivers, sensors, actuators, and protocol stacks suitable for communication in constrained environments where target hardware has limited resources, allowing them to gather data from the environment and act on it and providing an interface to the physical world. Users can wear these things or deploy them in the environment [6]. They are typically confined, with limited memory and energy storage, and they must meet severe low-cost requirements. Data storage, computing, and analytics are essential prerequisites for enriching and transforming raw IoT data into meaningful information. Introducing computer resources to the edge of access networks, according to the "Edge Computing" model, may provide many important benefits for IoT scenarios: low latency, real-time capabilities, and context awareness, which are all important features. Additional processes, such as machine and deep learning, can be used to process the big data, transforming raw data created by connected items into meaningful information [6]. The beneficial data

Figure 6.2   Stage of IoT

will then be broadcast to appropriate devices and users, or it will be kept for later processing and access.

6.3   SG

Technical advances and the ever-increasing demand for energy necessitate a significant role for SGs. The concept of SG exists for quite some time, but there are only countable SG installations mainly in the academic institutions and far remote locations. As shown in Figure 6.3, SG comprises grid monitoring and home network systems, which are an addition to the traditional grid systems. The grid monitoring system has different energy storage devices for storing energy during the period of intermittency. The advanced monitoring system used is for the two-way communication system. The distributed side network of SG is included with distributed energy resources (DERs). Even with these systems, the technology of SG was suffering from communication problems.

SG terminology was a niche technology until the IoT arrived. The advancement of communication technology has transformed the generation and distribution of electrical energy. SGs with a higher percentage of renewable energy sources are on the verge of becoming an integral part of today's power system. Consumers and service providers will benefit from new technological advancements that allow them to have complete cost control; reliability and energy sustainability also allow all stakeholders to participate actively. This was built on the top of communication infrastructure, which is made up of the combination of hardware and software communicated with reporting software. Consumers will have utility corporation abilities for responding to challenges in the SG era. The electricity transfer between utility and consumers becomes two way in this system. There will be greater transparency achieved.

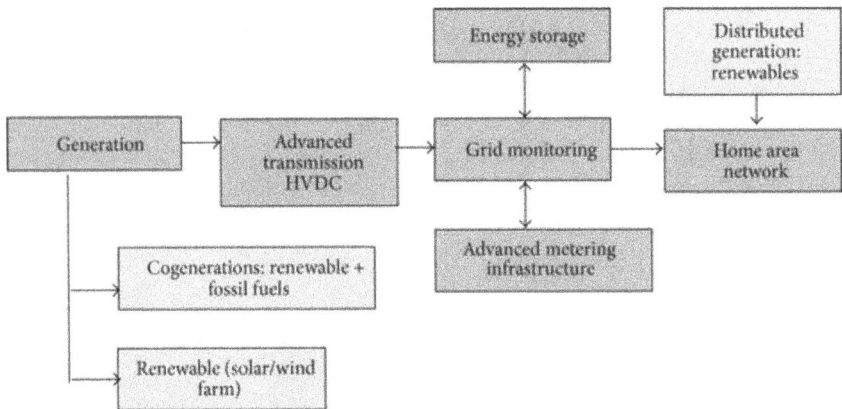

Figure 6.3   SG block diagram

6.4   Requirement of SGs

In today's world's scenario, stringent carbon emissions are mandated post COP21 agreement deployment across the globe [2]. For achieving super economic growth under these circumstances, the evolution of the energy industry is important. An SG system with a technology-intensive and superior communication network will enable locally controlled and highly reliable power. Depleting fossil fuel reserves will make SGs, with higher renewable energy sources, penetration increasingly cost-effective. With the combination of renewable and storage systems, the peak hour demand can be well managed, hence reducing the cost of energy during peak hours of consumption [7]. The SG will use storage and high-cost instantaneous power sources at the local level to control demand and supply in order to meet creatively at all times. This will lower the capital cost of installed capacity at the neighborhood, district, state, and national levels [6]. A recent study by *The Wall Street Journal* revealed that any assault on just nine key substations among the total of 55 000 could paralyze the entire US power system for many weeks to months. So in order to prevent such eventuality and improve the reliability of the power system, local-ized generation, distribution, and consumption would be the right pick. In nutshell, the SG solution will fulfill environmental, reliability, sustainable energy, and eco-nomic growth requirements. The SG has progressed swiftly, but unevenly, during the past few years. Recent events in Japan, such as the nuclear reactor meltdown at Fukushima, have made the energy equation even more complicated, driving the need for an SG with higher renewable penetration [7]. Figure 6.4 gives a brief idea about the need for SG.

6.5   Challenges of SGs

Companies and the public at large believe that current SG development is slow and that SGs have not delivered on the promises made to the public by the government. These promises include establishing two-way communication between users and

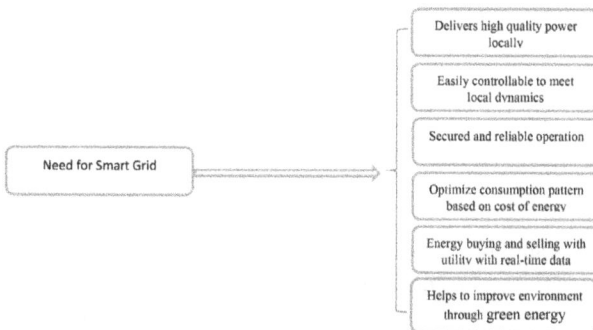

Figure 6.4   Requirements of SG

Table 6.1   Challenges and Enablers

Challenges	Enablers
Communication infrastructures	Development of communication infrastructure
Information technology (limited supervisory control and data acquisition (SCADA) capabilities, smart meters only for large consumers)	SCADA improvement, proliferation of smart meter
Noncompetitive electricity markets	Opening up electricity markets
Regulated government pricing	Policies and regulation dynamics pricing
Unsteady supply from renewable energy	Incentives for renewable energy

grids, and allowing users to control their own energy generation and consumption; increasing community job opportunities was not established as aimed.

There is a significant issue with how utilities and electricity companies communicate with customers who have already installed smart meters.

SG research and demonstration activities have progressed smoothly; however, investment excitement among power companies and other private firms is low, which has an impact on the future development of SGs to some extent.

In fact, many power firms' power asset ownership and administration are decentralized, and electrical equipment investments are substantial and have long life cycles. As a result, before making investment decisions, each power company must do a cost-effectiveness study. As a result, obtaining further finance is challenging Table 6.1.

6.6   IoT services to SG

IoT's broad sensing and processing capabilities can boost SG capabilities, such as processing, disaster recovery, self-healing, warning, and reliability. The integration of IoT and SG can lead to the improvement of communication devices, information equipment, and sensors. In several components of the SG, the IoT can be employed to achieve dependable data transfer in wire and wireless communication infrastructures as follows.

- IoT can be used to monitor power generation from various types of power plants (e.g. solar, wind, coal, and biomass), gas emissions, energy storage, and energy consumption, as well as estimate the amount of power required to serve consumers.
- An IoT can be used to track power usage and towers, monitor and protect transmission lines and substations, schedule, and administer and control components.
- IoT can be utilized in smart meters on the client side to measure various metrics, intelligent power consumption, network interoperability, charging and discharging of electric vehicles, and managing energy efficiency and power demand.

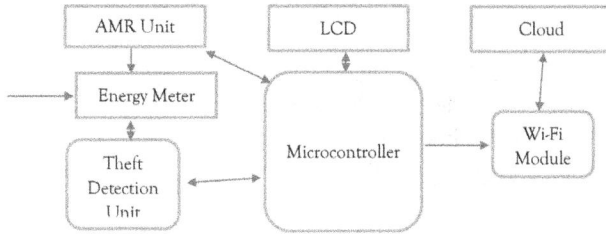

Figure 6.5   IoT-based smart energy meter

The application of IoT with SGs is as follows:

- Advanced metering infrastructure (AMI) with high reliability: SGs depend greatly on AMI. IoT can be utilized in AMI to collect data, monitor electricity quality and distributed energy, and analyze customer usage patterns [6].
- Smart home: A smart home can be used to communicate with users and SG, improve SG services, satisfy marketing demand, increase Quality of Service (QoS), manage smart appliances, collect power consumption data from smart meters, and manage renewable energy sources [6].
- Transmission line monitoring: The transmission cables can be monitored utilizing wireless broadband communication technology in order to detect and eliminate faults [6].
- Electric vehicle assistant management system: Integration of IoT-based data management system helps in assisting the charging station, electric vehicle, and monitoring center are all part of the electric vehicle assistant management system. Users can look up local charging stations and parking information using GPS. The indication for the management coordinates the management of automobile batteries, charging equipment, and charging stations, as well as the allocation of resources [6].

The block diagram of IoT-based smart energy meter is shown in Figure 6.5.

- Energy meter

An electric meter, often known as an energy meter, is a device used to measure how much electricity a building, tenant space, or electrically powered equipment consumes. Electric utilities assess the amount of electricity supplied to their customers using electric meters put on their premises for billing purposes [8].

- Theft detection unit

When a customer attempts to tamper with the meter, the theft detection device detects it using magnetic sensors. The microcontroller and relay circuit receive the signal generated by the sensors. The power source is disconnected via a relay circuit [8].

- Automatic meter reading (AMR) unit

AMR is a technique that collects consumption, diagnostic, and status information from energy metering devices (gas and electric) and sends it to a central database for invoicing, troubleshooting, and analysis [8].

- LCD

It is the display unit connected to measure the reading.

- Microcontroller

Microcontrollers are condensed microcomputers used to manage embedded systems in office equipment, robots, home appliances, automobiles, and a variety of other devices. Memory, peripherals, and, most crucially, a CPU are all included in a microcontroller. Here, this unit manages and monitors the work of other peripheral devices.

Microcontroller here is the main coordinating unit that sends signals to all other peripheral devices excluding the energy meter. The energy input reading is taken by the energy meter from the input supply and coordinated bidirectionally with a theft detection unit and single direction coordination with an AMR unit. AMR unit, in turn, establishes a bidirectional communication thereby controlling the energy meter reading. The microcontroller is connected to cloud through Wi-Fi module for storing and extracting the data [8].

6.6.1   Requirements for using IoT in SG

For the effective deployment of IoT-based sensors to SG, the following few technological requirements have to be satisfied:

- communication technologies
- data fusion techniques
- energy harvesting process
- operating in harsh environments
- reliability
- security
- sensors

A brief description of each requirement is listed below.

- Communication technologies: Information regarding the state of SG devices can be received and transmitted using communication technologies. We have communication technology standards for both short-range and long-range communication.
  (a) Short-range communication technology—Bluetooth, ultra-wideband technologies, and ZigBee.

    (b)   Long-range communication technology—Optical fiber, satellite commu-
          nication network, and wireless cellular networks, such as 3G and 4G

- Data fusion techniques: Because IoT terminals have limited resources (such as batteries, memory, and bandwidth), it is not possible to convey all information to the intended recipient. Thus, data fusion techniques can be used to gather and integrate information to boost the efficiency of data collecting.
- Energy harvesting process: Because most IoT devices rely on batteries as one of their key power sources, energy harvesting is critical for IoT applications, such as monitoring different portions of an SG with various sensors and cameras.
- Operating in harsh environments: High-voltage transmission lines and power stations require IoT devices that can withstand extreme conditions. As a result, we need to have sensors that are immune to high and low temperatures, anti-electromagnetic, or waterproof to increase their lifetime in these settings.
- Reliability: In diverse situations, IoT applications must meet varied needs, such as reliability, self-organization, and self-healing. As a result, an appropriate IoT device should be chosen depending on the actual environment in order to over-come environmental challenges. When some devices, for example, are unable to communicate data owing to a shortage of energy, an alternative path for the data must be established to ensure network resilience.
- Security: To transmit, store, and process information, prevent security breaches and losses, and preserve data, security mechanisms must be included at all IoT layers.
- Sensors: Sensors collect and send raw data such as current, voltage, frequency, temperature, power, light, and other signals for processing, transmission, and analysis. Nanotechnology has recently been employed to develop high-performance materials for a variety of sensor applications, boosting the sensor industry's growth. Nanotechnology has recently been utilized to develop ris-ing materials for a variety of sensor applications, boosting the sensor sector's growth.

6.7   Sources of data

The procedure for collecting, measuring, and evaluating correct insights for study using established approved procedures is known as data collection. On the basis of the data gathered, a researcher might evaluate their hypothesis [9]. Regardless of the subject of study, data collection is usually the first and most significant phase in the research process. Depending on the information needed, different approaches to data gathering are used in different disciplines of study.

    The data are broadly classified into the following categories, as shown in Figure 6.6.

1.   Primary data
    (a)   Survey and direct collection

2. Secondary data
    (a)   Internal data

        i.   Organization data

    (b)   External data

        i.   Government data

6.8   Procedure for IoT-based sensor data management in SGs

A three-layer structure with perception layer, network layer, and application layer is presented. The power layer (i.e., power sensor), tags, and readers (i.e. RFID tags/readers) or sensors (e.g., GPS devices/cameras) are used to collect information from the device layer [10]. The network layer contains various types of wired and wireless networking (e.g. 2G, 3G, 4G, wireless broadband, public telephone networks, private networking, WiFi, and ZigBee) and the Internet, which maps the information collected from sensors within the perception layer into communication protocols. It is used in processing, monitoring, and accessing the core network for the transmission of these mapped data to the application layer [10]. It comprises administration and information centers. The information received from the network layer is

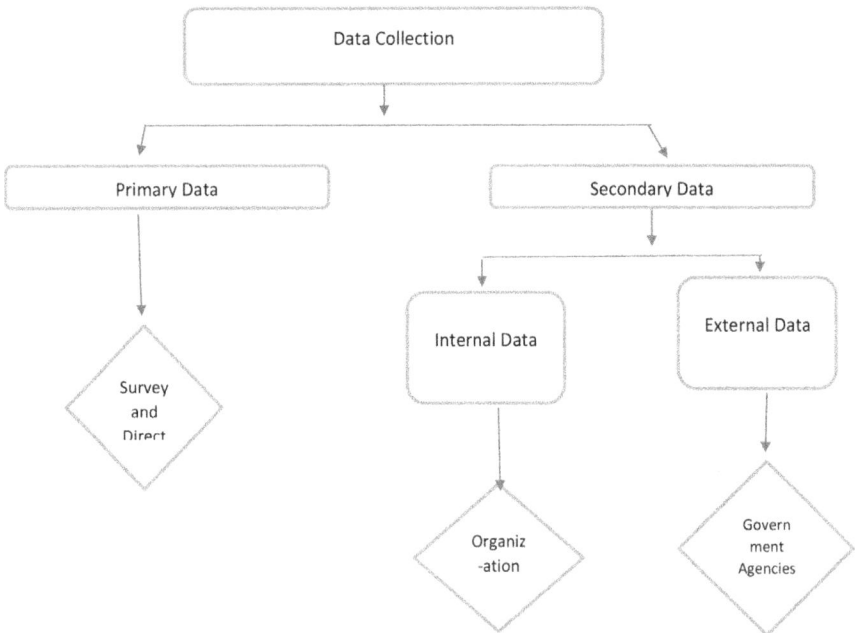

Figure 6.6   Classification of data

processed by the application layer, which allows IoT devices to be monitored in real time. It implements a comprehensive collection of IoT applications using a number of IoT technologies and includes application structure. The application layer of the IoT can facilitate the integration of information technologies.

There are four layers: cloud management layer, device layer, network layer, and application layer. Device layer contains two sublayers.

1. To detect the surroundings, collect data, and control home appliances, the IoT layer (which includes various types of sensors, smart meters, smart tags, and actuators) is used.
2. The gateway layer (which includes microcontrollers, communication modules, and local display and storage) governs how elements of the thin layer are connected.

Demand response management, dynamic pricing, and energy management are examples of services provided by the application layer to end users such as homes and utilities. The supporting layer is the fourth layer in the four-layer paradigm, and it combines certain common IoT technologies.

6.8.1 Requirements of data collection in SGs and the analyzing technique

The collection of data in SGs is done with help of sensors placed in the smart meters. The growing competition in the energy market needs the collection of data using sensors and other artificial intelligence techniques [11]. Data collected from IoT-based sensors at beginning of each feeder helps in the protection of the system. The data of current magnitude collected is required for maintaining voltage regulation by the automatic on-load tap changer.

The collected data has to be pre-processed for improving its quality. The stage of pre-processing has three major steps.

- Data integration

The process of merging data from several sources into a single, cohesive perspective is known as data integration. Ingestion is the first step in the integration process, which involves cleansing, Exact Transform Load (ETL) mapping, and transformation.

- Data cleansing

The process of detecting and correcting (or removing) corrupt or inaccurate records from a record set, table, or database is known as data cleansing, and it entails identifying incomplete, incorrect, inaccurate, or irrelevant parts of the data and then replacing, modifying, or deleting the dirty or coarse data.

• Data transformation

The process of modifying the format, structure, or values of data is known as data transformation. Data can be modified at two phases of the data pipeline in data analytics projects. Data transformation can be used in a variety of processes, including data integration, data migration, data warehousing, and data wrangling.

6.8.2 Process of IoT-based sensor data management in SGs

Even though with a great development in many techniques for processing big data, the data management of SG poses a big challenge. The real-time response and its proactive solutions and accurate prediction and forecasting raise critical issues. SGs are deployed with SCADA for core decision-making. This system is used in real-time monitoring and power grid control [12]. The sensors will be distributed all over the power grid and the SCADA system collects the data from these sensors. SCADA systems provide remote monitoring of the system by helping the flow of power with high reliability and efficient demand energy management. The increase in the size of the electrical grid with sensors causes a burden to manage data and upgrade it. The increasing use of cloud computing and its capabilities makes the system robust to host SCADA systems [3].

The model of cloud computing in the SCADA system helps in access sharing resources like storage, application, network, and server. The service to the end users by this enterprise of cloud computing is delivered in three models, namely software as a service (SaaS), infrastructure as a service (IaaS), and platform as a service (PaaS). The ready use of the application for end user is provided by SaaS, and it is capable of transferring the complete data of the SCADA system to the cloud and storing and processing it. The operating system and its storage, network, and database services for the end user's operation are provided by IaaS. For the end users, PaaS provided the ability to understand the programing that is available in the cloud [13].

The usage of cloud computing systems in SCADA has advantages like reduction in cost and higher collaboration. In the same pace with some disadvantages of network and security system, the problems will be caused due to data storage in the cloud. This causes the system more prone to cyberattacks. Even the limitation of bandwidth in certain areas reduces the effectiveness of this system [14]. Fog computing is a highly distributed model which can be used as a better replacement for cloud computing for the developing SG application. The fog computing helps to extend the cloud computing to network edges thereby data can be processed inside the devices on network edges and the need for shifting data to the cloud for processing can be avoided [15].

The most frequently used technique for analyzing huge historical data sets is the Hadoop MapReduce technique. In this method, the small data sets are divided from big data for processing. Then the developed small data sets are processed in machines using small codes. Static applications of SGs such as weather forecasting and these types of other similar applications can use this MapReduce fit technique [16]. But if the same technique is used for real-time application, then self-healing, online monitoring, and

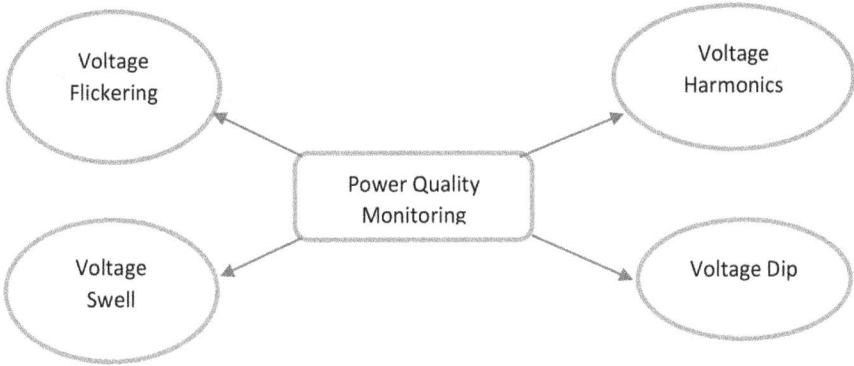

Figure 6.7   Power quality monitoring

fraud detection may be difficult and not efficient [17]. For processing the sensor data, stream processing is also considered a good platform. Stream processing technique design is done for handling the big data and fault-tolerant architecture. As applied to data analytics, stream processing has a greater scope.

6.9   Advantages of placing the sensors in SGs

The application of IoT-based sensors for data management helps in processing, warning, self-healing, disaster recovery, and reliability. Integrating SGs with IoT-based sensors can help in the effective usage of communication devices and smart terminals. For the different parts of SGs viz generation transmission and distribution, this IoT method of data management helps in reliable data transmission. The following are the parameters successfully overcome by placing sensors in SGs.

Power quality monitoring: The major concerning issue in the power grid is the maintenance of the quality of power. The quality of power refers to the frequency and the magnitude of voltage and current as referred to the safe operating area. The issue of power quality arises due to the increase in nonlinear loads, power electronic loads, and harmonic distortion owing to stability problems on the generation side.

The effect of bad power quality leads to many problems, as shown in Figure 6.7, such as

1. Voltage flickering
2. Voltage harmonics
3. Voltage swell
4. Voltage dip

1. Voltage flickering—Voltage flickering refers to a sudden change in voltage. Arc furnaces, any arcing condition in the power supply, constant starting/stopping of the motor, and oscillating loads all cause voltage flickers. This can be reduced in SGs with the placement of IoT and ensuring proper data management of loads.
2. Voltage harmonics—A harmonic of a voltage or current waveform in an electric power system is a sinusoidal wave with a frequency that is an integer multiple of the fundamental frequency. This leads to bad power quality.
3. Voltage swell—A voltage swell is a rise in voltage values over a short period of time. Overvoltages are defined as voltage swells that continue longer than two minutes. Large load shifts and power line switching are major causes of voltage swells and overvoltages. This leads to the deterioration in power quality, which can be overcome by the methods discussed in this book.
4. Voltage dip—Voltage dips (also known as "sags") occur when the voltage drops by 10% or more below normal or acceptable levels, which caused problems in the power system.

6.9.1   Predictive maintenance/condition-based maintenance

The concept of operation of SG at the distribution level is called distribution automation. This is fed from data of IoT-based sensors management. This successful implementation helps in localizing the faults and minimizes the restoration timings. For this process of distributed automation, large data has to be collected using advanced metering or supervisory data acquisition (SCADA). Pole-mounted autorecloser (PMAR) is the type of electronic device installed in the overhead line of distributed network for protection. References [2, 4] have stated novel methods for avoiding the problems of PMAR.

The development in the field of IoT technology and its implementation in power systems have improved the performance of the entire system. A huge volume of data is collected via IoTs. This power system data helps in avoiding possible threats.

6.9.2   Identification of topology

With the help of the information layer in SG, many challenges can be resolved in the distribution network. The system can be made sensible by using advanced sensors for applications like monitoring, communication and control, and measurement. The sensors used in SCADA and Wide area monitoring system (WAMS) provide huge real time-data [18] (Ghosh *et al.*, 2013). The dynamic solution approach has many solutions. The advantage of layers in SG is found to be the most suitable approach for the problems of RES in distributed networks. Low-carbon technologies are being pushed by the government through the use of heat pumps, electricity, electric vehicles, and other useful appliances in low-voltage distribution networks with the goal of creating a greener society.

6.9.3 Renewable energy forecasting

Wind and electrical phenomenon energies, which are abundant and environmentally beneficial, are expected to be the primary energy source for future facility generation. However, because of their randomness and intermittent nature, they constitute hurdles on a large scale, in an extremely stable manner. To address these obstacles and achieve better dispatching, maintenance programming, and regulation, a correct and trustworthy RES strategy is required. Figure 6.8 shows the data of US energy generation taken from environmental impact assessment (EIA). With the deployment of IoT-based sensor data management in SGs, the issue has been resolved.

The meteorological data found in historical records are employed in a cluster strategy to categorize times into distinct groups [12]. After that, a fabric algorithmic program-driven neural network is trained to predict wind energy outputs. Rather than using a neural network, the support vector regression methodology is used to predict wind speed using historical wind speed statistics that have been decomposed into multiple intrinsic mode functions and residues. In an exceedingly short-run probabilistic, wind power forecast methodology is bestowed supported the thin theorem classification and Dempster-Shafer theory as a statistic approach.

The types of renewable energy forecasting can be broadly classified into

- forecasting for long term
- forecasting for medium term
- forecasting for short term
- forecasting for very short term.

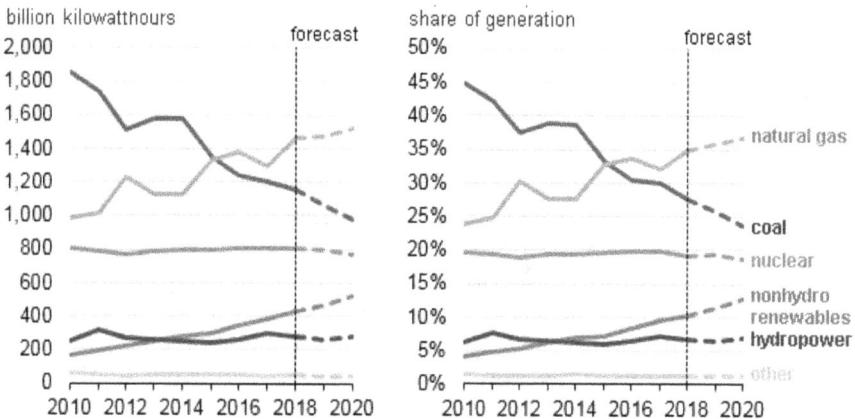

Figure 6.8   Energy generation data

6.9.4   Data management issues in SG

The SG with IoT-based sensor technology comes with storing and processing large data. Figure 6.9 shows various structured data. There will be data for every minute consisting of faults in the power line, the status of network components load demand at the user end, energy consumption scheduling, prediction of load demand, records of advance metering devices, and many more [19]. The developed integrated technology should be capable of managing, storing, handling, and processing the data effectively and safel [20].

There will be a collection of huge data characterized as big data. Big data is defined as the data with a large variety, having increased bidirectional flow, more volume, and high velocity. The high data velocity needs a higher processing speed, which is a challenging issue. However, there are some tasks that have to be performed all day, such as monitoring of real-time data of various power grid components. And some tasks require a speedy energy outrage for real time.

Among the various ways for the development of big data in SG is the implementation of the SCADA system. The implementation of phasor measurement units (PMUs) has the capability of scanning at a faster rate of 30–60 samples of data per second. Its installed with advanced meters having a capacity of reading the data in 15-minute intervals replacing the traditional monthly reading meters. The collection of data will be huge leading to 96 data per day (i.e., 24 hours × 4) and for a month 2 880 data from each power metering. This scenario is just an example of using just one advance meter for PMUs. If other advanced devices, such as intelligent electronic devices (IEDs), sequence of event recorder (SER), and digital fault recorder (DFR), are used, then a huge volume of data will be collected. And various events of data have to be stored for increasing the stability of the power grid.

The grid events can be broadly classified as voltage events, frequency events, and phase angle events. There is a requirement for storage events pertaining to each event, and each event is again subclassified as follows. In the power grid, the

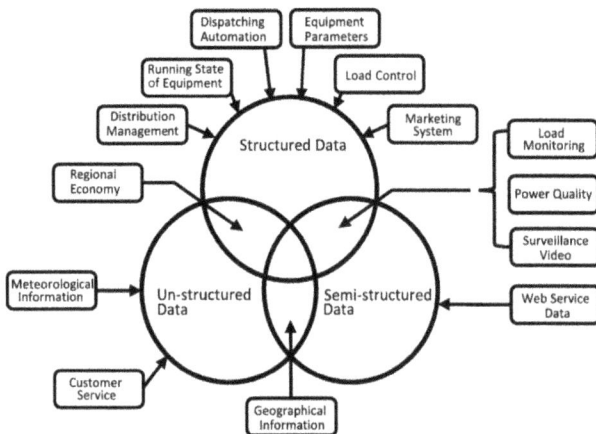

Figure 6.9   Structured data

three-phase fault in the system leads to a dip in the voltage. Even a small dip may lead to great disaster.

6.10   Benefits of sensor data management in SGs

The big data of SGs has numerous benefits to end users and generators.

1. Increase in efficiency: Big data analytics helps in better understanding physical limits and operating characteristics, thereby the system efficiency is increased.
2. Increase in asset utilization: The deployment of IoT-based sensors for data management helps calibrate and validate the physical model of SG, and the analysis necessary for a new model and effective utilization of the old model can be done.
3. Increase in stability and reliability of the system: The major concern of the power system is always the safety of the grid. And the safety is comprised of both stability and reliability. The stability and reliability of the system are achieved if the data of events like voltage stability, event detection and restoration, oscillation detection, post-analysis of faults, islanding detection, etc. are maintained properly.
4. Customer satisfaction: Introducing smart meters at residential and industrial premises helps in effective utilization of energy, reduces fault detection time, and hence great customer satisfaction can be achieved.
5. In generation sectors, the data collected from the sensors can be used for monitoring different types of power plants like coal, wind, and solar; biomass; gas emission; energy storage for its optimal usages; and prediction of required power to consumers.
6. Sensors placed help to know the systems about the consumption of electricity and its monitoring ability, the state of transmission lines, and their control equipment.
7. The different types of parameters like intelligent power consumption, interoperability of networks, and charging and discharging can be easily monitored, as shown in Figure 6.10.

6.11   Future trends and tendencies

From a policy standpoint, there is still a need to adapt and improve applicable laws and regulations for the long-term development of SGs in terms of controlling risks and benefits. They must do so while defining industry-acceptable technical standards and achieving the integration of various equipment manufactured by various companies. This would stimulate active engagement from a variety of private businesses and electricity firms, as well as attract more investment. Furthermore, improving user communication is critical, as it can assist SG developers in comprehending the drawbacks of the SG process. The growth of SGs must continue to enhance

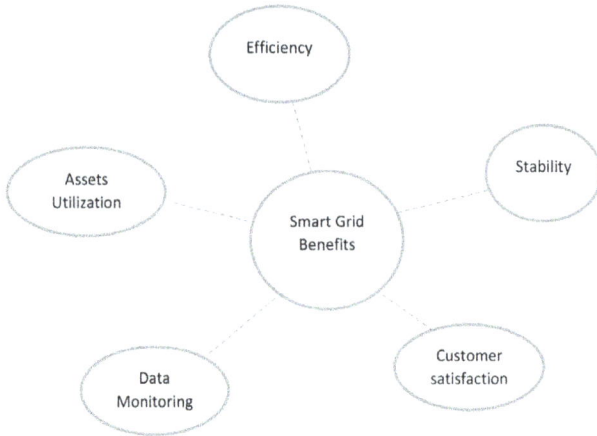

Figure 6.10   Data structures

the ties and interactions between the three institutions (administration, industry, and academic institutions) and seek to establish a new way of promoting SGs while providing new services in line with user needs. The growth of SGs in the future should include the following things, as shown in Figure 6.11.

6.12   Conclusion

Proper data management has the potential to increase the reliability of SGs. It also helps in enhancing economical gains. This chapter gives a review of the deployment

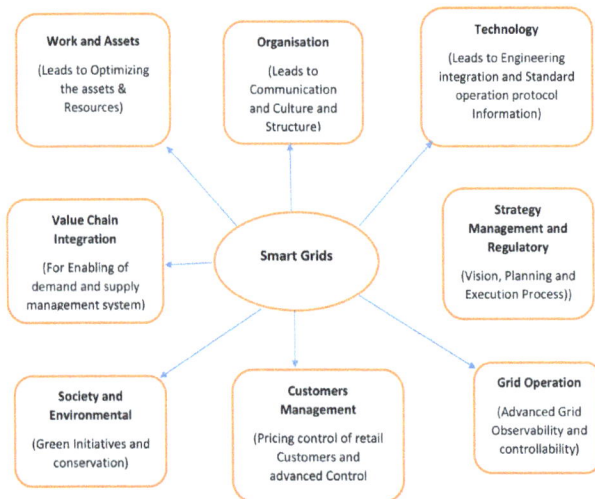

Figure 6.11   Data structures

of IoT-based sensors for management in SGs. The SG is found to be a technology that solves the problems of unidirectional flow of energy and information and wastage of energy. It is also found that SG helps in increasing the reliability of traditional power grids. Implementation and deployment of IoT-based sensors for the management of huge data are found to be effective in improving automation, connectivity, and tracking process. The study gives information about different ways of data management in SGs and their corresponding issues. Furthermore, the chapter discusses the interdependence of IoT data management in SGs.

References

[1]   Giusto D., Iera A., Morabito G., Atzori L. *The Internet of Things: 20th Tyrrhenian Workshop on Digital Communications*. Springer; 2010.

[2]   Atzori L., Iera A., Morabito G. 'The Internet of things: a survey'. *Computer Networks*. 2010, vol. 54(15), pp. 2787–805.

[3]   Xun L., Qing-wu G., Hui Q. 'The application of IOT in power systems'. *Power System Protection and Control*. 2010, vol. 38(22), pp. 232–6.

[4]   Shu-yong C., Shu-fang S., Lan-xin L. 'Power system technology'. *Survey on Smart Grid Technology*. 2009, vol. 33(8), pp. 1–5.

[5]   Morabito G., Iera A., Atzori L., Computer Networks. 'The Internet of things: a survey'. October 2010, vol. 54(15), pp. 2787–805.

[6]   Atzori L., Iera A., Morabito G. 'The Internet of things: a survey'. *Computer Networks*. 2010, vol. 54(15), pp. 2787–805.

[7]   'KNX association, Diegem, Belgium, version 2.0 of the KNX specification'. 2009.

[8]   Buettner M., Greenstein B., Sample A., Smith J.R., Wetherall D. 'Revisiting smart dust with RFID sensor networks'. *Proceedings of the 7th ACM Workshop on Hot Topics in Networks (Hotnets-VII)*; Calgary, Alberta, Canada; 2008.

[9]   Alqassem I., Svetinovic D. 'A taxonomy of security and privacy requirements for the Internet of things (IoT)'. *IEEE International Conference on Industrial Engineering and Engineering Management*; 2014.

[10] Shu-wen W. 'Research on the key technologies of IOT applied on smart grid'. *IEEE*. 2011.

[11] Romer K., Mattern F. 'The design space of wireless sensor networks'. *IEEE Wireless Communications*. 2004, vol. 11(2004), pp. 54–61.

[12] Trifa V., Wieland S., Guinard D., Bohnert T.M. Design and implementation of a gateway for web-based interaction and management of embedded devices. *Proceedings of the 5th IEEE International Conference on Distributed Computing in Sensor Systems (DCOSS Workshop)*; 2009.

[13] Kuzlu M., Rahman S., Pipattanasomporn M. 'An algorithm for intelligent home energy management and demand response analysis'. *IEEE Transactions on Smart Grid*. May 2012, vol. 3, pp. 1–8.

[14] Han D.-M., Lim J.-H. 'Design and implementation of smart home energy management systems based on zigbee'. *IEEE Transactions on Consumer Electronics*. August 2010, vol. 56(3), pp. 1417–25.

[15] Langhammer N., Kays R. 'Performance evaluation of wireless home automation networks in indoor scenarios'. *IEEE Transactions on Smart Grid*. December 2012, vol. 3(4), pp. 2252–61.

[16] 'International Electrotechnical Commission Std. 62056 Electricity metering - Data exchange for metre reading, pricing, and load control'.

[17] Euridis' essentials [online]. Available from http://www.euridis.org/solutions/details.html.

[18] 'Iso 16484-5 data communication protocol for building automation and control systems'. *International Organization for Standardization Std*.

[19] Kapar Z. 'Power-line communication - regulation introduction, PL modem implementation, and possible applications'. *12th International Conference on Power-Line Communication*; University of Johannesburg, Johannesburg, South Africa.

[20] Yashiro T., Kobayashi S., Koshizuka N., Sakamura K. 'An Internet of Things (IoT) architecture for embedded appliances'. *Humanitarian Technology Conference (R10- HTC), 2013 IEEE Region*; 2013. p. 10.

Chapter 7

Security and privacy of smart grid data and management using blockchain technology

B H Swathi[1], K S Anusha[1], S Gagana[1], and Hong Lin[2]

Through the confluence of the Internet of Things (IoT) and wireless sensors, the smart grid (SG) is proposed to fix the possible energy supply concerns. However, problems of reliability and safety in the trading of energy and the utilization of data cause significant obstacles to SG adoption. Blockchain technology is investigated to fix these issues in SGs. This chapter focuses on elaborating on the SG, blockchain technology, and some of the most critical SGs, which use blockchain applications. A systematic investigation is done in-depth, including recommendations for an appropriate blockchain design, a model block layout, and the blockchain technicalities that could be used. To address the problems of privacy and protection in the grid, data aggregation techniques that are efficient and based on blockchain technique are investigated. Energy delivery systems may use blockchain to potentially regulate electricity transfer to a specific location directly by tracking consumption statistics. In addition, blockchain-based systems will aid the process of diagnosis and upkeep of SG facilities. Ultimately, different obstacles that must be overcome in order to integrate these two systems are explored.

7.1   Introduction

Electricity is critical to the contemporary economy and society. The majority of the globe depends on inadequate electrical infrastructure that was built many years ago (more than 20% loss). Supply is continually modified to fit the needs of these systems, resulting in demand-driven control. Production, transport, and distribution are all one-way transfer processes in conventional grids [1]. These grids aren't safe enough against security flaws. These systems, on the other hand, might be made more dependable and long-lasting [2].

[1]Vidyavardhaka College of Engineering, Mysuru, India
[2]University of Houston-Downtown, Houston, TX, USA

Resource and
system efficiency

High Efficiency

**Traditinal Energy
Production**

Exhaustible fuels that
burden the environment

**Advanced enegy
Production**

Energy efficient and or
low emission production

Solar Economy

Solar based production with
high overall efficiency

waste to energy
CHP

Wind

Gas

Bio CHP

Sun

Coal

Conventional
CHP

Ocean

Oil

Nuclear today

Hydro

Carbon capture

Geothermal

Low Efficiency

High Emissions

Emission Free

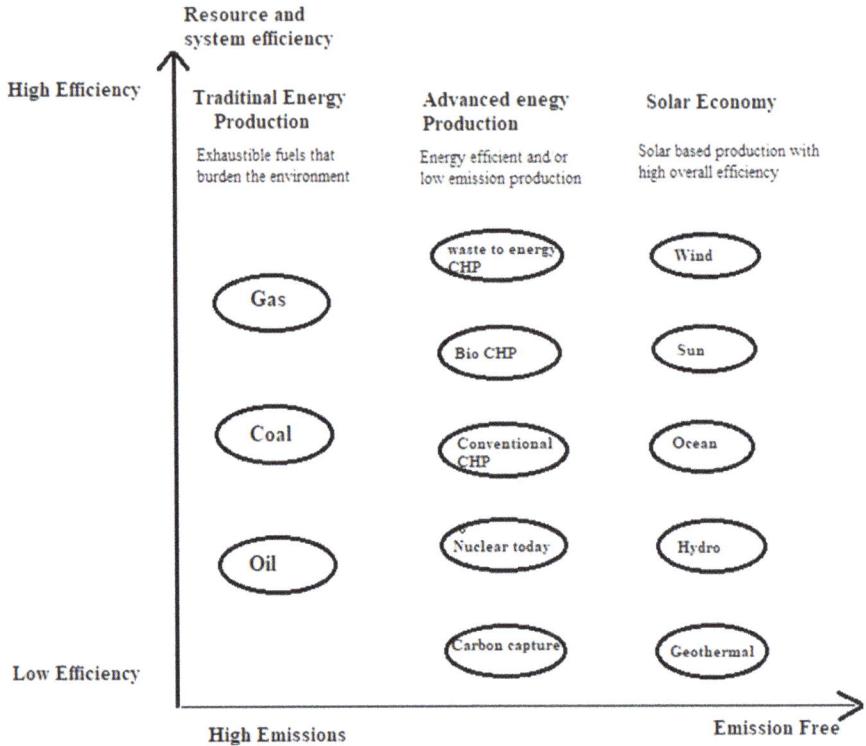

Figure 7.1   Energy system framework in future

The SG was defined in the United States by the Energy Independence and Security Act in January 2007. It's a digitally managed intelligent network of energy microgrids that can observe, regulate, and self-heal in the event of a power outage. SG provides a solution for clients to make the most efficient use of their power. SG's main advantages are increased electricity supply efficiency and dependability, enhanced cybersecurity of control and monitoring infrastructure, and the incorporation of distributed renewable energy sources into the existing grid. It should result in a decrease in carbon footprint due to increased efficiency. Energy will be generated by solar panels, rechargeable fuel cells, a vast chamber of pumped hydroelectric power, wind turbines, and other sources in the next generation of energy infrastructure. These systems should be able to manage simultaneous energy flows, and SGs will be required for better optimization. SG can modify demand by responding to dynamic changes in energy supply [3]. Figure 7.1 shows the future energy system framework.

The SG, called the intelligent grid, is the future production power grid environment that was created for more systematic and dependable energy management. The SG employs smart information communication technology to overcome the shortcomings of conventional unidirectional grid systems.

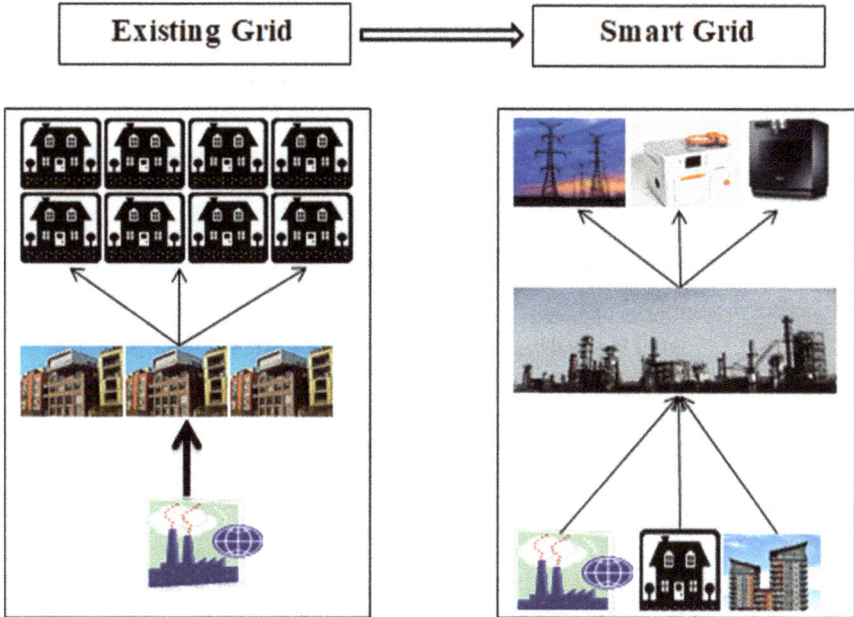

Figure 7.2   Transition from an existing grid system to an SG [5]

The SG has the benefit of being able to function as an ideal energy management system by linking electricity generation and delivery facilities, as well as power users, to the network and exchanging information in real time. However, because there are insecurity issues with data created by grid members and the implications of these issues can be fatal, efforts to detect and handle these threats are required [4].

7.2   SG

From the 1970s through the 1990s, as demand for electrical energy expanded, the number of power plants has increased since the first grid was erected in 1886. This development can be due to the world's urban growth and infrastructure development. Power plants have been at the top of the existing grid system's hierarchy, as shown on the left of Figure 7.2. Due to the hierarchy, the existing grid system is unable to understand and collect instantiations data on the assistance provided to consumers and a massive waste of power loss and surplus power is considered [6].

Using communication technologies and smart information in the existing grid, SG in connection to condition the grid gives enhanced "situational awareness" [7]. The SG can alternatively be defined as a system that provides a secure two-way communication channel that generates power from integrated information, dissemination, distribution, substation, and consumption using Information Communications Technology (ICT) and cybersecurity communication technology [8]. Table 7.1 shows a comparison of the properties of the existing grid and the SG.

Table 7.1   Comparison of the properties of the SG and the existing grid

Existing grid	SG
Centralized generation	Distributed generation
Blind	Self-monitoring
Electromechanical	Digital
Few sensors	Sensors throughout
One-way communication	Two-way communication
Manual check/test	Remote check/test
Few customer choices	Many customer choices
Failures and blackouts	Adaptive and islanding
Manual restoration	Self-healing
Limited control	Pervasive control
Hierarchical	Network

7.3   Blockchain technology

Blockchain technology is a distributed ledger consisting of a series of blocks. It stores various types of information or all the transaction data. This technology was initially described as a sequence of blocks. The blocks are connected together like a chain, with each block referencing the cryptographic hash of the data contained in the preceding block in the chain. With the blockchain network, blocks are continuously created and attached to the chain in a periodical manner, and the chain is copied across the participants in the network. Figure 7.3 shows the sequential structure and directed acyclic graph (DAG)-based logical structure.

Every block might even contain other information such as a nonce, a timestamp, a hash tree known as the Merkle tree, smart contract code, and many more. The

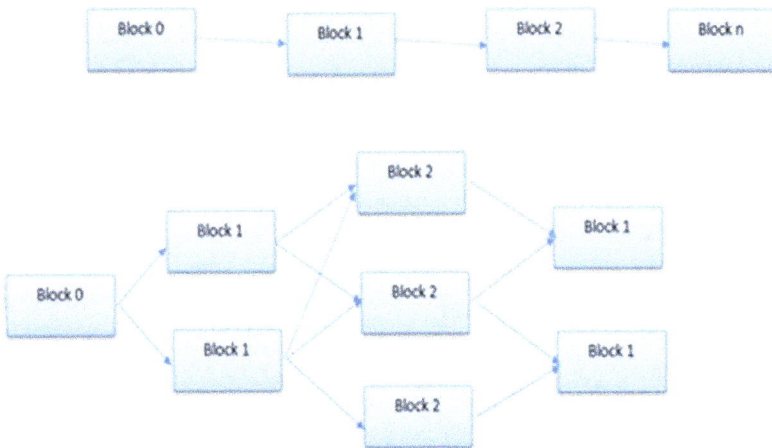

Figure 7.3   Sequential structure and DAG-based structure

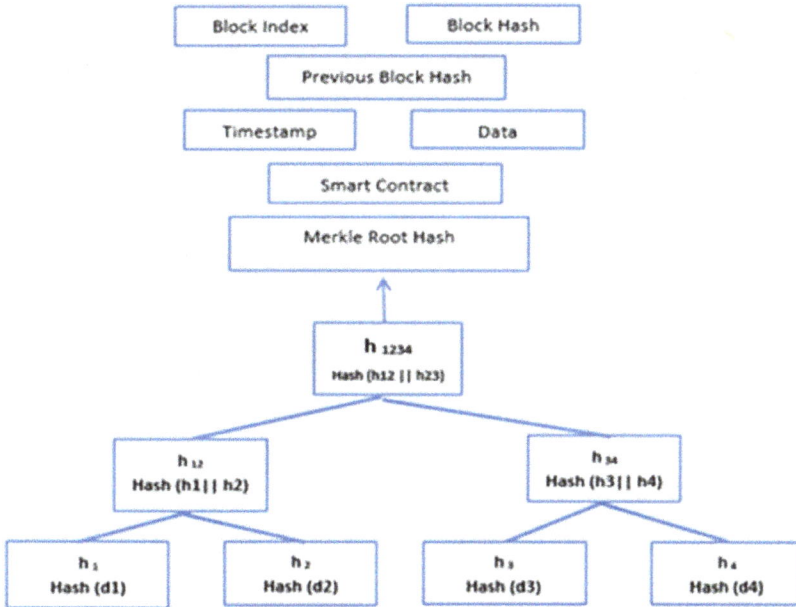

Figure 7.4   A classic block structure and Merkle hash tree

use of a hash and a Merkle tree enables the verification that the material contained within the block has not been altered, hence assuring integrity. They use lightweight devices to look for the addition of information rapidly and check the information. In an attempt to transform the content of any block, it is necessary to change the content of all the blocks, because the hash of a block changes invariably if any of its content changes, and because each block contains the hash of the previous block, it is virtually impossible to maliciously modify the chain. Furthermore, in addition to this linear chain-based structure, a DAG structure is also provided, whereby each block refers to many preceding blocks. Figure 7.4 shows the Merkle tree hash.

The smart contract was invented by Nick Szabo in the 1990s as a computerized protocol that may be used to execute the conditions of a contract without the need for human involvement. In the context of blockchain, a smart contract [9] is a computer program and it is stored and executed on the blockchain network. Rather than using legalese, the smart contract gathers events and conditions such as an asset's goal value, finish dates, or transaction details and then stores that data [10]. In the smart contract, if the condition is satisfied, the contract can be instantly performed as per the rules written for it. Because a smart contract is implemented on the blockchain, it is capable of operating independently. Ethereum is the most widely used smart contract technology that is built on the blockchain [11]. Figure 7.5 shows the working principle of a smart contract [12–15].

In other words, the chain of immutable blocks-based consensus algorithm is the blockchain which is used for secure block transactions and to secure the integrity [16]. The transactional legitimacy of each node is checked before adding a

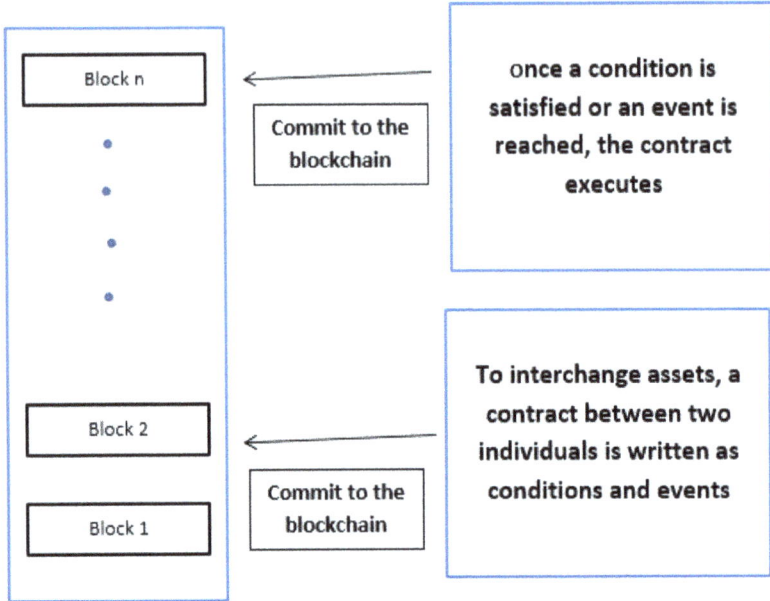

Figure 7.5   Working of smart contract

block to the blockchain to achieve an agreement, ensuring a high level of account-ability and security, as shown in Figure 7.6. A blockchain's blocks are linked together in a blockchain hash value. Block hash is the hash value of the current block; prior block hash is the previous block's hash value; nonce is an incidental integer used to check the authentication of the transaction; to get the updates about the transactions, a timestamp is used and the hash of transactions in a block is the Merkle root.

A block's data/transactions are unchangeable (i.e., data are tamperproof). Using a consensus algorithm, the blockchain transactions are validated by consid-ering each node's participation and if the nodes agree on the transaction's authen-ticity, the data are stored in structures known as blocks. Because every participating node of the network has a replica of the complete blockchain, malpractices like the change of a block in the blockchain may be easily identified. A blockchain is a distributed ledger in which each member of the network owns a complete copy of the chain. It could be public, private, or consortium. Table 7.2 summarizes a com-parison of these three categories based on several parameters. A public blockchain has no permissions and may be joined by anybody, but a consortium blockchain is a semi-decentralized method that is owned by different firms. However, given its public reactivity, the privacy issue in each of these blockchain variants is particu-larly difficult. As a result, a private blockchain is a possible alternative for making the Electronic Transaction (ET) secure and ensuring the anonymity of ET partici-pants [17].

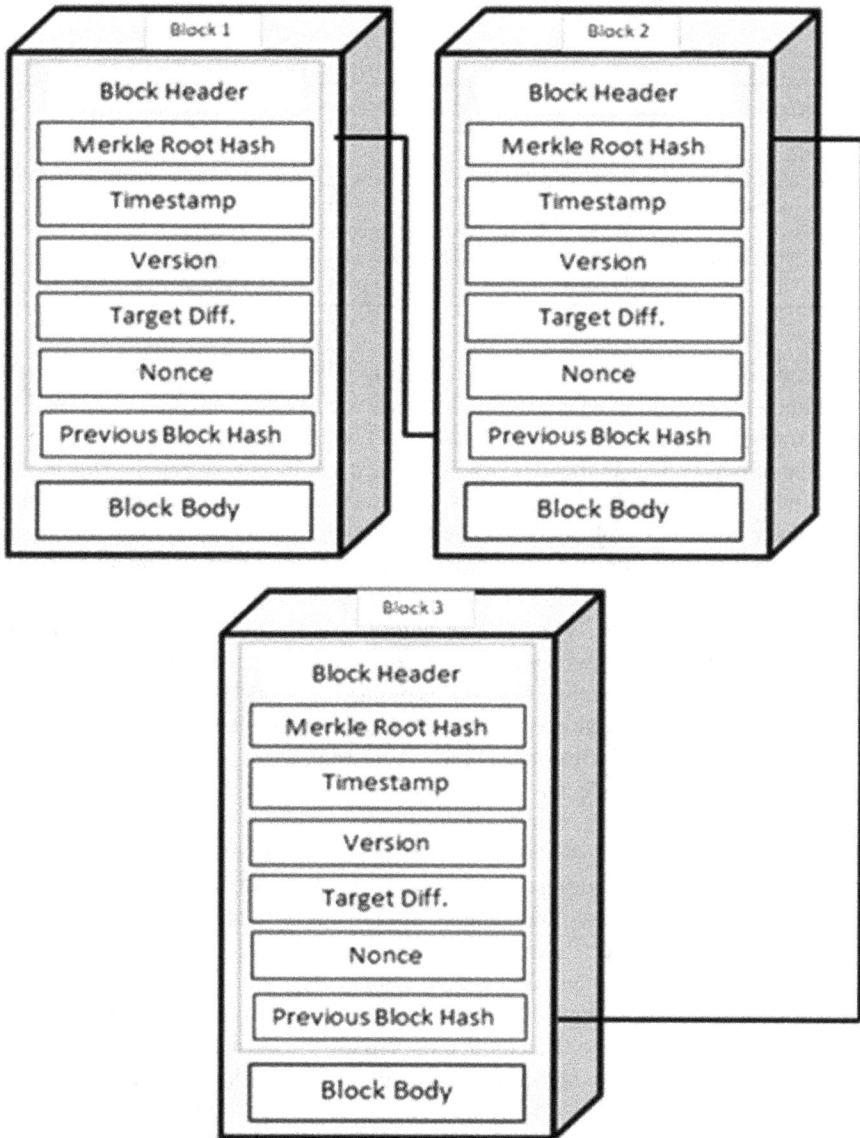

Figure 7.6   Blockchain structure

7.4   Consensus mechanisms

Consensus mechanisms are one of the most essential aspects of blockchain technology since they allow for the inclusion of freshly released blocks into the network. The public and private blockchains use a consensus process to maintain confidentiality in

Table 7.2   Comparison of public, private, and consortium on several parameters

Parameter	Public blockchain	Private blockchain	Consortium blockchain
Obscurity	More	Less	Less
Decentralization	Fully decentralized	Less decentralized	Less decentralized
Access to read	Anyone	Open to the public	Anyone
Access to write	Anyone	Internally controlled	Specific nodes
Speed of transaction	Low	Extremely high	High
Receptivity	Fully open	Open to a person	Open to some nodes

the network. This mechanism comprises a group of validators/miners who come to a common agreement on whether a block is genuine or not.

Anyone can participate in consensus on public blockchains without relying on the trust of other nodes in the network, which is a significant advantage. Due to the trust and a limited number of validators or miners, the consensus processes of private blockchains are significantly simpler than those of public blockchains, allowing for better bandwidth utilization than those of public blockchains. Following that, we will provide a quick explanation of some popular consensus procedures.

The proof of work (PoW) consensus algorithm was presented in Bitcoin as the first public blockchain consensus system. The underlying concept of this is to generate a new block. Here the consensus node is required to work out the puzzle which is conceptually expensive and is called a PoW problem. After obtaining the solution, the same will be attached to the latest block. By attaching this solution to a new block published by a certain node, any other nodes can check that the block is valid [18]. The process is referred to as mining in this context, and it is typically motivated by financial incentives. The PoW, however, is subject to a 51% attack, in which a single or multiple malevolent nodes gain control of 51% of the total processing power in the network, despite the fact that the goal of the PoW is to attempt to escape other types of attacks. Moreover, the mining method for PoW has a number of limitations, such as minimum throughput, delay, and more energy utilization, which make it inappropriate for a wide range of many other blockchain applications [19, 20].

The proof of stake (PoS) technique is perhaps the most widely used replacement for the PoW method, and it seeks to address some of the shortcomings of the PoW technique [9]. For PoS blockchains, the word mining is substituted by validating, which means that blocks are frequently validated rather than mined. It is the underlying principle of PoS to haphazardly select the validators to construct brand-new blocks. In other words, in PoS, instead of executing computational complexity puzzle-solving algorithms, the validators must demonstrate their contribution to the network in accordance with the current chain [21–23].

The Delegated Proof of Stake (DPoS) is another category of the PoS; however, in DPoS, only a few designated representatives can give rise to and authenticate the blocks. The DPoS is brisker than the PoS. The consensus process of the BitShares blockchain is based on the distributed PoS [24, 25].

This is made possible by the leased PoS (LPoS) [26], which permits the nodes to lease out their equity to others. One of the primary goals of this method is to enhance the likelihood of becoming validators. This would lead to the growth of participants who are eligible to vote, as well as a decrease in the possibility that the blockchain network will be dominated by a single set of nodes. In most cases, the rewards are distributed in a standardized way.

The proof of activity consensus mechanism is created on the basis of the PoW and PoS consensus mechanisms. First, the block makers of the succeeding blocks in the blockchain act as miners, employing the PoW method to defend against safety assaults, and, as a result, they begin to get rewards. After miners have accumulated a sufficient amount of coins (asset), they can go on to utilize the PoS process to publish new blocks [27, 28].

The proof of burn (PoB) is a middle-of-the-road option between PoW and PoS. By delivering the verified, publicly accessible and unspendable addresses, the validators are able to produce a new block and receive a reward after they have burned their own coins/assets. This spending coin is seen as a long-term investment. Following their investment, a user can place their holdings on the blockchain and gain the status of the approved validator. The PoB, in contrast to the PoW and PoS, does not necessitate the use of energy. Slimcoin is a cryptocurrency that was created on the basis of PoB [29].

A proof-of-inclusion for Merkle trees is extremely effective for establishing the presence of information. We need not disclose or compute most of the data in the tree, and simply essential bits of content are required to travel from the root to the required leaf [30].

Intel designed a proof of elapsed time for permission in blockchain applications [31]. This is to lower the high cost of energy in PoW. It applies a trusted election model for the population of validator, and it aimlessly selects the next leader in order to publish the block. Validators in the network demand a random wait time from their enclaves, which they receive. The validator who has the least waiting period for a specific block is voted as the leader, and it is required to wait until the waiting time is passed before publishing the next block in the chain. The equipment that generates the time is recognized as a source of confidence.

The proof of authority (PoA) is specifically built for permissioned blockchains. As specified by the process, before a participant can be designated an authority to publish a block, the participant must first validate its identification in the network. In contrast to PoS, where a stake is represented by a collection of coins or other assets, PoA considers a participant's identification to be a stake. Furthermore, it is considered that the authorities who would publish a block have been pre-selected and can be trusted. Another advantage is that it is simple to detect malicious authorities and notify other nodes about their dangerous behavior. Parity Ethereum is developed based on PoA [27, 28].

The Byzantine Generals Problems can be solved using practical Byzantine fault tolerance (PBFT) [32, 33]. It is presumed that at least two-thirds of the total number of nodes are trustworthy when PBFT is implemented. It is divided into the following stages [34].

a.   In order to build and validate a block, a primary node is chosen to serve as the leader. In addition, more than two-thirds of all nodes support the selection of a new primary node, which is supported by the rest of the network as a whole.

b.   Following the receipt of a request from the user, the leader generates a new block, which is then considered to be a candidate block for the request.

c.   Other nodes that are capable of participating in consensus are notified of the block by the leader, which allows for verification.

d.   Upon getting the block data, every node verifies this and publishes the outcomes to other nodes together with a hash value. A comparison is made between the audit findings including other nodes.

e.   They achieve an agreement on the candidate block and send a reply to the leader, which contains audit and comparative findings as well as other information.

f.   Again, when the leader has received the results from at least two-thirds of the nodes that have approved a candidate block, the leader can finalize the block and include that in the chain, if the results are favorable.

7.5   Blockchain applications to provide security in SG

Figure 7.7 depicts some of the most prominent blockchain applications in the context of SGs.

Here we emphasize the five major applications in SGs with IoT wherein blockchain technology is used [6, 35, 36].

7.5.1   SGs equipment maintenance

The SG system includes features, such as maintaining the health of equipment, problem detection, and monitoring remotely. However, in traditional techniques of diagnosis, technicians are required to physically reach the field for detecting and managing. It also entails a financial commitment in terms of labor and other costs, with the risk that the client may be dissatisfied with the services. This necessitates the development of technologies that may minimize maintenance time while remaining unaffected by geographical constraints. SGs include a wide range of devices, from substations to smart meters (SMs) put in houses. To maintain high efficiency and dependability, such a complicated smart system necessitates smart equipment maintenance procedures [37, 38].

7.5.2   Power generation and distribution

Hackers use different tactics to modify data and get control of SGs in the past. Regional power disruptions and even total blackouts have happened as a result of this. Because one of the primary qualities given by the blockchain system is its capacity to assure data unchangeableness, using blockchain technology in power generation and power supply system helps to avoid data tampering [39].

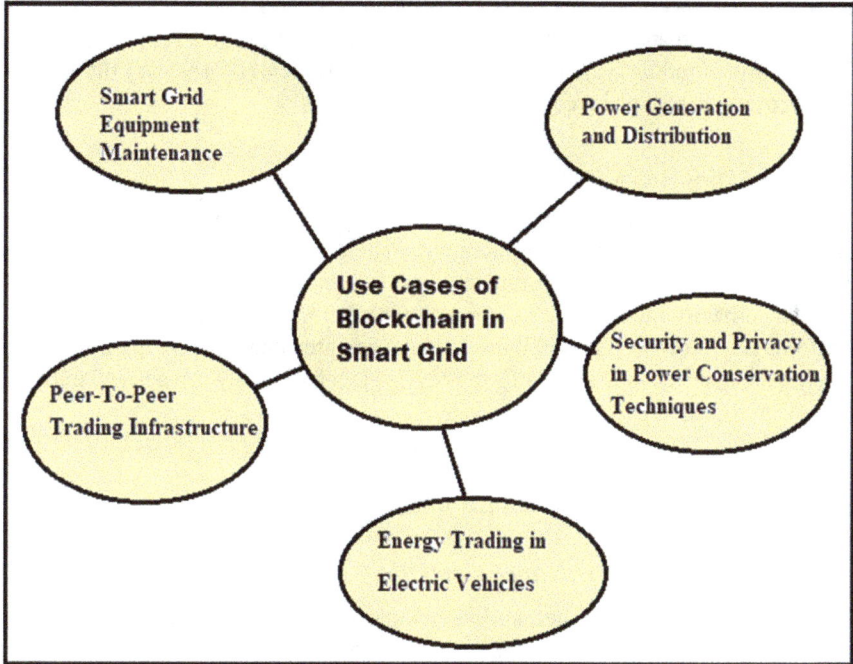

Figure 7.7   Blockchain applications in SG

7.5.3   Security and privacy in power conservation techniques

SMs are installed in every home as the real-time data will be provided by an SG on the consumption of energy, which is utilized for a variety of reasons. Malicious entities can follow the power usage behavior of customers by evaluating their electricity consumption profiles, revealing the consumers' sensitive information. Hence it is suggested a blockchain-based approach for fast data aggregation while maintaining anonymity. They separated the users into several groups, and in each of which the data of the users was recorded using a blockchain. The technique employs the Bloom filter for rapid authentication, allowing for fast verification of the validity of the user ID. Users utilize pseudonyms to keep their identities private within the community. While blockchain technology cannot guarantee data privacy in and of itself, advanced cryptographic techniques can be used to enable data privacy [5, 40].

7.5.4   Energy trading in electric vehicles

In distributed renewable energy, electric vehicles play a critical role. Vehicle-to-vehicle and vehicle-to-grid are two options for charging electric vehicles. Electric vehicles use electric hotspots in parking lots or in vehicle charging stations in peer-to-peer (P2P) techniques. To meet the demand for electricity supply, vehicles will discharge the energy. However, owing to security concerns, discharging electricity is hesitant to engage in the energy trading market, resulting in an uneven electricity supply [41]. Hence there are some serious issues with typical centralized power-trading schemes that rely on intermediate companies. As a result, a secure, decentralized electricity trading system that protects electric vehicle (EV) privacy throughout the electricity exchange is required.

7.5.5   Peer-to-peer trading infrastructure

The insecurity for transactions induced by the presence of intermediaries is the fundamental flaw in present grid networks. The grid's hierarchical organizational trading structure results in high running expenses and low operational efficiency. A blockchain-based trading infrastructure, on the other hand, provides a decentralized platform that allows for safe P2P trading between prosumers. In comparison to the old system, the decentralized platform provides greater identity privacy and transaction security [42, 43].

7.6   Case study of blockchain in SGs

7.6.1   Blockchain in advanced metering infrastructure

The SG network, which connects utility companies, consumers, and producers, will be able to interact even more through SMs that permit automatic and two-way communication, thanks to the introduction of advanced metering infrastructure (AMI). SMs, as opposed to traditional meters, are advanced meters that can collect detailed data on energy usage and production, as well as status information [44]. The service includes advertising, user contraption administration, keeping track, and fixing errors. This varied transforming information, on other hand, is carried out via a wide area network from a standard centralized storage system or the cloud.

The presence of a centralized organization may be related to the perils of change, data leakage, and there is only one point of failure. Additionally, increasing the count of connections to a centralized system may cause adaptability, accessibility, and concerns with delayed responses [45]. SMs and vehicles that run on electricity, on the other hand, produce significant volumes of information on electricity usage and payment records in SG systems, which are frequently exchanged with other factors for observing, invoicing, and dealing. Even so, in such a difficult structure, broad exchange of information presents intentional security concerns, as middlemen, intermediaries, and trusted third parties may reveal sensitive information and personal information. Sensitive information includes things like identity and location. Trust among centralized parties, producers, and consumers is also an issue. As a result, producers and consumers may find it difficult to recognize centralized parties' fairness and openness.

The goal of secure, privacy-protecting, and dependable decentralized AMI structure must be met. We share a variety of related blockchain studies on AMI in this section. The compress will function as a middleman between producers and consumers, lowering prices and increasing transaction rates while also strengthening security. When a transaction occurs, the connected SM sends the information

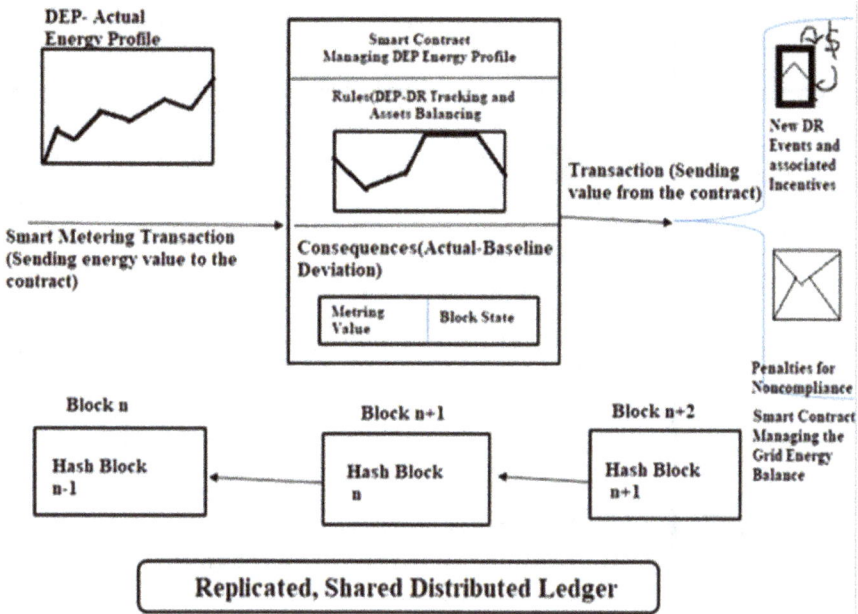

DEP- Actual
Energy Profile

Smart Contract
Managing DEP Energy Profile

Rules(DEP-DR Tracking and
Assets Balancing

Transaction (Sending
value from the contract)

New DR
Events and
associated
Incentives

Smart Metering Transaction
(Sending energy value to the
contract)

Consequences(Actual-Baseline
Deviation)

| Metring Value | Block State |

Penalties for
Noncompliance

Block n

Block n+1

Block n+2

Smart Contract
Managing the
Grid Energy
Balance

| Hash Block n-1 | Hash Block n | Hash Block n+1 |

Replicated, Shared Distributed Ledger

Figure 7.8   The self-enforcing contract structure [45]

to the distributed ledger, which generates a new block with a timestamp that can be checked later [46]. The consumer can then be charged depending on the ledger's information. However, a major critique of this work is the lack of substantial technical explanation.

To accomplish decentralization and autonomy, management of demand paradigm for dynamic networks has been presented. The blockchain is employed in this design to create a decentralized energy network that is safe and automated in which every vertex operates independently without the need for centralized monitoring and Digital Storage oscilloscope (DSO) control [47]. It's also used to store energy usage data gathered from SMs in tamper-resistant blocks. Figure 7.8 depicts the use of a smart contract to provide decentralized control and calculate rewards.

Finally, through energy usage and production traces from UK building databases, this notion is tested and evaluated by creating a prototype on the Ethereal blockchain platform. According to the findings, this model is capable of adjusting demand in near real-time by executing energy flexibility levels and validating all demand response agreements [48]. On this public blockchain, however, it is unclear how energy profile privacy has been protected. There's a chance that looking at publicly available transactions will yield some fascinating results.

An SG network's functionality and energy security can be improved with blockchain and edge computing. To combat criminal behavior within communication channels and central data centers/clouds the blockchain is mostly used to maintain participant privacy and decentralized data storage [49]. Super nodes (SNs), edge

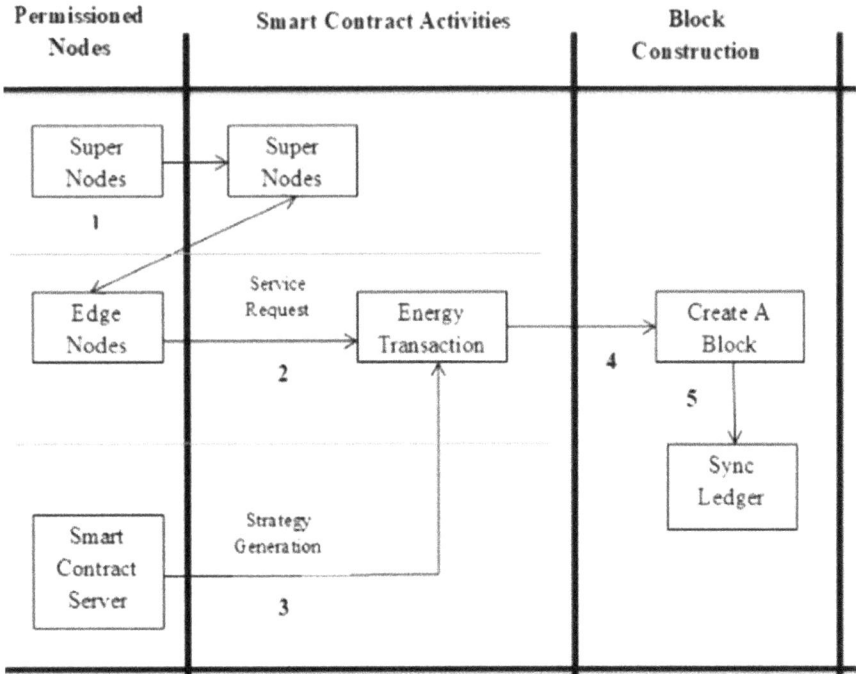

Figure 7.9   Three primary components of the suggested strategy, as well as their interrelationships [46]

devices, and smart contract servers are the three permissioned entities that ensure the blockchain network's integrity and trustworthiness (as depicted in Figure 7.9). In this case, edge devices are treated as ordinary nodes, just as they would be in a modern blockchain system.

SNs, on the other hand, are a form of the node that may choose which network devices to connect to. To participate in the consensus and voting procedures, edge devices are used. Before requesting to join in the selection procedure, super vertices, identity authorization, and covert channel authorization techniques must be used to authenticate the identities of edge nodes. Nodes aren't bad guys. Hence, they're less vulnerable to the 51% attack. Smart contract server nodes, on the other hand, are in charge of executing and connecting the script for the contract to the blocks [50]. This contract in smart assigns energy resources to electricity customers in the most efficient way possible, as determined by edge devices, while accounting for energy usage, latency, and security. However, there will be a point of failure, if the SN is under attack.

7.6.2   Decentralized energy trading and market using blockchain

Consumers can function as producers and vice versa due to the bidirectional energy and information flow characteristics. A growing number of consumers, producers, and prosumers (producer + consumer) are projected to use the SG. According to distributed energy trading scenarios, they should be able to exchange surplus energy from distributed sources such as microgrids and energy storage units with one another to achieve some benefits such as lowering load peaks and minimizing power loss in transmission. As a result, the energy trade, along with the associated formalities, must be integrated for managing bids, negotiating contracts, and executing contracts among parties [51].

Furthermore, energy can be traded directly and effectively between consumers and producers. This direct energy transaction can benefit all parties and is excellent for renewable energy projects because it eliminates the need for intermediaries traditional methods, on the other hand, only allows consumers and producers to engage in such trade operations indirectly through a slew of third parties and retailers, all of whom profit. Significant operational and regulatory costs are imposed as a result, and monopoly motivations, rewards, and penalties all contribute to an uncompetitive market.

The core characteristics of blockchain make it a perfect instrument for creating a more decentralized and open energy market. To assure the security and decentralization of energy trade in a variety of scenarios, introduce an energy token and P2P energy trading system based on consortium blockchain technology, credit-based payment mechanisms, and Stackelberg game theory. Figure 7.10 depicts the proposed system, which consists of four elements and processing steps. The credit-based payment system was created to address the problem of transaction confirmation delay, which is common in Bitcoin and other PoW currencies. Peer nodes can use their credit scores to apply for energy coin loans from credit institutions, allowing them to make quick and easy judgments.

Its payment system is speedier and more efficient than Bitcoin's. In order to maximize the economic benefits of credit banks, the game theory of Stackelberg is implemented to develop loan pricing, which is an approach to this system. However, no formal evidence of the double-spending hazard is provided in this paper, and the anticipated energy blockchain prototype is not implemented. The goal of this PriWatt is to make SG energy trading systems more secure by addressing transaction security and user identity privacy. A blockchain-insisted multi-signatures, smart contracts, and anonymity encrypted communication paths make up this system. Smart contract agreements are used by PriWatt to help buyers and sellers complete complicated transactions. The flood of anonymous communications is approaching [52, 53].

The multi-signature mechanism is employed to protect against theft, while the validation scheme ensures accuracy. For a transaction to be valid, it must be signed by at least two participants. To minimize Byzantine failures and the risk of double-spending, they use PoW for consensus, similar to Bitcoin. However, it is unclear whether nodes are PoW miners, how miners will do PoW, or how much mining will pay out if it is successful. The Ethereum smart contract was created with the goal of enforcing data access rights and transaction transparency. Pseudonyms were also employed to preserve users' privacy by concealing their true identities.

Ethereum is described a blockchain-based crowdsourced energy system (CES) with P2P energy sharing at the distribution level, as well as its framework

Energy Block chain

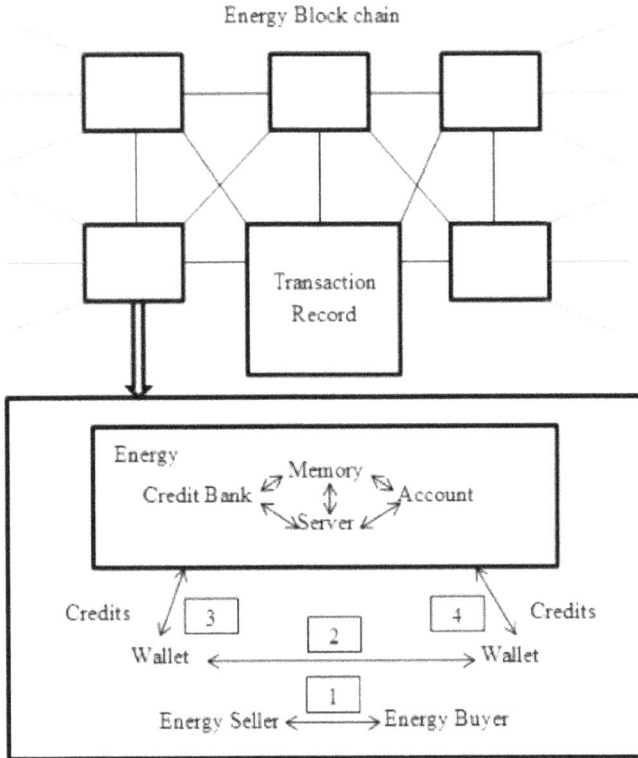

*Figure 7.10   The consortium's energy trading system demonstrated in detail,
based on blockchain technology [45]*

and operational algorithms. Several functions are included in this architecture, including charging and discharging electric vehicles, delaying electric consumption, and connecting to renewable energy sources [54]. These chores are automated using distributed blockchain implementations and SMs. The ultimate goal of this CES architecture is to manage network difficulties such as real-time demand shortages and surpluses. The operational algorithm, on the other hand, contains two phases.

To handle the majority of grid operations, the first stage focuses on day-ahead production scheduling and regulated Distributed Energy resources (DERs), while the second stage uses incentives to balance hour-ahead energy shortages and surpluses. These algorithms enable P2P energy sharing, resulting in a logical approach to distribution network regulation. Simultaneously, it encourages and inspires crowdsources to contribute to dispersing network ecosystems like microgrids. Finally, the hyper ledger fabric platform is used to prototype the CES framework and algorithms. However, there is no indication of how to cope with malicious crowdsources, market stakeholders, or outsiders.

7.6.3   *Monitoring, measuring, and controlling using blockchain*

Present Cyber-Physical System (CPS) uses various sensors to monitor the system. Supervisory Control and Data Acquisition (SCADA) system is used to regulate the electrical networks. This internet-connected SCADA system improves the controlling and monitoring of the SG system. Power device status information is often collected by many IoT sensors and other smart devices. These fine measurements are used by CPS, suppliers, consumers, and grid operators to provide more control, stability, grid safety, and dependability. Phasor Measurement Unit (PMU), Master terminal Units (MTU), and Remote Terminal Units (RTU) are also used by CPS. Malicious attackers or insiders can use a variety of methods to initiate cyberattacks including changing data, initiating availability assaults, and introducing unwanted information via PMUs and various sensors. The attackers can pass the bad data by accessing the control channels and they can announce malicious data. The blockchain provides additional controlling, measuring, and monitoring capabilities in the decentralized SG.

ICS-BlockOpS, a blockchain-based architecture, has been deployed to enhance the security of operational data. This scheme is primarily aimed to address two primary concerns in Industrial Control System (ICS): immutability and redundancy, by utilizing blockchain technology. The tamperproof characteristic of blockchain ensures the immutability of data. On the other hand, a blockchain-assisted efficient replication approach based on the Hadoop Distributed File System is offered to assure data redundancy. However, no mention is made of how to deal with malicious or hacked nodes pumping false data, or how resource-constrained sensors and actuators will operate in a blockchain network. The three sorts of nodes in this blockchain network are SMs, consensus nodes, and utility corporations, each with its own set of tasks. Consumers' digitally measured energy data is frequently fed into the blockchain network through SMs. Consensus nodes, on the other hand, are in charge of maintaining energy usage records, validating new blocks, and broadcasting them to the main chain.

Before constructing new blocks, these nodes, on the other hand, produce temporary forms for specific users, which include meter IDs and other consumer information. These forms will be turned into blocks after they have been audited and accepted by consensus nodes. The blockchain is utilized to create an immutable data record system that secures SM data from both consumer and utility company manipulation in this situation. Furthermore, by setting criteria for identifying unauthorized electric power usage, detecting fraudulent manipulation of usage figures, and implementing penalties, the smart contract promotes transparency. This work, however, is lacking in practicality.

This architecture has five layers, as seen in Figure 7.11. Several sensors and computing equipment make up the first layer. Computing devices are in charge of gathering and pre-processing sensor data. Before producing blocks and putting them in the distributed ledger, the second layer performs cryptographic operations on the data. The distribution and synchronization of the whole block are taken care of by the third layer. A good storage methodology and efficient distributed algorithms are required to join all the layers. In the end, the last layer serves the functionalities to recognize the failure and monitor the system in real time.

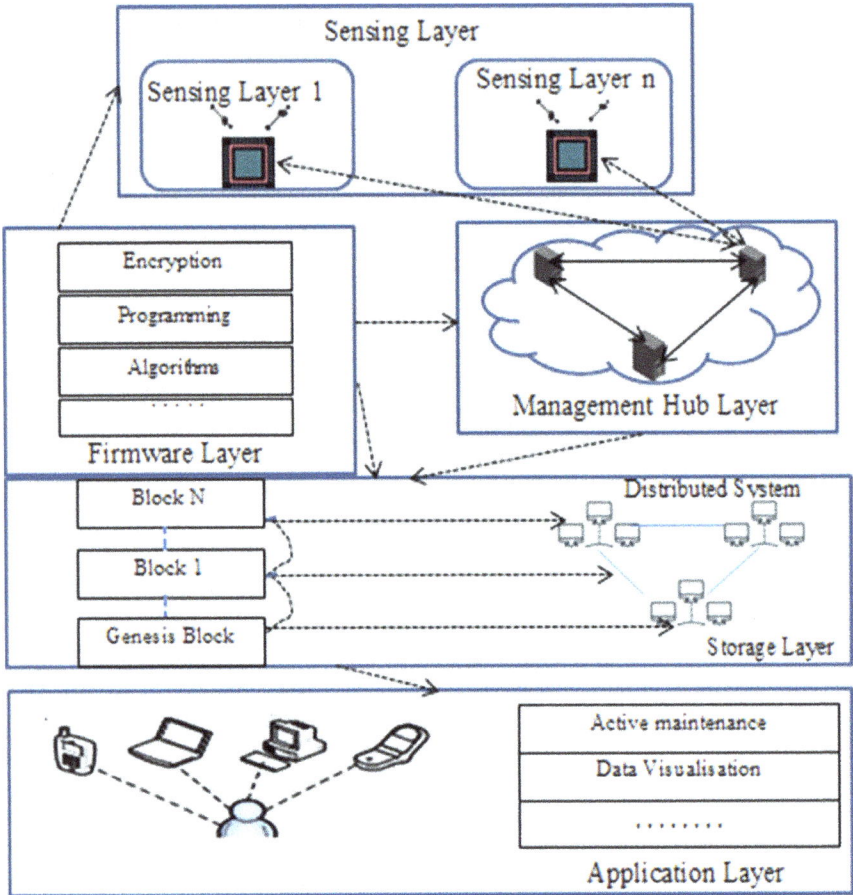

Figure 7.11   The blockchain-assisted architecture

Finally, the architecture is deployed in a fully automated production platform. According to the findings of the investigations, the proposed architecture provides greater security and privacy than traditional architecture. The same is shown in Figure 7.11.

7.6.4   Blockchain-based access control protocol in IoT-enabled SG system

Several organizations make up a blockchain-based SG system, including the trusted service providers (SPs), registration authority (RA), IoT-enabled SMs, and SM users. SPs are in charge of the system for allocating and exchanging electricity. SMs are in charge of keeping track of how much power is used and keeping wages the same for the customers. The transmission between SPs and SMs has to be safeguarded to prevent passive and active attacks. A private blockchain can be used to

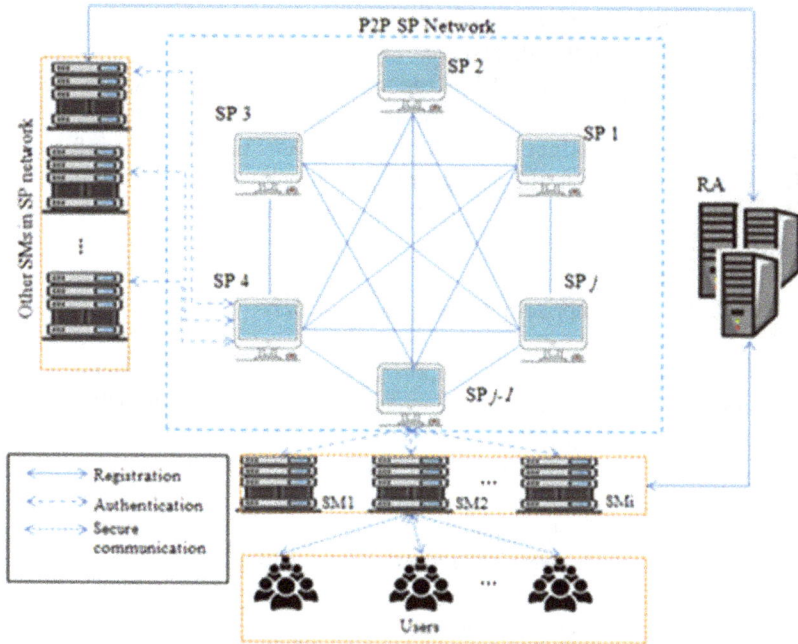

Figure 7.12 *Third-party untrusted IoT-enabled SG architecture based on blockchain*

store and secure data by means of blocks. The P2P SP network's SPs are in charge of validating new blocks before they are added to the blockchain through the consensus algorithm. To alleviate the discussed issues, a decentralized blockchain-based access control protocol in IoT-enabled SG (DBACP-IoTSG) system is designed. Before constructing the blocks, the data from the SMs are safely collected by their respective providers of services and the P2P SP network uses a voting-based consensus process to add those blocks to the blockchain. Because the blockchain imparts immutability and transparency, once a block is added to the blockchain, it cannot be tampered with by anybody, including legal professional associations of the network of SGs, and anyone can view the information saved in the block. The work specifically focuses on private blockchain because the data collected from SMs by SPs is both confidential and private.

The network model used in this study is depicted in Figure 7.12, wherein a smart meter SMi is linked to several users, and a set of SMs is linked to a service provider SPj. A P2P network operator, also known as the P2P SP network, will be formed by a group of service suppliers. In offline mode, a trusted RA is in charge of registering all installed smart meters SMi and service providers SPj. The registration process is carried out by the RA in a secure manner. Secured communication is used between users and a smart meter SMi, whereas a smart meter SMi and a service provider SPj, on the other hand, communicate securely and then use a session id generated between them via a system for restricting access. Also, the SPs develop confidential pairwise identities among themselves for authentication protocols. The data is surreptitiously collected by SMi from its associated users and then secretly delivered to the service provider SPj, under which the smart meters SMi are registered. SPj then produces a block of transactions based on the information gathered. Following then, the newly produced block can be added to the existing blockchain if the SP network's SPs reach an agreement. When the block is put into the blockchain, it cannot be modified or deleted in order to maintain the "immutability" attribute.

DBACP-IoTSG comprises multiple phases, namely

a. Setup of the system: The trustworthy RA is in charge of selecting system parameters like selecting a "non-singular elliptic curve, identity IDRA, chooses a master key as the private key, as well as its public key and pseudo-identity, picks a one-way cryptographic hash function". Finally, the RA preserves the master key as its private key and releases additional parameters that are publicly available to all network entities.
b. SM and SP's registration: The RA runs this step in offline mode to register all of the SMs that have been deployed.
c. Access control: Using the node authentication job, a smart meter SMi will be able to authenticate with its respective service provider SPj during this phase. Then, with the support of the key establishment task, they would construct a common session key after the mutual authentication.
d. SP's essential key management: Imply that two SPs, SPj and SPl, agree to form a partnership. A pairwise symmetric key is generated between them.
e. Block formation and insertion of blocks in the blockchain: The P2P SP network maintains a pool of transactions, and when it hits the transaction threshold, the secure leader selection algorithm [55] selects a leader (L). Following that, in the voting-based consensus for block verification and addition to the blockchain algorithm [55], the leader L will construct a block and initiate the consensus algorithm in order to add this block to the blockchain.
f. After the initial deployment in the SG environment, new SMs will be added: Some service providers SPj may become faulty nodes at times, or some SMs may be physically hacked by an adversary. As a result, certain new SPs or SMs must be added to the existing IoT-enabled SG system.

The DBACP-IoTSG presented here operates without the use of a trusted third party. The blocks are then carefully validated in the P2P SP network by a leader selection procedure. After that, the leader is in charge of safely conducting the consensus procedure to validate the blocks by its peer nodes using the PBFT approach, and the blocks are added to the blockchain after successful validation. The transactions are kept secure with the Elliptic Curve Cryptography (ECC) encryption technique and the public key of the network operator SPj which creates the blocks containing the transactions so that only SPj can decrypt them. DBACP-IoTSG was

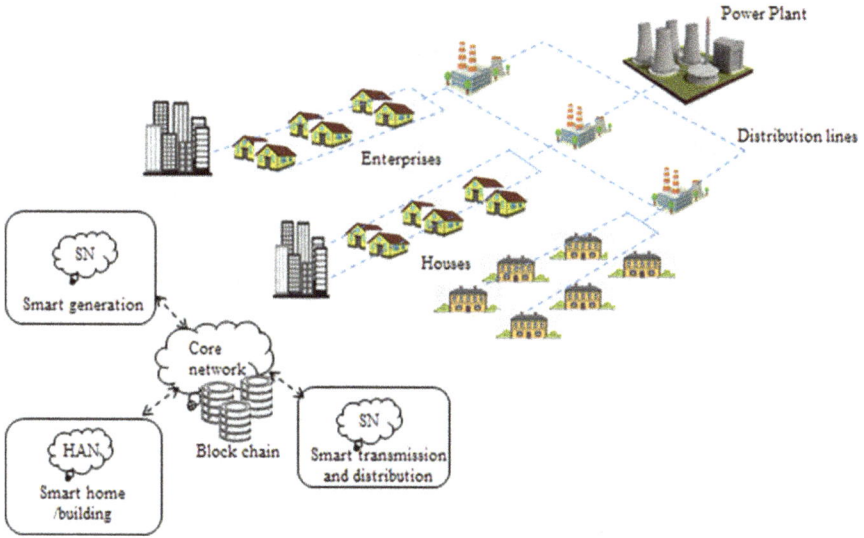

Figure 7.13   System architecture of blockchain-based SG

used to calculate the computing time for different numbers of blocks in the block-chain, as well as different numbers of transactions per block.

7.6.5   Blockchain-based cybersecurity and advanced distribution in SG

Cybersecurity can be increased using blockchain in SG cyber layers. This will, in turn, provide privacy, integrity, scalability, and governance. Field measurements, electricity bill payment, data administration, data aggregation, and system operation are the five primary areas where blockchain is being used. Figure 7.13 shows the system architecture of blockchain-based SG. It includes many layers such as sensor networks (SN), Neighbor Area Networks (NAN), and Home Area Networks (HAN). These can be used to provide efficient communication between the components of SG.

Managing authorities like maintenance companies and operators are connected to the core network. General Packet Radio Services (GPRS), Worldwide Interoperability for Microwave Access (WiMAX), and Transmission Control Protocol/Internet Protocol (TCP/IP) are used to communicate in the core network. Clients within a core network have complete control over monitoring and sending commands to the SN and HAN. Local industrial workers are directly connected to NAN and HAN, and blockchains can be utilized to enable a variety of applications including local energy exchange and electric vehicle charging.

Transactive energy is related to the financial and surveillance system that allows a perfect interplay of supply and demand across the entire electrical grid. Using value as a primary operating criterion, blockchain gives operators more transparency over

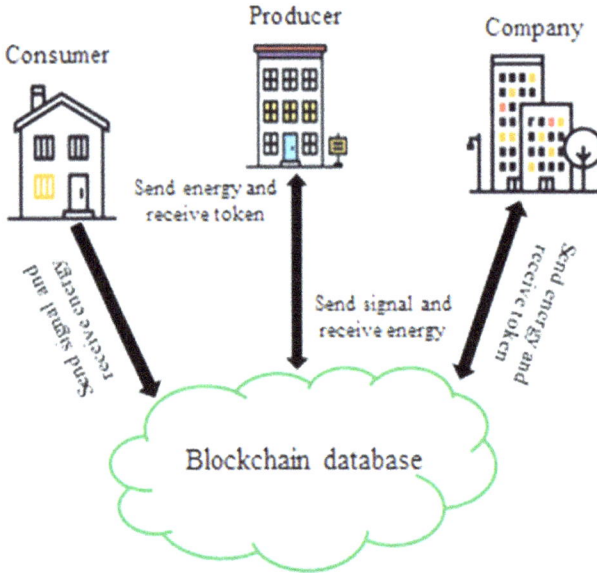

Figure 7.14   Electricity market

pricing mechanisms, device control, and a variety of other shared data that is entered into the blockchain ledger, as illustrated in Figure 7.14.

In this SG system, private blockchains will be used. Here the flow of data via the network is recorded on the blockchain and to exchange the data in the network, only the published data will be used. The data movement inside and between the network is validated and stored by blockchain.

Figure 7.15 shows an energy transmission diagram for handling sensors, data, and judgment in a smart structural system. The houses that have already been set up with these smart devices allow people to engage and modify energy loads. To put these power electronics management systems in place, we need to detect user authentication, smart contracts, energy trading and distribution, load control, and P2P transactions.

SG metering uses developed metering infrastructure (AMI) and automatic meter reading (AMR) devices to communicate data bidirectionally between users and the provider. To improve network security, such adaptive measurement would be con-nected to the Bitcoin platform and used to access and trade data across the system. Power transmission grid operators are able to estimate the usual load profile with adequate safeguards using trustworthy electricity usage data collected from their cli-ents. Smart buildings, such as their process automation and transportation systems, may all be monitored with today's IoT device-enabled AMIs. The key functional-ities of these AMIs are remote meter checking, procurement, order fulfillment, and prepaid power grids.

The upgrading of the energy system has been considerably aided by the decen-tralization of global energy markets. As a result, clients can now purchase energy at

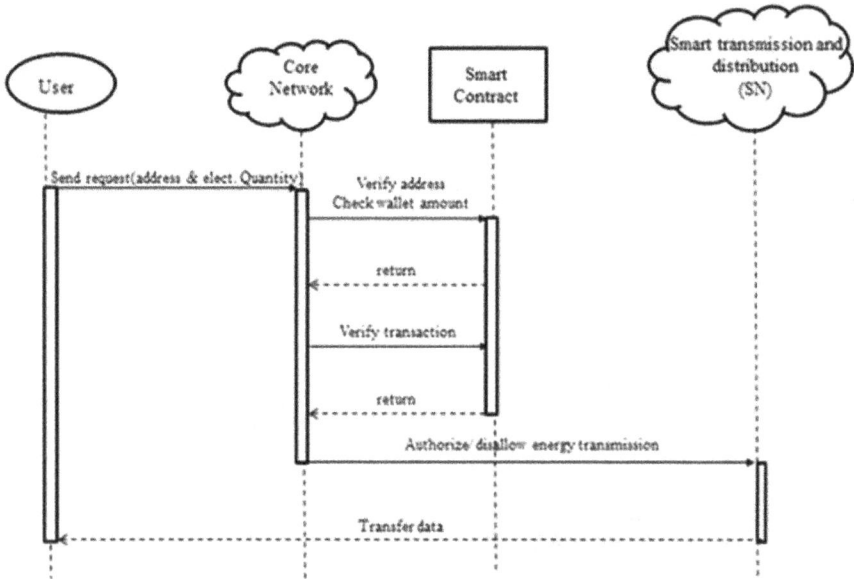

Figure 7.15   Energy transmission

favorable rates on the site and from potential marketplaces. Despite the variety of power sources, which might range from a small private DER to a huge industrial-scale power plant, the forms of power trading platforms vary.

In the power grid, the quantity of consumers is rapidly increasing. Blockchain technology is used to decentralize present energy trading networks, which allows prosumers to carry out direct P2P resource sharing on the system with customers and other prosumers. Customers can choose unique energy sources through their preferred prosumers, thanks to blockchain. Prosumer, consumer, and service reputation structures are expected to decentralize as a result of blockchain platforms. The energy trading, as illustrated in Figure 7.16, platforms that allow consumers and prosumers can trade electrical energy in a safe system at competitive rates during peak and off peak demand periods.

An SG system has the potential to increase energy distribution dependability, flexibility, and quality. In SG architecture based on edge computing, the ability to achieve securely encrypted communication between different users and edge server is critical. To improve SG cyber resilience and secure transitive renewable energy, the use of blockchain and smart contracts is discussed in this study. One of the most pressing issues in SG networks is server stability, and although this fosters reliable network enforceability, blockchain has been proved to be a method that is reliable and used as part of data mining.

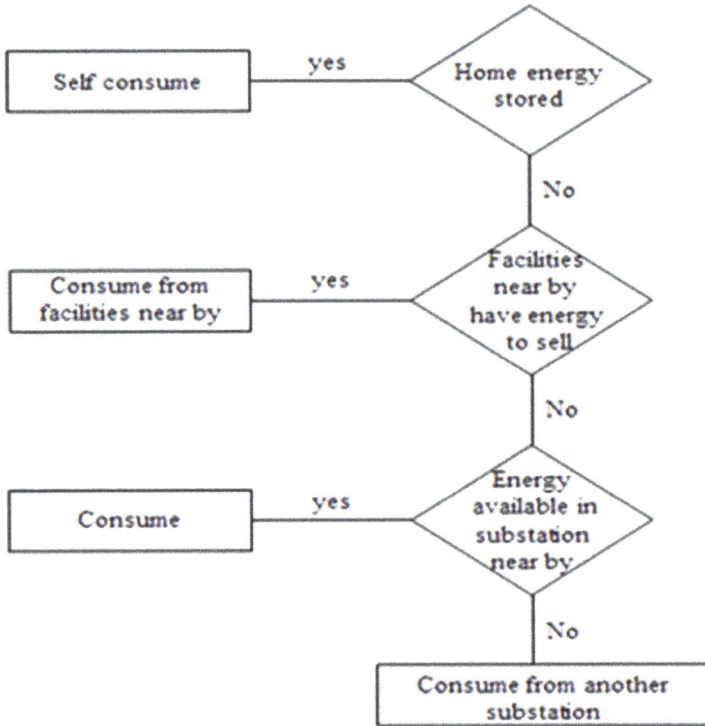

Figure 7.16   Energy trading

7.7   Conclusion

The rapidly growing technology in the energy sector is the SG and it requires a stable and secure operating environment. This chapter examines SG, blockchain technology, and the applications of blockchain in SG to provide security and privacy to the SG. Furthermore, two case studies Blockchain in Advanced Metering Infrastructure and Decentralized Energy Trading and Market using Blockchain are described wherein the SG is integrated with the blockchain to widen up a slew of latest possibilities. Though blockchain technology does not directly guarantee advanced cryptographic techniques and privacy, it can be used to ensure privacy is maintained. However, in order for blockchain to be widely adopted in the SG, scientific communities and the industry will need to collaborate to overcome the substantial hurdles that lie ahead.

References

[1] Gharavi H., Ghafurians R., Rodrigues J.J.P.C., Kozlov S.A. 'Blockchain in smart grids: A review on different use cases'. *Sensors*. 2019, vol. 19(22), p. 4862.

[2] Kumari A., Tanwar S., Tyagi S., Kumar N., Obaidat M.S., Rodrigues J.J.P.C. 'Fog computing for smart grid systems in the 5G environment: Challenges and solutions'. *IEEE Wireless Communications*. 2019, vol. 26(3), pp. 47–53.

[3] Bayindir R., Colak I., Fulli G., Demirtas K. 'Smart grid technologies and applications'. *Renewable and Sustainable Energy Reviews*. 2016, vol. 66(1), pp. 499–516.

[4] Mylrea M., Gourisetti S. 'Blockchain for smart grid resilience: Exchanging distributed energy at speed, scale and security'. Resilience Week (RWS); Wilmington, DE; Sep 2017.

[5] Mo Y., Kim T.H.J., Brancik K., *et al.* 'Cyber–physical security of a smart grid infrastructure'. *Proceedings of IEEE*. 2011, vol. 100, pp. 195–209.

[6] Kim S.M., Lee T., Kim S., Park L.W., Park S. 'Security issues on smart grid and Blockchain-based secure smart energy management system'. *MATEC Web of Conferences 260*, 25 Jan 2019; 2019.

[7] Mo Y., Kim T.H.J., Brancik K., *et al.* 'Cyber-physical security of a smart grid infrastructure'. *Proceeding of IEEE*. 2012, vol. 100(195), pp. 991–1003.

[8] Gharavi H., Ghafurians R. 'Smart grid communications architecture: A survey and challenges'. *Proceedings of the IEEE*. 2011, vol. 99, pp. 83–9.

[9] Proof of stack [Online]. Available from https://en.bitcoin.it/wiki/Proof of Stake [Accessed Nov 2019].

[10] Duong T., Fan L., Zhou H.-S. *2-hop blockchain: Combining proof-of-work and proof-of-stake securely* [Online]. 2016. Available from https://eprint. iacr. org/2016/716.

[11] Parity Ethereum [Online]. Available from https://wiki.parity.io/Parity-Ethereum [Accessed Nov 2019].

[12] Merkle R.C. A digital signature based on a conventional encryption function. *'Conference on the Theory and Application of Cryptographic Techniques'*; Springer, Lundy Place, San Jose; 1987. pp. 369–78.

[13] Szabo N. *Smart contracts: Building blocks for digital markets* [Online]. 1997. Available from http://www.fon.hum.uva.nl/rob/Courses/InformationInSpeech/ CDROM/Literature/LOTwinterschool2006/szabo.best.vwh.net/smart contracts 2.html [Accessed Nov 2019].

[14] Smart contracts [Online]. 1994. Available from http://www.fon.hum.uva.nl/rob/ Courses/InformationInSpeech/CDROM/Literature/LOTwinterschool2006/ szabo.best.vwh.net/smart.contracts.html [Accessed Nov 2019].

[15] Buterin V. 'A next-generation smart contract and decentralized application platform'. *White Paper*. 2014, vol. 3, p. 37.

[16]   Wang S., Taha A.F., Wang J., Kvaternik K., Hahn A. 'Energy crowdsourcing and peer-to-peer energy trading in Blockchain-enabled smart grids'. *IEEE Transactions on Systems, Man, and Cybernetics: Systems*. 2019, vol. 49(8), pp. 1612–23.

[17]   Jindal A., Aujla G.S., Kumar N., Villari M. 'Guardian: Blockchain-based secure demand response management in smart grid system'. *IEEE Transactions on Services Computing*. 2019, vol. 13(4), pp. 613–24.

[18]   Chepurnoy A., Duong T., Fan L., Zhou H.-S. 'Twins coin: A cryptocurrency via proof-of-work and proof-of-stake'. *IACR Cryptology ePrint Archive*. 2017, pp. 232–17.

[19]   Swathi B.H., Meghana M.S., Lokamathe P. 'An analysis on Blockchain consensus protocols for fault tolerance'. *2nd International Conference for Emerging Technology (INCET)*; 2021. pp. 1–4.

[20]   Chen G., He M., Gao J., Liu C., Yin Y., Li Q. 'Blockchain-based cyber security and advanced distribution in smart grid'. *IEEE 4th International Conference on Electronics Technology (ICET)*; 2021.

[21]   Nguyen C.T., Hoang D.T., Nguyen D.N., Niyato D., Nguyen H.T., Dutkiewicz E. 'Proof-of-Stake consensus mechanisms for future blockchain networks: fundamentals, applications and opportunities'. *IEEE Access*. 2019, vol. 7, pp. 85727–45.

[22]   King S., Nadal S. 'PPcoin: Peer-to-peer crypto-currency with Proof-of-Stake'. *Self-published Paper*. 2012, vol. 19.

[23]   Peer Coin, a peer-to-peer cryptocurrency [Online]. Available from https://peercoin.net/ [Accessed Nov 2019].

[24]   Delegated Proof of Stack (DPoS) [Online]. Available from https://en.bitcoinwiki.org/wiki/DPoS [Accessed Nov 2019].

[25]   The white paper of Bitshares blockchain [Online]. Available from https://bitshares.org/technology/delegated-proof-of-stake-consensus/ [Accessed Nov 2019].

[26]   Leased Proof-of-Stake (LPoS) [Online]. Available from https://coinstelegram.com/2018/10/30/what-is-leased-proof-of-stake-lpos/ [Accessed Nov 2019].

[27]   Bentov I., Lee C., Mizrahi A., Rosenfeld M. 'Proof of activity: Extending bitcoin's proof of work via proof of stake'. *IACR Cryptology*. 2014, vol. 2014, p. 452.

[28]   De Angelis S., Aniello L., Baldoni R., Lombardi F., Margheri A., Sassone V. 'PBFT vs Proof-of-Authority: Applying the cap theorem to permissioned blockchain'. 2018.

[29]   Stewart I. *Proof of burn* [Online]. Available from https: //en.bitcoin.it/wiki/Proof of burn. [Accessed Nov 2019].

[30]   Slimcoin a peer-to-peer crypto-currency with Proof-of-Burn [Online]. Available from https://github.com/slimcoin-project/slimcoin project.github.io/raw/master/whitepaperSLM.pdf [Accessed Nov 2019].

[31]   Chen L., Xu L., Shah N., Gao Z., Lu Y., Shi W. On security analysis of Proof-of-Elapsed-time (PoET). *International Symposium on Stabilization, Safety,*

and Security of Distributed Systems, SSS 2017; Springer, Boston, United States; 2017. pp. 282–97.

[32] Castro M., Liskov B. 'Practical Byzantine fault tolerance'. *OSDI.* 1999, vol. 99(1999), pp. 173–86.

[33] Lamport L., Shostak R., Pease M. 'The Byzantine generals problem'. *ACM Transactions on Programming Languages and Systems.* 1982, vol. 4(3), pp. 382–401.

[34] Mollah M.B., Zhao J., Niyato D., *et al.* 'Blockchain for future smart grid: A comprehensive survey'. *IEEE Internet of Things Journal.* 2021, vol. 8(1), pp. 18–43.

[35] Abdella J., Shuaib K. 'Peer to peer distributed energy trading in smart grids: A survey'. *Energies.* 2018, vol. 11(6), p. 1560.

[36] Li Z., Kang J., Yu R., Ye D., Deng Q., Zhang Y. 'Consortium blockchain for secure energy trading in industrial Internet of Things'. *IEEE Trans. Ind. Inform.* 2017, vol. 14, pp. 3690–700.

[37] Mengelkamp E., Notheisen B., Beer C., Dauer D., Weinhardt C. 'A blockchain-based smart grid: Towards sustainable local energy markets'. *Computer Science - Research and Development.* 2018, vol. 33(1-2), pp. 207–14.

[38] Miller D. 'Blockchain and the Internet of things in the industrial sector'. *IT Professional.* 2018, vol. 20(3), pp. 15–18.

[39] Huh S., Cho S., Kim S. 'Managing IoT devices using blockchain plat-form'. *Proceedings of the 2017 19th International Conference on Advanced Communication Technology (ICACT), Bongpyeong, Korea;* Bongpyeong, Korea, Feb; 2017. pp. 464–7.

[40] Münsing E., Mather J., Moura S. Blockchains for decentralized optimization of energy resources in micro grid networks. *Proceedings of the 2017 IEEE Conference on Control Technology and Applications (CCTA);* Mauna Lani, HI, Aug; 20. pp. 2164–71.

[41] Ipakchi A., Albuyeh F. 'Grid of the future'. *IEEE Power and Energy Magazine.* 2009, vol. 7(2), pp. 52–62.

[42] Avancini D.B., Rodrigues J.J.P.C., Martins S.G.B., Rabêlo R.A.L., Al-Muhtadi J., Solic P. 'Energy meters evolution in smart grids: A review'. *Journal of Cleaner Production.* 2019, vol. 217(10), pp. 702–15.

[43] Kosba A., Miller A., Shi E., Wen Z., Papamanthou C. The blockchain model of cryptography and privacy-preserving smart contracts. *Proceedings of the 2016 IEEE Symposium on Security and Privacy (SP);* San Jose, CA, 22–26 May; 2016. pp. 839–58.

[44] Alladi T., Chamola V., Rodrigues J.J.P.C., Kozlov S.A. 'Blockchain in smart grids: A review on different use cases'. *Sensors.* 2019, vol. 19(22), p. 4862.

[45] Mollah M.B., Zhao J., Member D.N., Lam K.-Y., Zhang X., Ghias A.M.Y.M. 'IEEE Internet of Things Journal'. *Blockchain for future smart grid: A comprehensive survey.* 2020, vol. 5.

[46] Erturk E., Lopez D., Yu W.Y, Wei Yang Y. 'Benefits and risks of using Blockchain in smart energy: a literature review'. *Contemporary Management Research.* 2019, vol. 15(3), pp. 205–25.

[47] Winter T.M.G.L. 'The advantages and challenges of the Blockchain for smart grids'. *IEEE*. 2018, vol. 5.

[48] Tim van Genderen V.M.J.J. 'Blockchain in smart grids'. *IEEE*. 2018, vol. 6, pp. 1–48.

[49] Koyunoglu A.S. 'Blockchain applications on smart grid - A review'. *IEEE*. 2019, vol. 5, pp. 1–81.

[50] Kotsiuba I., Velykzhanin A., Biloborodov O., Skarga-Bandurova I., Biloborodova T., *et al.*. 'Using blockchain for smart electrical grids'. IEEE International Conference on Big Data (Big Data); 2018.

[51] Livingston D., Sivaram V., Freeman M., Fiege M. *Applying blockchain technology to electric power systems*. New York; 2018. pp. 2–3.

[52] Fu X. 'Design and implementation of a smart grid system based on Blockchain smart contract technology'. *IEEE*, pp. 1–44.

[53] Sawtooth Hyperledger [Online]. Available from https://sawtooth.hyperledger. org/ [Accessed Nov 2019].

[54] Malik H., Manzoor A., Ylianttila M., Liyanage M. 'Performance analysis of blockchain based smart grids with ethereum and hyperledger implementations'. *IEEE Conference*. 2018.

[55] Bera B., Saha S., Das A.K., Vasilakos A.V, Athanasios V. 'Designing Blockchain-Based access control protocol in IoT-Enabled Smart-Grid system'. *IEEE Internet of Things Journal*. 2021, vol. 8(7), pp. 5744–61.

Chapter 8

Machine learning and deep-learning algorithms for blockchain and IoT-driven smart grids

K C Suhas[1], Mahesh N[1], Asifullakhan[1], and D Kotresh Naik[1]

Over the last few years, the development of a blockchain mechanism for Internet of Things (IoT)-driven applications has emerged as a solitary, troubleshooting as well as a leading mechanism. The distributed database in blockchain technology gives more importance to information security and privacy. It also has a consensus system that assures information security and legitimacy [1]. This mechanism is widely used in the financial economy, IoT, cloud computing, large data, and edge computing. Nonetheless, it introduces novel security concerns, such as majority assault and double-spending, and many more. Hence, there is a need to address these novel security concerns as this technology is being used in many critical and edge-cutting applications. Hence, this chapter introduces mechanisms for the amalgamation of data analytics with blockchain to secure the information. The value of new technologies, machine learning (ML) and deep learning (DL), is highlighted through analytics on securing the information on the blockchain [2]. To improve the accuracy of outcomes, data reliability and exchange are critical. When these two technologies (ML and DL) are combined, they can produce extremely exact outcomes. In the next section, we focus on the broad knowledge of ML and DL. After that, we emphasize ML and DL mechanisms for smart grids that are to be followed by them for blockchain in smart grids.

8.1 Introduction

DL and ML on the blockchain has several advantages, but each has its own set of drawbacks. ML and DL algorithms have design and efficacy concerns, but blockchain has challenges with energy demand, privacy, confidentiality, and effectiveness. They can

[1]CIT, Gubbi, India

be linked as two separate study directions that benefit from natural integration [1]. These two technologies are similar in terms of gathering, data security, and trustworthiness as they can operate together. Algorithms, computer power, and data are all crucial components in ML and DL.

Establishing a shared ledger that is simultaneously shared and distributed blockchain, for example, can enhance the trustworthiness, integrity, protection, and confidentiality of business processes [3]. A blockchain, or a distributed database in general, could be used as a register to record any form of resource.

One of the major drawbacks of IoT today is the capacity to capture and organize massive amount of information. A convergence of technologies employing blockchain technology could make data management more scalable. The sheer lack of sustainability of blockchain systems is connected to the usage of something that requires a lot of energy procedures to verify transactions, such as contracts on evidence, about ML as well as DL of blockchain technology.

However, scalability can be enhanced by implementing more energy-efficient consensus procedures, such as proof-of-stake or proof-of-authority. The Bitcoin network's tremendous energy usage will soon become an artifact. Aggregation issuer (AI), in conjunction with blockchain, can help to boost scalability, even more, including a methodology for improving the productivity of blockchain-enabled IoT applications [3]. This method might be useful for deep reinforcement learning (DRL), a variety of ML, to achieve a larger level of throughput. To improve performance, the paper proposed a "DRL-based strategy" to dynamically select/adjust the block producers, blockchain network, block size, and block interval.

To put it neatly, blockchain technology [4] helps IoT device data management because of its immutability, confidence, legitimacy, data integrity, safety, and privacy features. It can overcome the present limitations of IoT data when combined with ML and DL.

The term "smart grid" refers to software that operates the discharge of power as well as operational data across a private system. Power grid systems that use the information and digital communication systems are commonly used in construction grids. Smart grid has emerged in recent years, causing a change in electric grid systems [5]. Utilities can take a proactive approach to end-to-end electricity generation, with a smart grid providing energy. In comparison to the traditional grid, the smart grid electromechanical grid, there is a lot to improve in terms of transaction systems and electricity generation, storage, and transmission. As a result, smart grid is seen as the electrical grid for the future. The way electricity is generated, distributed, and monitored is supposed to change. According to researchers, a smart grid will make human life a little easier, safe, and sustainable. As a result, countries all around the world are installing smart grids to ensure that the advantages reach out now to their communities.

A new system generates huge amounts of data in a variety of sources. a variety of intelligent devices, for a variety of reasons, incorporates energy flow optimization, operations management, system monitoring, and production decisions. Client statistics, business metrics, metrics for the power system, location, and weather are all examples of data. There could also include personal information on private clients and partners, information on a country's central power grid, etc.

Attackers could be capable of manipulating such information when they have accessibility to it, the grid might be severely harmed, and civilization could be thrown into chaos. As a result, appropriate protection methods are critical for ensuring the reliable and secure operation of a smart grid, as well as the confidentiality of such sensitive data. Any attack on a smart grid system must be prevented by a protection mechanism. Among many of the numerous sectors are privacy and security protection systems. Where ML methods are increasingly routinely deployed ML algorithms make a model by providing it with data that has been labeled with specific labels [2], and features that refer to the data characteristics required by the system to make predictions, and also label referring to the "true response" that the model should be able to predict on its own. After that, a new model has been created and it had been instructed based on particular selected features, it may be capable of predicting labels.

Reinforcement learning, support vector machines, and decision trees are among the others. In recent years, smart grid industries have begun monitoring power data by employing ML and DL models, fuzzy logic, convolution neural networks (ANN), and genetic algorithms to improve results for anticipating specific power consumption. Such algorithms have been used in the industry for energy efficiency planning and control. Security and privacy issues in the power grid, on the other hand, have received far less attention [2].

Multiple concerns must be examined to fully benefit from smart grid technology. Smart grids have been subject to threats including denial of service attacks, data theft, and data corruption injection, to name a few. The property of non-tampering with records is the driving force underlying blockchain development [2]. The mining method that calculates the hash value and adds the block and the future block to the network is just not controlled by the central structure.

The additional clarification can be used by member nodes to arbitrate transactions, and public-key cryptography ensures transaction authenticity and integrity. In a safety system, use blockchain. Data are very important in smart grid systems.

These devices produce massive amounts of data, which are maintained in cloud storage and then sent on numerous networks. A cyber-assault on the smart grid power system could be dangerous, according to researchers specifically to generate electricity as well as supply chain networks [2]. Cyber-assault on a smart grid may have serious implications for system failures, power failures, data theft, and total blackouts, all of which can result in financial and other damages. It's simple to see everything that passes through a power system to identify cyber threats and attacks. While ML has risen in popularity in the field of cybersecurity, it necessarily requires the use of certain multiple techniques and theoretical perspectives to interact with the massive amounts of information generated as well as transfer across multiple networks in a smart grid environment.

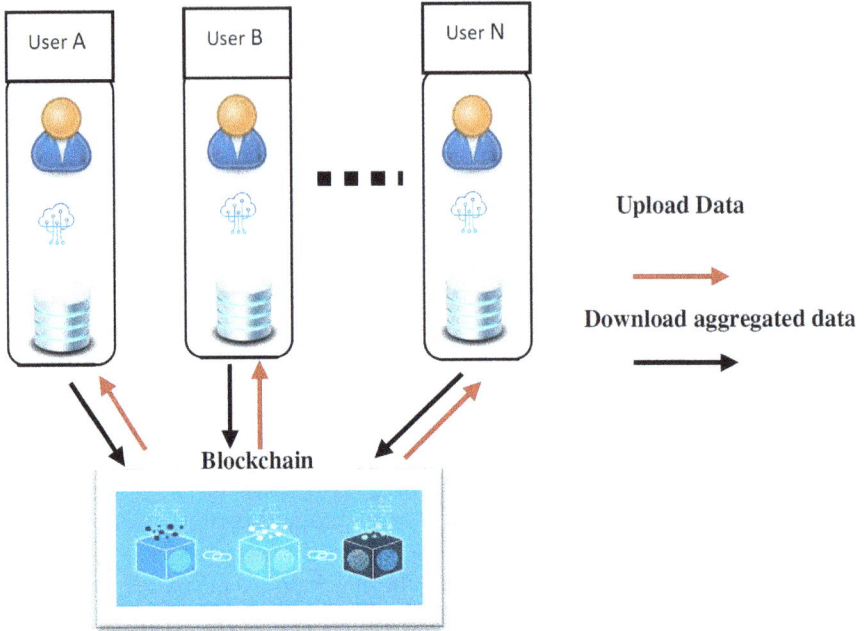

Figure 8.1   An overview of the system at a high level

8.2   Privacy-preserving data aggregation mechanism based on blockchain and homomorphic encryption for smart grid

The federated training technique has two units—users as well as servers—that are the basis for privateers' protection in a gadget model. After that each user has consented, a merged deep neural network (DNN) is presented. The topology of the classified feature is employed for the criterion function in the unified DNN, and neighboring data are ciphered in the user level and transferred to servers on the cloud throughout the training technique. The outputs are transferred to every users when the data have been combined and the encrypted text has been generated. Lastly, the neighboring DNN variables are modernized by each user using the decoded encrypted information. This looped-up procedure is repeated till the DNN is unable to get the optimal situation. Figure 8.1 depicts a top-level representation of the suggested scheme's device model.

This study looks at the critical issue of abnormal power usage with the help of various customers. During our review of the literature, we observed that most researchers assumed that users in each Home Area Network (HAN) of the clever grid use it regularly, which is practically impossible. Regular users of an HAN provide high-quality information units, while occasional users generate low-quality information units. The records uploaded by irregular clients can also affect the training phase of the version, which affects the model's forecast accuracy. The inverse

Generation of Transactions

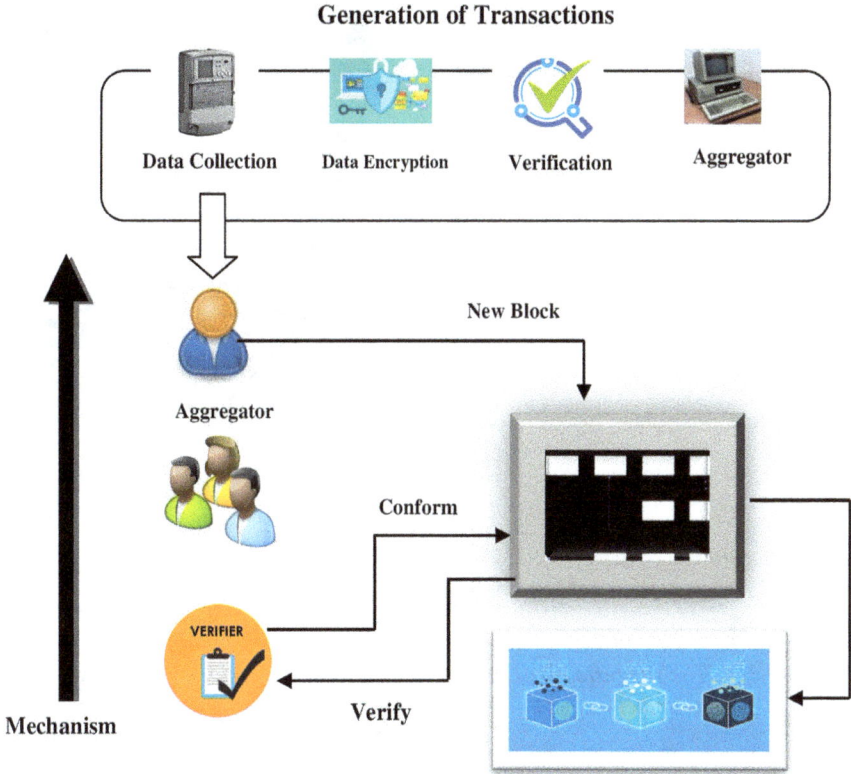

Figure 8.2   Block introduction method in blockchain

effect of an irregular workload on training results is reduced using an enhanced information aggregation technique proposed in this paper [6]. Some researchers are comfortable with the mechanism's authentication method, while others report the transaction using blockchain. Our machine aims to provide a robust technique that allows data aggregation while maintaining power usage privacy.

8.3   Evaluation of the scheme

8.3.1   System version

The suggested Blockchain-Based Hierarchical Data Access Model (BHDA) paradigm consists of three primary components: (1) a HAN with intelligent meters (SM), (2) an AI, and (3) a power middle (computer) or grid issuer. The proposed system version is represented in Figure 8.2 at an excessively high level. Smart meter (SM): computers set up the SM in every HAN for recording the usage of electricity. Within the version, the symmetric homomorphic key is an exclusive entity and isn't discovered by any third-celebration for authentication purposes [7].

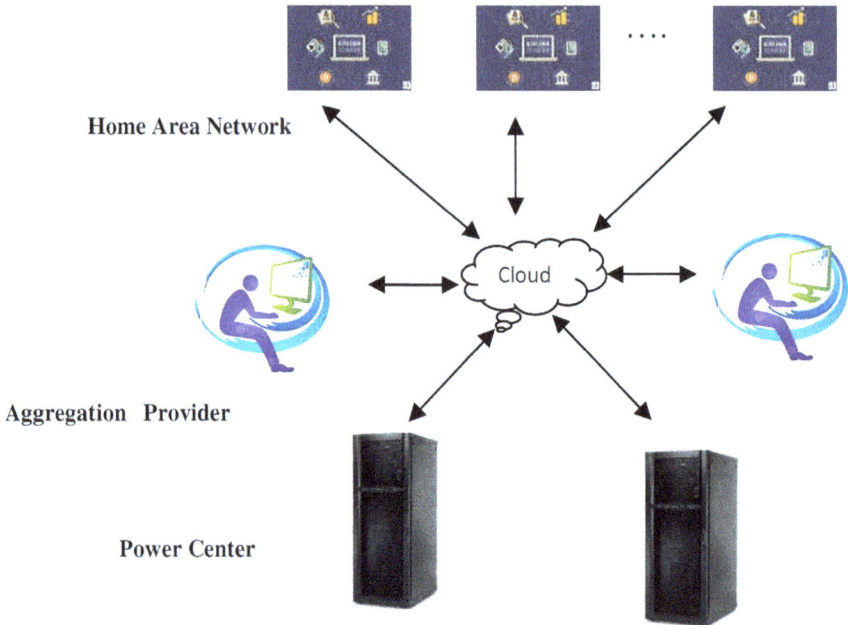

Figure 8.3   *The suggested BHDA data aggregation approach has a layered architecture*

1. AI: The AI's primary task is to validate the gathered facts and executes information aggregation before forwarding the facts to the computer. It effectively serves as a pseudo-identity for each SM's series request.
2. Power center (PC): The computer is critical in distributing electricity to each HAN. The power use information from SMs is collected regularly to optimize the power distribution. Another function of the computer is to secure connections by ensuring that the correct authentication technique and session keys are used.
3. The symmetric homomorphic key is forwarded to SMs when the consultation key and authentication technique have been appropriately authenticated (Figure 8.3).

8.3.2   Models of assault

The suggested BHDA model takes into account the following types of attacks:

1. Collusion attack: The goal of a collision attack on encrypted data is to find two inputs that have a similar hash price.
2. Internal attack: This attack is carried out by a single person or a group of individuals from the same company. The enterprise's technical users or professionals

utilize its computer systems to launch the attack and subsequently remove the evidence.

3. An outer assault occurs when anybody from outside the company attempts to get into the setup using social media [6], hacking, harmful set of programs. Hence, it is considerably harder to find rather than an assault from within.

8.4 The scheme that has been proposed

The suggested BHDA model is discussed in this segment. A sophisticated grid system has several layers. We provide privacy-based total records collection of consumers as well as servers inside a smart grid environment in the ciphertext area [8]. The technique which has been proposed is divided into three phases, as follows.

8.4.1 Setting up the machine

The proposed BHDA version's first step is to load the gadget with the necessary additives. With high order (HO), the cyclic organizations' (CO) and multiplicative cyclic group of prime order p (CGT) are multiplicative. When CO is used to create co, the bi-linear map is CGT = CO*CO. For this purpose, the SHA-256 hash function is used, and each SMi, AP, and machine generates a unique identity. In our approach, the gadget initialization is provided by a trusted third party.

8.4.2 Authentication system

Bellowin and Merritt [9] authentication system password-authenticated key agreement (PAKE) is a protocol which is based on the password that was proposed. If each entity has the same password, the consultation secret is the same; however, if the passwords are distinct, no entity can access the statistics. PAKE has been the subject of a lot of research to enhance its ability and authenticity. In this research, work has been carried out on encryption based on homomorphism and presented Abdalla and Pointchevel's [10] chance analysis of key leaking. In this essay, we will discuss how to protect against symmetric encryption attacks. The main entity is the computer, and the authentication method is as follows. To collect the records, the computer produces the SM list and assigns a random number V random. Each SM's key is generated and forwarded to all other SMs in the organization.

1. Each SM creates a session key entirely based on the acquired key. Those session keys are also used to generate the time-stamped verification code. The computer receives this verification code.
2. Based on the time stamp and verification code, the computer produces the session key. Authentication is successful if the verification code is correct; otherwise, authentication is unsuccessful.

8.4.3   Blockchain transaction report

The risk of tampering with the facts in between is the number one difficulty in information evaluation of intelligent grids. To guarantee digital integrity, an appropriate technique for signing records is required. To achieve this, the aggregation module is used to combine the information and store it in the block. A consensus mechanism aids in the creation of the blockchain [11].

- Transaction generation: At a given time slot, there are n SMs in the subarea. The data Dij are gathered from the smart meter SMij. Eq.1 is used to construct the ciphertext CT.

$$CT = N3\%(Rn * GDij) \tag{8.1}$$

  The % modulus is defined by the percent. It is used to defer the processing cost and large results of cryptography operations such as encryption and decryption.

- Statistics and verification aggregation: The Accounts Payable (AP) module collects documents from the SMs and sends the aggregated records to the computer. The session secret is given to the computer and is used to decrypt HKi. HK = decrypt is the symmetric homomorphic key (HKi, HKps). Validation is also reported using the time stamp. The aggregate ciphertext CTj is obtained from the subarea as consistent once the SM verification is completed.

- Block creation: The AP module records the transaction details for each block. After that, this block is forwarded to the subarea for authentication purposes. Figure 8.2 illustrates the block introduction approach. This block has three additives: current hash, previous hash, and is used to compute the modern block's hash price.

$$Hct-bk = SHA256(\Sigma ijTij + index + Hpv-bk + Pi + \Delta s + CTj) \tag{8.2}$$

  The BHDA technique is based on a hybrid approach for garbed circuits and homomorphic encryption.

- Blockchain generation [33]: After the aggregation process, a new block is formed and disseminated to the subareas. To meet the requirements of the intelligent grid, the facts are verified with each node inside the subareas. A signature is used to secure the transmission of records. If the verification fails, the partition approach is used to locate the illegitimate signature. When the verification is successful, the encrypted records are gathered using CTsum= Σ CTi. SGsum = H is used to produce signatures in a similar way (CTsum, Kpa, R).

8.5   Units of low-excellent statistics are processed

The federated training method gathers data from a variety of HANs, and low-quality data stored in the analytic system may have an impact on prediction accuracy. The data from the SMs are uploaded to the cloud through blockchain, and the local HAN uses DNN for training purposes. Truth discovery is a notion with various applications in crowdsourcing, transportation, and health care. The user's trustworthiness is vital in the aggregation process, and unusual usage can have an impact on each computer's burden needs.

- Consumer reliability: The reliability of customers is represented by Ri, and the combined value is defined by An.

  $Ri = C/\sum_{i=0}^{n} \mathrm{d}(An)$
- Contribution aggregation: Each SM generates records regarding electricity usage; however, for the aggregated records, we keep in mind the most reliable meters.

8.6 Evaluation of overall performance as well as safety

8.6.1 Overhead in computation

The AP module's processing overhead was used for the aggregation [7] and verification strategies. As a result, we will start by talking about the computing cost of the suggested model [10]. For the aggregate, the gadget used the strength data. The batch and single verifications were completed successfully, indicating that the computational overhead increased as the quality of the records improved. The batch verification was tested using a 1-to-1 comparison, and the results show that the suggested model enables batch verification at the highest level.

8.6.2 Cost of computation

The computing cost of the authentication operation is evaluated for the entire device in this section. In the simulated environment, we varied the number of cloud servers from one to ten. For the computation value, the suggested BHDA strategy was compared to a lightweight privateness-preserving records aggregation technique for the edge computing system (LPDA-EC) [12] and the security-better records aggregation scheme (SEDA) [11]. The proposed BHDA version is more value-green than the LPDA-EC and SEDA information aggregation systems, according to the test findings. Adding extra SM records improved the performance of the records aggregation procedures. The suggested BHDA system employed the blockchain to dispose of the verification method, with authentication at the time of initialization, unlike the current LPDA-EC and SEDA methods that used bilinear pairing, which makes the process more complex. As a result, we were able to reduce the computational cost of the suggested BHDA model (see Figure 8.4).

8.6.3 Protection of personal information

This section examines the advantages and disadvantages of our proposed BHDA version for aggregating [7] privateness-keeping records. Authentication, integrity,

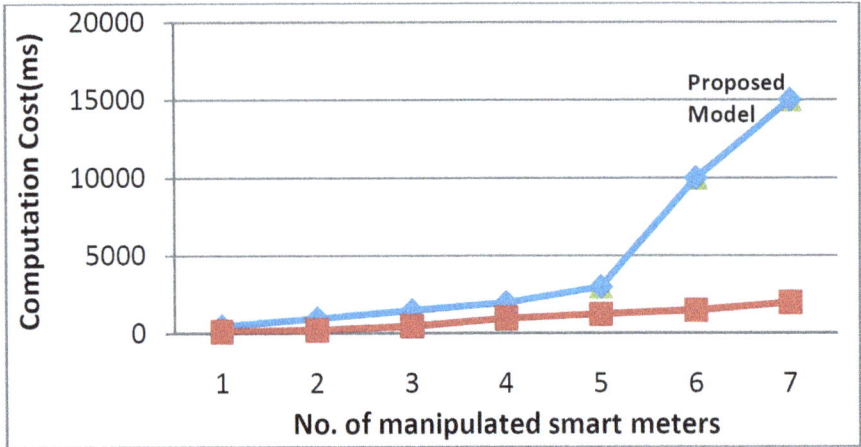

Figure 8.4   The overall calculation cost of the proposed BHDA and current fashions are compared

privacy, and aggregated records secrecy are used to validate the proposed BHDA version's performance.

With homomorphic encryption and an appropriate authentication method inside the intelligent grid environment, all communications are kept private and secret. Within the suggested framework, there is no straightforward textual content communication. Furthermore, because the records aggregator no longer acquires clear data, it is unable to view the facts. The aggregator, on the other hand, plays a vital function, but the attackers must leave the aggregators. We can provide clear facts to the PCs using homomorphic encryption [11]. To ensure that the records are coming from legitimate resources, an authentication mechanism is used, and the proposed model's authentication mechanism assures that contributors inside the facts aggregation module are the most effective SMs from legitimate HANs.

In addition, the proposed version can handle false-data-injection attacks, man-in-the-middle attacks, and stem replay attacks. The attackers inject harmful information into the SMs during the first assault, resulting in horrible information. In the planned BHDA version, the person's trustworthiness is taken into account to avoid horrifying records. In a guy-in-the-middle attack, the attacker pretends to be a different sender to enter into the discussion of the participants. The authentication [11] and blockchain consensus mechanisms address this onslaught. The data packets are replayed after the interception in the one-third attack, and the hash characteristic employed in blockchain makes it easier to avoid this type of attack. As previously stated, the suggested BHDA technique can prevent all types of potential attacks within the intelligent grid environment.

Because authentication and blockchain are used within the process, the suggested BHDA version has a larger communication overhead than the privacy-preserving data aggregation scheme with fault tolerance (PDAFT) and Efficient

Privacy-Preserving Power Requirement and Distribution (EPPDR) techniques. The overall quantity of aggregated data, however, is the same for all strategies. As a result, the proposed BHDA model outperforms the other styles in terms of overall performance.

8.7   DL mechanisms

The smart city's manufacturing data have increased as a result of the ongoing operation of sensors and the IoT. In the world sensing facts, factories are rapidly becoming more networked. Big data is defined by five operators: a massive amount, a massive rate, a tremendous veracity, a massive diversity, and a massive velocity. The key obstacles for sophisticated city applications such as stylishly developed, and others include latency, scalability, centralization, dependability, security, and privacy [13]. Meanwhile, in recent years, blockchain has been used to reduce central authority control and provide a safe environment.

A DL-based blockchain-driven scheme for a secure smart city, in which blockchain is employed in a circulated approach at the smog layer to assure manufacturing facts truth, decentralization, and defense. DL is used at the cloud deposit to boost making, and computerize data processing, with boost statement bandwidth in the industrial IoT and smart manufacturing systems.

We give a case study of vehicle production with the most up-to-date provision scenario for the future format and compare it to an active explore study utilizing key characteristics like security and privacy tools. Finally, based on findings, research directions problems are presented.

8.7.1   Identification of smart grid equipment

In managing the life cycle of intelligent grid machines, encoding intelligent grid assets play an important role. It is a good asset management website, intelligent online monitoring, power testing, bug fixing, and other subprograms. The smart grid machine code is the only indicator of recognizing the various functions of a website. The establishment of an integrated web-based grid, a smart grid device coding, and identification will be important, as shown in Figure 8.5.

8.8   Blockchain layers

The application layer and connectivity of blockchain and other programs are handled by the smart application layer in Figure 8.6. Application background: It is a semantic description of the blockchain system, and its key task is to develop blockchain solutions for various applications and industries. An example of a semantic translation is to explain cryptocurrency and then create a rule of thumb on how to exchange money between different businesses.

Background of consensus: It strongly supports a widely agreed consensus for a blockchain system. This layer is used to ensure the suitability of each blockchain,

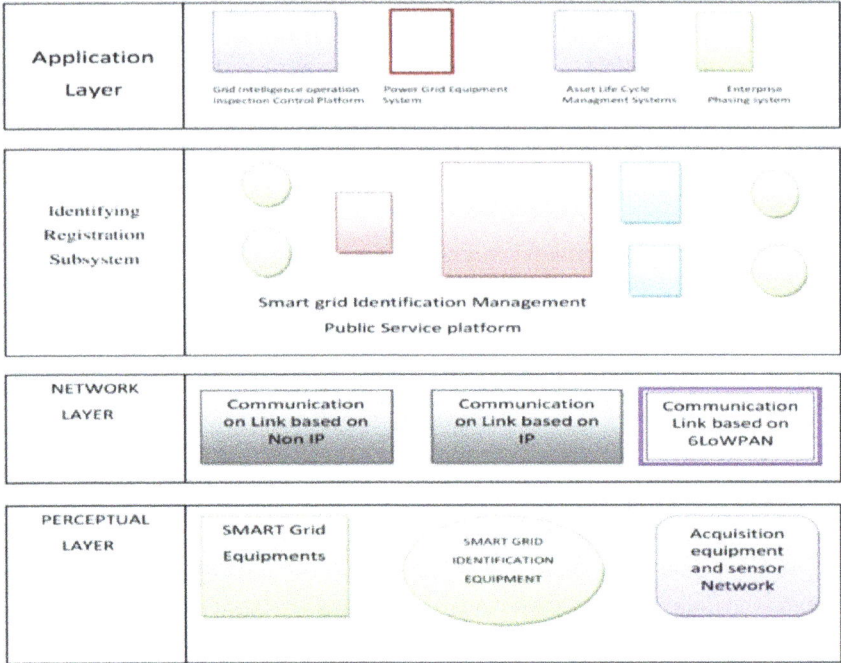

Figure 8.5   Smart grid equipment identification platform: instruction for smart grid equipment identification platform

to determine the order of the blockchain, and finally to achieve the required consistency by the system. Network coverage: It is primarily responsible for adding blockchain chains, updates as well exchanging information on blockchain chains and other activities [1].

8.9   Collection algorithms

Collection strategies be model collected of several weak models to be separately skilled with whose estimates are combined within a few methods in order to create the general calculation. a great contract attempt is positioned interested in pardon? type of vulnerable beginners in the direction of mix in addition to the methods inside which in the direction of mixing them. This is an extremely effective magnificence of technique, as well as such, could be extremely famous.

Like:
Boost.
Bootstrapped Aggregation (bag).
AdaBoost.
Subjective standard (combination).
Stack simplification (stack).

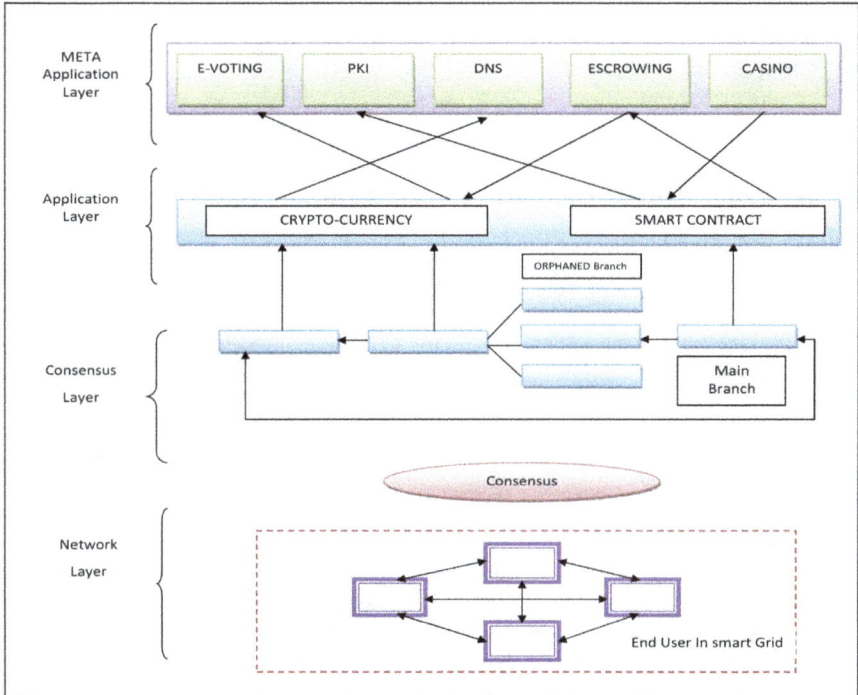

Figure 8.6   Blockchain layers

Incline boost equipment (IBE).
Incline boost failure timber (IBFT).
An accidental wooded area.

8.9.1   *Previous mechanism knowledge algorithms*

Numerous algorithms have been no longer included. I did no longer cover algorithms from uniqueness duties within the procedure of gadget mastering, along with

- function collection algorithms
- algorithm correctness assessment
- overall presentation method
- optimization algorithms

I additionally do now not wrap algorithms as of subject subfields of the system getting to know, which includes

- computational intelligence (evolutionary algorithms, and many others)
- laptop idea (CV)

- talking dispensation
- recommender system
- strengthening mastering
- graphical fashions

8.9.2 Difference between mechanism knowledge and profound knowledge algorithms

- Profound learning is a kind of system gaining knowledge, which is a subset of artificial intelligence.
- Deep learning is set computers studying to think using systems patterned on the human mind. Device training is ready networks capable of supposing and reacting with less human intervention; device learning is set computers studying to think using systems patterned on the human brain.
- Device mastering calls for much less computing energy; deep gaining knowledge of commonly wishes less ongoing human intervention.
- DL can analyze images, movies, and unstructured facts in approaches device studying cannot effortlessly do. Every industry could have professional paths that contain systems and deep getting to know.

8.10 Smart grids

The elegant network is a brand original era control device with a lot of traits, consisting of relative informatization, mechanization, as well as communication, and so on [4]. It could recover the dependability of network energy delivery, sell power-saving as well as production discounts, and recognize the maximization of network advantage as well as common advantage. It represents the destiny progress path of the electricity organization. The production of elegant networks emphasizes the interconnection of huge IoT gadgets as well as the communication of data. As elegant network communication, 5G Multi-access Edge Computing (MEC) is responsible for collecting data as well as community within every component of the luxurious network, as well as ensuring a safe, dependable, and competent data transmission channel designed for a variety of power provider structures in order to sell the overall well while correlating process of the power system.

8.11 IoT machine identification

IoT tool identifier is capable of exclusively recognizing a unit otherwise a category of the unit as well as is used in the direction of discovering the tool within the machine. In a similar instance, the variety of statistics within the IoT gadget is additionally required in the direction of broadening a combined identifier toward comprehending the records communication as well as contribution. The use of blockchain knowledge, encryption generation, as well as protection algorithms can be used to shield virtual identity, in that way structure, an additional protection with handy digital characteristics verification device designed for IoT. The responsibility of IoT tool identifiers is not simply in the direction of signifying their individuality traits other than additionally in the direction of helping the look for as well as detection of IoT gadgets. Consequently, while formulating a combined classification scheme designed for IoT gadgets, it is far essential to take into account the big quantity of IoT gadgets within pass-device as well as pass-platform situation, since glowing for the reason to the variety of kinds, as well as make sure with the intention of the strategy be able to live rapidly as well as correctly searched.

8.12   Method

This segment generally introduces the classification structural design as well as the organization process of 5G MEC blockchain-based elegant networks IoT.

8.12.1   *Structure block designed for potential IoT machine organization: blockchain as well as 5G MEC*

5G-based clever network software is presently within its immaturity. Within the potential, electricity organizations have to occupy in the company of telecom operator as well as communiqué system producers in the direction of guiding the consistency of technology inside the control communiqué fled, selling the simplification of electricity conversation fatal module, as well as manufacturing and assisting stage designed for communiqué commerce organization toward assisting clever grid sustainable improvement. Thinking about the profound combination through the 5G system, we can focus on the study of combined code as well as classification generation designed for clever network gadgets. On the identical instance, blend by blockchain along with portable aspect compute, we resolve to build a community control stage in the direction of sign in as well as manipulate the advantage statistics of a clever network system with a way of the usage of the combined code as well as classification principles. Within different phrases, we recommend transporting the gadget toward the blockchain designed for added dependability as well as safety. Elegant network is a brand original improvement drift of conventional electricity gadget, which fuse collectively cutting-edge superior generation [1], consisting of IoT, blockchain, as well as 5G cellular border compute (MBC). Moreover, everywhere clever energy IoT may be found to be able to completely observe the gadget's popularity along with economically development statistics. Blockchain along with 5G MEC is a mutually original technology, with the big capability of the aggregate of the 2 has ended up increasingly distinguished.

8.12.2   *A mixture blockchain organization designed for the elegant network*

5G drives the huge implementation of elegant gadgets within the clever network, which resources to the blockchain that may contain greater statistics than constantly previous to, as well as that information determination significantly encourages the globalization of elegant network skill. Blockchain skill can now build awake designed for the shortcoming of 5G's deprived isolation as well as protection, require of consideration within digital communication, as well as insufficient belongings privilege safety. The devolution of blockchain, the operation data, the privateer's safety, as well as the anti-tampering of chronological information within the elegant network determination efficiently encourage the improvement of the 5G network within the clever network. Abort prices and transaction time. As an instance, IBM makes use of blockchain generation to resolve the issues confronted with the aid of a brief hard work contract as well as have advanced equivalent statement understanding goods in the direction of addressing bill issues due to temporary exertions. Reconciliation via the virtual ledger cannot most effectively ensure the compliance of fee phrases, put off disputes bobbing up from invoices, however, additionally lessen the fee of integration invoice as well as cut down the employment sequence. Since a mutual digital ledger designed for footage communication, the blockchain era has introduced a clean appearance in the direction of the production global [3].

8.12.3   *Blockchain container subsists separated into three types: community, confidential, along with association blockchain*

- Community blockchain: The community blockchain is a blockchain that each person within the globe can liberally study, post communication, as well as contribute within the removal manner. But, every one node be capable of taking part inside the agreement procedure, along with there's a process verification in the making instance, which results in short operation throughput. Not as good as, its operation facts are obvious toward the general community, which is not beneficial toward protective confidentiality.
- Confidential blockchain: The appropriate method of inscribing being inside a private blockchain be engaged in by someone or else a company. Let's in them realize their non-public wishes whilst maintaining sure transaction data is personal. It can exist carefully a central system by the shortcoming.
- Association blockchain: The association blockchain is partly decentralized and mutually recognized by a couple of companies. Just a few legal nodes contribute to the agreement, which improves the effectiveness of the agreement as well as the whole transaction. The charge node adds the information with the purpose of requirements in the direction of existing demonstration in the direction of the equivalent blockchain as well as shops it temporarily toward maintaining questions next to several instances. Circumstances that affect the choice of blockchain nodes to be comfortable hardware assets or else an improved

running atmosphere, that be able to get better the implementation effectiveness of the gadget also make sure the safety of the organization [4].

8.12.4 Learn on top of mixture blockchain disagreement algorithms

Blockchain devices are an allotted organization. When the conventional unmarried join structural design evolves interest in the disbursed machine, the primary hassle is in the direction of making certain the reliability. Stipulation of the allotted machine cannot undertake the reliability of dispensation outcomes, the enterprise system constructed on top of it going to not employ usually. Constancy is the on the whole essential as well as essential trouble of a blockchain machine. Stipulation of the dispensed device is able to attain "constancy," and it is capable of gifting a faultless as well as scalable "effective join," which have higher presentation as well as constancy than bodily nodes. Compromise describes the procedure of achieving a contract on top of a certain kingdom among more than one node in a dispensed gadget. A dissimilar compromise mechanism can assemble the requirements of dissimilar tiers of compromise [5].

8.13 Blockchain and IoT for smart grids

Using IoT devices to capture useful information on energy usage in the house, for example, recommending the best approaches to save energy might help people use energy more efficiently and save money. In addition, data from IoT sensors can be utilized to automatically send all important information about multiple energy providers to consumers, allowing them to choose the best option.

In addition, data from IoT sensors can be utilized to automatically send all important information about multiple energy providers to consumers, allowing them to choose the best option.

Smart grids are a concept that brings intelligence to the voltage control cycle from the provider to the user. Users can benefit from this form of intelligence by being more conscious of energy usage and price optimization. A smart meter, which gathers, records, and evaluates electricity use at various times of the day, is also one of the smart grid's key applications. Consumers can utilize this information to save costs by adjusting their energy usage and changing their habits.

The IoT now has 5 billion related devices. This figure is assumed to expand to 29 billion by 2025 [14]. On the web, each tool creates and markets data. It is hard to mention the underlying security issues for large information infrastructure. The obstacles that IoT implementations confront are discussed in this section. The distributed architecture of the IoT is a significant challenge. Each node in this network is a potential cause of weakness that can be used to get going cyberattacks like Distributed Denial-of-Service [7]. A setup of networks with numerous contaminated devices working at the same time can suddenly collapse. Another point of contention is its monolithic configuration.

A commonly weak thing like this is a risk that has to be handled. Information security and authentication are another constant and possibly one of the most serious threats [15]. Data are generated that may be hacked and exploited unless the security of data is not given. Furthermore, as new technologies emerge in which devices may exchange things such as data, computing capability, or electricity on their own, data security becomes increasingly important.

Data reliability is another issue for IoT [7]. Smart grids are one of the most important IoT applications. Data gathered from the collection of sensors can be used to make quick judgments. As a result, protecting the system against injection assaults, which attempt to insert bogus parameters and so impact making decisions, is serious. For self-driven systems that analyze real-time data, such as transportation networks, production industries, as well as smart grids, availability is crucial. Sensor outages can lead to losses that range from financial to existence. With the rise of the machine economy, where data-generating sensors have the capacity of selling information in information exchanges and operating an end-to-end intelligent machine, establishing trust among participating entities has become a key hurdle [8]. The inclusion of a checkable audit trail alone without the involvement of a trusted intermediary is desired, as it solves the non-repudiation problem.

8.13.1   *Blockchain principles and strength*

The BC's main purpose lies to liberate humans from any kind of confidence that we are today required to place in the middle that controls and "handle" a big half of people's lives. The blockchain (BC) is a mechanism that was first utilized to facilitate financial activity (trades) using a new monetary called Bitcoin [16], which is decentralized and autonomous of governments and banks. The advantage of this monetary method, or even more precisely cryptocurrency, is it does not require middlemen like the government. What's more intriguing, however, is figuring out how to materialize and apply this new coin.

It will be interesting to see how this new cryptocurrency is realized and implemented. Several technologies, as well as privacy and cryptographic operations, have been used to attain this purpose. The BC is made up of the synergy of all of these technologies. However, this technology is beginning to be used in a variety of applications other than Bitcoin. It is the most intriguing aspect of this groundbreaking novel methodology. People frequently confuse BC with Bitcoin, although Bitcoin refers to a cryptocurrency that uses BC technology to allow it to travel independently and internationally without the banks' oversight. To put it another way, Bitcoin is merely a financial application of this tremendous methodology. Before we go into the BC technology and see what problems it tries to answer, we must first establish a crucial condition. The BC is nothing more than a database system, which is distributed and built on well-known regulations that allow value to be transferred among entities. The BC is the only one that has all three qualities at the same time. (1) Untrustworthy: It is not necessary to have a digital identity that is certified. The parties engaged need not recognize each other, and they can also nevertheless share data without knowing each other's identities; (2) permissionless: No one decides who is allowed to operate on the BC system and who is not. There are no permits or controllers, and (3) the system is resistant to censorship: Anyone can transact on BC since it is a system with no controllers, wherein entities rely upon only the integrity of the cryptographic algorithms that control the conduct. A transaction cannot be prevented or censored once it has been submitted and approved. BC can also be classified into dual sorts and classified on how it works: without permission and with permission. The agents who can engage in the system state's unanimity are limited by a permission BC. Only a small number of users have permission to check transactions in a permission BC.

It may also limit who can build smart contracts to those who have been approved. Permissionless BC, on the other hand, permits anybody to connect to a network, contribute to the block validation process to reach a consensus, and even construct smart contracts.

The four key concepts that underpin BC technology are as follows: (1) a peer-to-peer network: this approach eliminates the need for a centralized TTP, meaning that all network nodes have had the same rights. Nodes in this network can communicate with anyone using private/public key pairs. The private key is being used to validate transactions, while the public key serves as an address for reaching out to others in the network. (2) Open and distributed ledger: Consider a ledger as an entity's balance sheet that records all of the network's transactions in numerical order. This database model is not a centralized item; instead, each node has a replica of it. Everyone can see the ledger because it is open and transparent. Everybody on the network can visualize the presence of assets as well as how much each person has in their account. Three primary processes are required to achieve this goal: (1) openly broadcast fresh transactions made to the network, (2) verify transactions, (3) add verified purchases to the ledgers, and (4) mining. There are network latencies in a distributed network, so not all nodes get transactions at the same time.

Miners are one-of-a-kind nodes capable of adding transactions to the chain. Miners will compete between themselves to see who can be the early one to accept a new transaction, confirm it, and enter it into the ledger. The early miner to achieve so will be rewarded financially. A miner must verify the transaction and complete a numerical prediction game to be the earlier one. Just one miner at a moment will be allowed to offer operations to the BC in this manner. Furthermore, to prevent system attacks such as the better-known "double spending attack" [33], a means of making the "game" difficult for unethical miners is required. This technique simply costs a lot for enemies to add transactions. Proof-of-work (PoW) [17] is a reverse hashing procedure that determines a value (nonce) so that the SHA-256 hash pair "set of data," indicative of the block and the selected "nonce," is within a specified constraint. The procedures to add an extra block to the BC are shown in Figure 8.7.

Understanding that even a ledger is a linear sequence data structure of operations, it is important to understand that it accumulates things known as blocks. Each block consists of a collection of transactions. Each block in the chain, in particular, has two items: (1) Header: It contains a timestamp, the PoW's difficulty goal, and the previous block's hash value. Merkle tree root, which encrypts the operations inside the block in a singular checksum with leaves representing data blocks as well

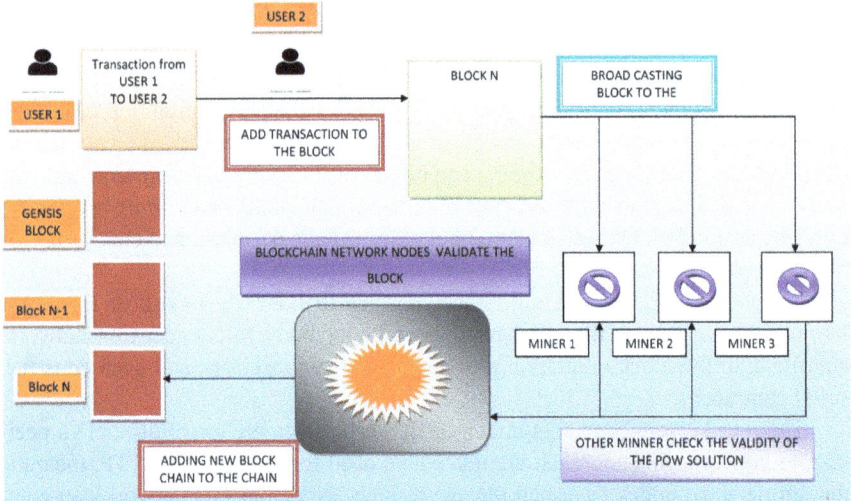

Figure 8.7   Transaction block validation and addition flow

as the nonce, which is essential to addressing the PoW while also avoiding a replay attack. (2) Block content: This includes all of the transaction's inputs and outputs. The output of prior transactions is included in the inputs, as well as a field carrying the verification only with the private key of the owner. The item to be transmitted is included in the outputs, as well as the recipient's location. Since only this private key will establish asset ownership, the recipient will be the only one who can transfer the asset.

In other words, each time a set of transactions is accepted, it is linked to the prior block via a hash, a distinct and unchangeable stamp that ensures that nobody can alter the data obtained. The only manner to alter the BC is to obtain 51% of the entire network's computational power acting on it. As a result, the individual is unable to make modifications to the ledger. This is a critical component: the BC's decentralization is indeed what makes it safe and distributed; also, this decentralization allows for the removal of any centralized system, instead of relying on the "vote of computing power," that is ensured by the BC's thousands of members, as shown in Figure 8.8.

8.14   DL insists reaction within the elegant network

In the direction of upholding the call for in addition to providing swapping inside the stack control structures, the Demand Resources (DR) model comes interested in lifestyles. DR can be an individual of the crucial techniques accompanied by utilizing means of strength groups for the reduction of strength intake through the customers in eight hours. In this method, clients can pick out toward drop nonimportant load with themselves, or else it can exist accomplished routinely with using

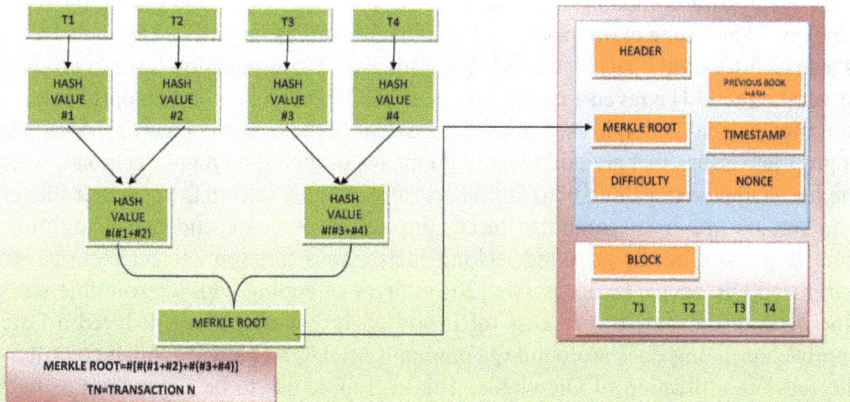

Figure 8.8   Block structure

the application at some stage in tip hours of the date [18]. DR can be a settlement involving the clients with the application primarily based on top of positive situations approximating instant durations, rate, as well as cargo. Via DR, strength intake will be able to exist compact at some stage within the tip hours of the date with the aim of the result within the discount of effective expenses, equipment fees, along with additionally the improvement of capacity Singapore (SG) disasters. DR ends in reducing the prices of strength. To be capable of guiding in the direction of a typical decrease in vend price, application businesses can inspire DR via supplying decreased energy prices at some stage in nonpeak hours toward inspiring the clients in the direction of dropping noncritical hundreds throughout top time. DR model utilizes purchaser as well as their stack intake information intended for preserving network sustainability with means of stack losing at some stage in height needs and managing the entire intake statistics styles correctly. As a result, developing an effective DR device is based on prevailing instance modifications of stack utilization facts styles that require exixting knowledge to gain particular stack utilization pattern. Within direct toward plot a green DR machine, a valuable forecast procedure is in the direction of existing carried out so that device can endorse appropriate solution sparkly utilization behaviors next to predict impending patron utilization pattern [18]. Designed for SG as well as clever built-up condominium area, around be enormous require designed for devise a unique power utilization forecast machine. A profound Recurrent Neural Network (RNN) utilizing a gated persistent component (GRC) machine can be urbanized in support of predicting power deliver-call for within suburban apartments designed for a tiny toward the average duration of instance [19]. The incorporated DRNN-GRU version is a 5-layered neural community such as optimized hyperparameters by normalizing enter. The primary level is a contribution level, fed through a day-by-day hourly stack utilization facts sample. The next level referred toward the primary GRC level in the direction of providing products designed for every direct instance. The third level stated the instant one

GRC level toward creating an advanced size production than the preceding level. This level has a tune extra quantity of weights as well as partiality. The fourth level is a straightforward concealed level. The fifth level is a production level to produce forecast effects. The reverse circulation is used intended for limitation optimization, min-max normalization is used toward the extent of the teaching records. More than appropriate issues that are addressed with means of including regularization toward the fee characteristic along with failure are implemented toward the concealed level. The version preserves combined forces through instance dependency of information, it is too efficient in fixing lacking facts issues through ancient records. An optimized DR device be accessible [20] in favor of coping with interruptible stack (IS) beneath the time-of-use cost for bendy energy stack pattern. It hired a DRL approach utilizing duel profound Q community (DDQN) pattern, which completes the sensible utilization of DR viable. The IS trouble has to be carried out using a Markov selection procedure toward advantage the maximum proceeds designed for a longer length, which define the nation, process, along with praise characteristic of the DR gadget. The Double Deep-Q Network (DDQN) company done concentrated sound along with more suitable the stableness of the representation, which indicates to the insecurity inside the junction of DDQN has to be decreased in the direction of acquiring the suitable fee of the defeat feature. Eventually, the version realizes the discount within the height energy intake call for along with the value of procedure of the parameter of energy surrounded by the maximum value lacking compromise protection.

The difficulty of profound strength utilization within the manufacturing sector is addressed utilizing Nabil *et al.* [19]. A version is urbanized with multiagent DRL designed for enforcing the DR method intended for dealing with an awesome production scheme. The economically developed gadget has been formulated like Multi Agent Deep Deterministic Policy Gradient (POMG) primarily, afterwards, the MADDPG set of rules changed into implemented and a most advantageous agenda of stack utilization used for each mechanism has to be intended. After that the representation changed into deploy on top of a sequence production gadget designed to evaluate its efficiency. The consequences of the replication prove to the DR device might present the slightest value of electricity intake as well as maintain smaller quantity manufacture fees in contrast to the non-DR standard machine. The replica permits the gadget in the direction of obtaining condensed energy utilization along with insecurity of the network. A resourceful short-term load forecasting (STLF) structure is evolved in the direction of dealing with the demanding situations offered by using energetic along with stochastic patron behavioral patterns within the suburban campus. The structure predicts individuals' strength utilization patterns the use of aggregate stack statistics of the appliance as well as the association between them with means of employing DL strategies.

Figure 8.9 represents flexible services and shared network services, similar processing, and universal access are some of the most popular cloud computing features in smart grid systems. Although the cloud computing model is considered to work well on smart grids, it has some drawbacks such as security and reliability. In this chapter, the design of the smart grid and its use is focused on it first. The structure

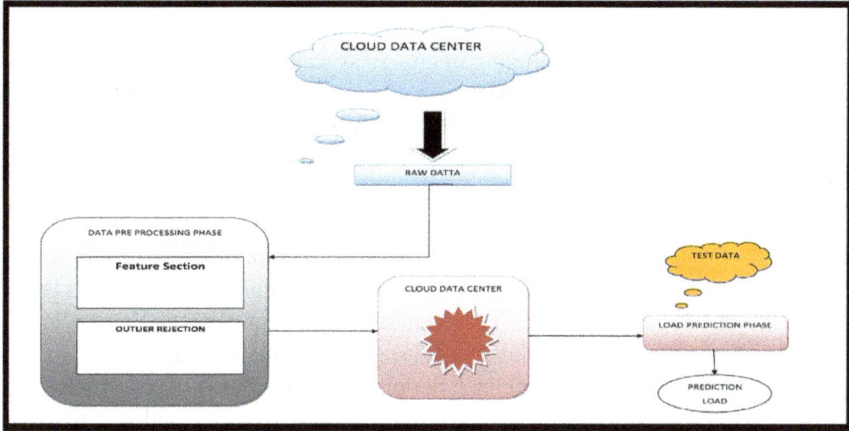

Figure 8.9   Load forecasting strategy in Smart Grid

of cloud computing is well defined. Then, cloud computing for smart grid apps was also introduced for efficiency, security, and usability. The technical and security issues of the cloud platforms are being analyzed. Finally, smart grid projects based on cloud service and open research issues are being introduced.

8.15   State-of-the-art DL procedures for call for reaction and smart grids

In this phase, numerous modern-day mechanism scheduled programs of DL intended for DR as well as SG approximating stimulating powered stack forecasting, nation evaluation within SG, power pilfering revealing within SG, electricity contribution as well as buying and selling within SG are mentioned. A stimulating stack forecasting is a method of predicting electricity; otherwise, electricity is essential toward convening the expectations call for. It is of excessive importance inside the power along with the control sector. Exact stack forecasting is vital and designed for efficient development along with the operation of the control system. It can exist separated addicted to two classes is immediate stack forecasting (ISF) as well as lengthy-time period stack forecasting. Normally, in ISF, a stack designed for the subsequent hour as much as the following two weeks is anticipated [21]. Taking into consideration forecasting, the electrical automobiles (EV) charge stack like their huge saturation causes excessive improbability within the energy to require of the strength classification. The document makes use of Artificial Neural Network (ANN) along with Long Short Term Memory (LSTM) primarily base DL tactics designed for hooking up EVs stack forecasting. A charge situation corporation of Plug-In Electric Vehicles (PEV) is used toward evaluating each conventional ANN move along with LSTM move. It became determined to the LSTM version has decreased mistakes as well as superior accurateness within a short period of EVs stack forecasting.

8.15.1   Power pilfering recognition

Electricity thievery refers to pilfering of energy as well as changing of one's stimulating powered strength statistics within an effort toward reducing his otherwise her influence receipt. It is far taken into consideration as one every of the most important crime within the unified state. Convenient is numerous behavior of power pilfering within an SG. Various of the strategies be erase logged procedures, tamper store requirements, tamper indicator, disconnection of the meter, and many others. Like, depicted in Figure 8.5 before, energy utilities used to ship organizations to investigate strength systems depending on scheduled community reviews. The boom within the development of metering communications completes it simpler in the direction of stumble on power robbery the usage of information accessible as of elegant meter. This gives an increase in superior metering infrastructures (AMI) [22]. But, convenient be numerous negative aspects to facilitate being added with AMIs together with having the ability to govern indicator reading. This brought about the involvement toward characteristic manufacturing structures, especially toward discovering pilfering within clever strength grid. Smith *et al.* [23] anticipated a configuration to facilitate mixing an inherent training set of rules along with a limited aggregate version cluster designed for purchaser segmentation. This is intended toward producing function situation which conveys the importance of call for more instance. Additionally, the evaluation of comparable households prepared it reliable to detect extraordinary performance as well as deception. An extensive variety of ML algorithms had been used. The conclusion is superb since this approach is computationally extremely realistic. In regards to the typeset of rules, the incline boost equipment surpassed the entire different ML organization fashions which be used previously. This has critical sensible symptoms designed for stimulating utilities and may be their survey higher within addition studies. Non-technical defeat within SGs is distinct from power to be transported, with the exception that it is not billable unless power pilfering is covered. This has come to be a first-rate difficulty within the energy delivery enterprise global, so that you can locate bypass of the indicator along with NTL of indicator manipulate by the identical instance. Kim *et al.* [24] projected an energy allocation community base totally version, referred to as the transitional reveal indicator (IMI). This version divides the set of connections hooked on best with a free network on the way to take a look at the control stream correctly as well as correctly discover the NTL. An NTL recognition set of rules is projected so that it will solve the linear structure of equations. This is evolved with inspecting the equilibrium of power through IMMs along with the investor. The author too said the hardware structural design of IMMs. This structure changed into green in phrases of time and turned into capable to face up to correct detection of at least 95% accuracy. It changed into additionally established to facilitate it reveals the ethics of clients with the defeat of power since bypass to be complex through conventional recognition methodologies toward conquering the loophole of active ML-based detection techniques. Chen *et al.* [25] projected a brand innovative recognition technique referred to as power robbery detection the usage of deep bidirectional RNN (ETD-DBRNN), that's used in taking pictures of the inner traits and the outside association by using inspecting the power intake files. Experiment at the actual globe records indicates the evidence of this process. In comparison to previously active strategies, it is determined to this approach capture a sequence of the power procedure statistics with the inner functions among everyday as well as odd energy utilization styles.

8.15.2   Energy sharing and trading

Strength control is essential to trouble involved with power structures. Power buying and selling are stronger organization procedures used in the direction of progress and the performance of the electricity system [26]. Its miles converting beginning central in the direction of dispersed way except through electricity contribution along with trade, come protection as well as dependability issue. Consequently, it is too essential with the purpose of the SG attack be averted. A structure with equally DL as well as DeepCoin, a blockchain-based power structure be projected using ref. [27]. A dependable peer-to-peer power gadget base scheduled Byzantine error acceptance algorithm is incorporated inside the blockchain proposal. The structure additionally makes use of short signatures and hash features to take advantage of the energy get right of entry as well as avoid the clever-grid assaults. The DL proposal is an interference recognition gadget that makes use of RNNs intended to detect community assaults inside the blockchain base strength association. The overall presentation of the structure is evaluated the usage of three datasets along with an elevated throughput is determined representative of the design is green. An iterative set of rules designed for limited electricity buying and selling between the distribution community's gamers is supplied via Gazafroudi *et al.* [28]. Outcomes of the simulations primarily base scheduled bendable activities of the quit-users to depict to move intelligent stop-users comprise additional energetic suppleness than to of self-consumption give up-customers. In addition, advanced proceeds intended for the allocation employer along with the aggregators be achieved through move intelligent give up-customers. It is what's more visible to the ensuing distribution community will become a sustainable electricity structure. A strengthening getting to know (RL)-primarily base micro-network (MN) strength trade design be projected through using Xiaozhen *et al.* [27]. Based totally on the expected destiny renewable power era, the approximate opportunity control call for, along with the MG sequence intensity, the trade coverage may exist selected as a consequence. The paper additionally provides an overall performance certain at the MG software. Reproduction effects base scheduled practical renewable power era with electricity call for facts display to this proposal perform higher than the standard proposal. During an elegant network such as three MGs, its miles found to the projected proposal increase the MG application through the way of 22.3% compared to the benchmark scheme.

Energy theft refers to the theft of power and also varying of one's electric authority information in command to reduce his or her control invoice. It is measured as a single of the major crime in the United States. readily available are quite a few habits of electricity theft in an SG. A number of techniques erase logged

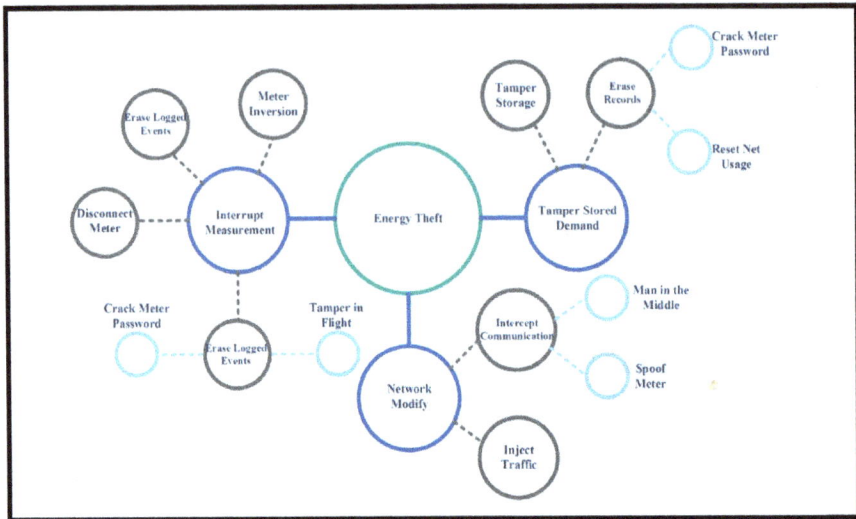

Figure 8.10   Various techniques for energy theft

proceedings, tamper stored demand, tampered meter, disconnection of the meter, etc., as depicted in Figure 8.10.

8.15.3   Inducement base immediate DR algorithm designed for elegant grid

The essential hobby essential intended for comforting the strength grid is the comparison of the production of energy along with its intake. Convenient resolve exists a growth within the expenditure designed for the manufacturer along with purchaser condition to hand may be some difference within linking deliver along with a call for. Numerous fashions were urbanized thus distant toward stability the strength version with improving the network dependability. The primary mission is an inducement base genuine instance DR set of rules projected intended for SG electricity structures the usage of DNN algorithm and reinforcement getting to know. This set of rules intends to offer support in the direction of the carrier issuer toward acquiring electricity possessions since its register clients if you want towards steady the power fluctuations as well as advance the dependability of the network [29]. A DNN set of rules be used to estimate the private costs along with the insist used for power. Strengthening mastering is functional to accumulate the most excellent inducement fee designed for diverse clients with a way of inspecting the net earnings of equal clients as well as providers. The carrier companies are capable of gathering the value starting the energy sell along with the call for power since numerous clients simply within the want of the hour since the inherent actions of the hierarchical strength sell toward explaining those varieties of expectations changeability, DNN is implemented to estimate the private fee along with call intended for power. The DNN collects innovative time with electricity needs as well as predicts the destiny charge along with insisting intended for strength. This action is recurring and frequently cultivated the conclusion of the exacting date. In addition, alongside the anticipated opportunity charge along with power difficulty, reinforcement mastering is implemented toward gathering the quality inducement quotes designed for numerous clients with means of analyzing the proceeds of deal vendors along with clients. The blessings of application of reinforcement gaining knowledge of have several blessings along with model-unfastened, the adaptive character with precision. The source issuer finds the satisfactory inducement fee through gaining knowledge of since clients immediately with it want now not need several preceding understanding otherwise determined regulation on a choice of inducement charges. The carriers preserve attain the inducement costs separately within a web technique remodeled to diverse clients. Similarly, it considers the flexibility with reservations of the power structure. The complete computation of an element of the strengthening algorithm be depending scheduled a map chart with its far extremely clean toward putting in force this within numerous genuine globe applications. The effects show to the projected set of rules instigate the contribution as of the call for aspect and improve the income of examination provider along with clients with complements the dependability of the machine through stabilizing the power assets. During opportunity, the inducement base set of rules preserves is there multiplied toward big competence useful source promote structure such as the network operator along with numerous facility carriers.

8.16 Challenge with sufficient methodological solutions be accessible now

8.16.1 Active price designed for call for reaction within the microgrid

A microgrid is a nearby power supply related in the direction of the SG surroundings that could perform autonomously. Microgrid adopts the active price method anywhere the facility issuer determination proceeds since a mediator (brokerage) among the end-user along with the application organization through using trade the control purchased as of the software agency toward the end-user. Still, even if the provider uses active price toward manipulating the microgrid, the reservations inside the microgrid with misguided client information (purchaser load call for stage, purchaser acquire model along with procedure styles) make it extra demanding toward establishing the price base scheduled client's expectations performance. Scheduled the previous dispense, clients countenance the difficulty of power utilization preparation incurred via reservations within power fee with their utilization. RL base multi-agent version changed into projected within [30] to study each the patron's power intake with the carrier company's active price base scheduled destiny performance. And, they contain got use the post-decision kingdom learning algorithm to beautify the client's gaining knowledge of fee of minimizing their power value. Stack give evaluation scheduled the call for the surface.

8.16.2   *Stack forecasting within the clever grid*

The electricity gives study that may subsist finished successfully via precise forecast of opportunity stack, i.e., the expectations power call for of clients primarily base scheduled their preceding procedure. This can make sure a sustainable power system through minimum misuse with an honest price. This electricity forecast relies upon lying on different factors like climatic change and evolution opportunity. Moreover, a miniature cargo forecasting blunder could cause huge economic pasting consequently a powerful load forecasting in SG is hard and has to be addressed correctly. DNN locate of policy (i.e. to utter lengthy speedy name memories construction) and ANN-based whole instance album model can live worn for exact calculation of power in inhabited and trade home [29].

8.16.3   *Cyberattacks in bidirectional strength buying and selling*

SG era permits bidirectional message flanked by the customer, and the submission venture consequently specializes in presumed in the first choice to the buyer. The bidirectional message handles unique kinds of aware accounts between the two entities. An SG makes exercise of facts furthermore oral swap age for a minimum of its games which makes it vulnerable to many cyberattacks. As well, the intruder to souse have a loan of the vigor or to lie to the supporter power outline can set off a pair of cyberattacks.

8.17   Conclusion

The current development in blockchain and ML has opened a new era in technical evolution. The various prediction algorithms in ML and DL in conjunction with the ledger (distributed) concept add strength to many applications such as smart grids

The new model is mainly determined by quality and security analyses, which revealed that the given BHDA version was 80% more powerful than the conventional method in identifying SM updating. Furthermore, if compared to existing technologies, the computational value of the competent BHDA model decreased by 20% from 80%. As a result, the suggested BHDA scheme implies a large update in total performance and privacy preservation for data aggregation in smart grids while suffering no computational overhead.

In this chapter, we have tried to give out exhaust information on various blockchain technologies, ML, and DL approaches to smart grids. This mechanism can be used to plan, implement, and ML, DL-BT-based solution for smart grid efficiency. The chapter also discusses various data hazards that encounter and gives a broad approach to efficiently tackle those hazards. The chapter also sheds light on various data aggregation techniques, privacy issues concerning the integration and accuracy of the records.

References

[1] Hang L., Kim D.-H. 'Design and implementation of an integrated IoT Blockchain platform for sensing data integrity'. *Sensors*. 2019, vol. 19(10), p. 2228.

[2] Karimipour H., Dehghantanha A., Parizi R.M., Choo K.-K.R., Leung H. 'A deep and scalable unsupervised machine learning system for cyber-attack detection in large-scale smart grids'. *IEEE Access*. 2019, vol. 7, pp. 80778–88.

[3] Singh P., Masud M., Hossain M.S., Kaur A. 'Blockchain and homomorphic encryption-based privacy-preserving data aggregation model in smart grid'. *Computers & Electrical Engineering*. 2021, vol. 93(1), p. 107209.

[4] Bodkhe U., Bhattacharya P., Tanwar S., Tyagi S., Kumar N., Obaidat M.S. 'Blohost: Blockchain-enabled smart tourism and hospitality management'. *Proceedings of International Conference Computer, Information Telecommunication System (CITS)*; 2019. pp. 1–5.

[5] Zheng Z., Xie S., Dai H.N., Wang H. *Blockchain challenges and opportunities: A survey [online]*. Available from http://inpluslab.sysu.edu.cn/files/blockchain/blockchain.pdf [Accessed 10 Feb 2019].

[6] Hossain M.S., Rahman M.A., Muhammad G. 'Cyber-physical cloud-oriented multi-sensory smart home framework for elderly people: An energy efficiency perspective'. *Journal of Parallel and Distributed Computing*. 2020, vol. 103, pp. 34–45.

[7] Li S., Xue K., Yang Q., Hong P. 'Ppma: privacy-preserving multisubset data aggregation in smart grid'. *IEEE Transactions on Industrial Informatics*. 2017, vol. 14(2), pp. 462–71.

[8] Lynn B. *Java pairing based cryptography library (IPBC) [online]*. 2020. Available from http://libeccio.di.unisa.it/projects/jpbc/ [Accessed 20 Sep 2020].

[9] Bellovin S.M., Merritt M. 'Encrypted key exchange: Password-based protocols secure against dictionary attacks'. Proceedings of the IEEE Symposium on Research in Security and Privacy, Oakland, CA; 1992.

[10] Abdalla M., Pointcheval D. *Simple Password-Based Encrypted Key Exchange Protocols Cryptographer's Track at the RSA Conference*. Springer; 2005. pp. 191–208.

[11] Zhang S., Lee J.-H. 'A group signature and authentication scheme for blockchain-based mobile-edge computing'. *IEEE Internet of Things Journal*. 2020, vol. 7(5), pp. 4557–65.

[12] Fan M., Zhang X. 'Consortium Blockchain based data aggregation and regulation mechanism for smart grid'. *IEEE Access*. 2019, vol. 7, pp. 35929–40.

[13] Tanwar S., Bhatia Q., Patel P., Kumari A., Singh P.K., Hong W.-C. 'Machine learning adoption in blockchain-based smart applications: The challenges, and a way forward'. *IEEE Access*. 2019, vol. 8, pp. 474–88.

[14] Barboutov K., Furuskär A., Inam R.. Available from https://www.ericsson. com/assets/local/mobility-report/documents/2017/ericsson-mobility-report- june-2017.pdf [Accessed 30 Mar 2018].

[15] Gubbi J., Buyya R., Marusic S., Palaniswami M. 'Internet of things (IoT): A vision, architectural elements, and future directions'. *Future Generation Computer Systems*. 2013, vol. 29(7), pp. 1645–60.

[16] Hawlitschek F., Notheisen B., Teubner T. 'The limits of trust-free systems: A literature review on blockchain technology and trust in the sharing economy'. *Electronic Commerce Research and Applications*. 2018, vol. 29(31), pp. 50–63.

[17] Karame G.O., Androulaki E., Capkun S. 'Double-spending fast payments in bitcoin'. Proceedings of the 2012 ACM Conference on Computer and Communications Security (CCS 2012); Raleigh, NC, Oct; 2012.

[18] Jindal A., Schaeffer-Filho A.K., Liu M.Y., *et al.* 'Hidden electricity theft by exploiting multiple-pricing scheme in smart grids'. *IEEE Transactions on Information Forensics and Security*. 2020, vol. 15, pp. 2453–68.

[19] Nabil M., Ismail M., Mahmoud M.M.E.A., Alasmary W., Serpedin E. 'Ppetd: Privacy-preserving electricity theft detection scheme with load monitoring and billing for AMI networks'. *IEEE Access*. 2019, vol. 7, pp. 96334–48.

[20] Buzau M.-M., Tejedor-Aguilera J., Cruz-Romero P., Gomez-Exposito A. 'Hybrid deep neural networks for detection of non-technical losses in electricity smart meters'. *IEEE Transactions on Power Systems*. 2020, vol. 35(2), pp. 1254–63.

[21] Ighravwe D.E., Mashao D. 'Predicting energy theft under uncertainty conditions: A Fuzzy cognitive maps approach'. *6th International Conference on Soft Computing & Machine Intelligence (ISCMI)*; 2019. pp. 83–9.

[22] Razavi R., Fleury M. 'Socio-Economic predictors of electricity theft in developing countries: An Indian case study'. *Energy for Sustainable Development*. 2019, vol. 49(5), pp. 1–10.

[23] Smith P.M., Granville L. 'Tackling energy theft in smart grids through data-driven analysis'. *International Conference on Computing Networking, and Communications (ICNC)*; 2020. pp. 410–4.

[24] Hock D., Kappes M., Ghita B. 'Using multiple data sources to detect manipulated electricity meter by an entropy-inspired metric'. *Sustainable Energy, Grids and Networks*. 2020, vol. 21(1), p. 100290.

[25] Han D., Zhang C., Ping J., Yan Z. 'Smart contract architecture for decentralized energy trading and management based on blockchains'. *Energy*. 2020, vol. 199(3),117417.

[26] Ferrag M.A., Maglaras L. 'DeepCoin: A novel deep learning and blockchain-based energy exchange framework for smart grids'. *IEEE Transactions on Engineering Management*. 2019, vol. 67(4), pp. 1285–97.

[27] Xiaozhen L., Xiao X., Xiao L., Dai C., Peng M., Poor H.V. 'Reinforcement learning-based microgrid energy trading with a reduced power plant schedule'. *IEEE Internet of Things*. 2019, vol. 6(6), pp. 10728–37.

[28] Gazafroudi S., Shafie-khah M., Lotfi M. 'Iterative algorithm for local electricity trading'. *IEEE Milan PowerTech*; Milan, Italy; 2019.

[29] Gungor V.C., Sahin D., Kocak T., *et al.* 'Smart grid technologies: Communication technologies and standards'. *IEEE Transactions on Industrial Informatics*. 2011, vol. 7(4), pp. 529–39.

[30] Lu R., Hong S.H. 'Incentive-based demand response for smart grid with reinforcement learning and deep neural network'. *Applied Energy*. 2019, vol. 236(8), pp. 937–49.

Chapter 9

Cloud computing-based techniques for blockchain-based smart grids

Manoj B O[1], Sindhu Shiranthadka[1], and Annaiah H[2]

Smart grids (SG) based on blockchain presents emerging employments of block-chain in SG structure and looks for future headways in the usage of blockchain development in the energy market. Quick development of sustainable power assets in power frameworks and critical improvements in the media transmission frame-works have brought about new market plans being utilized to cover flighty and con-veyed age of power.

The capability of blockchain innovation is seen by multiple businesses. This capability is sensible to theorize energy business, particularly SG, which uses or exploits frameworks based on blockchain, which controls itself and deals with legally binding data and exchanges.

To put it plainly, we can define blockchain as decentralized framework com-prising all the records of exchanges, which consistently occurs among the members, all in all, a public record as [24]. An ordinary blockchain exchange works in the accompanying manner. The exchange is addressed as a "block" with data put away inside. The square is communicated to each hub in the organization and sits tight for endorsement from the hubs that this exchange is legitimate. After the endorsement, the square is added to the current blockchain and the exchange is finished [3]. As indicated by "Shen and Pena-Mora (2018), the three 'columns' of blockchain stages are brilliant agreements, circulated agreement, and crypto token. This addresses the business rationale, information base, and framework of the stage individually. The utilization of blockchain has developed throughout the last decade, from blockchain 1.0 to 3.0 (Swan, 2015)".

Being "smart" has been attracting consideration from both the government and industry for ongoing years. This idea "encapsulates ideals of efficiency, security and utilitarian control in a technologically mediated and enabled environment." The "smart" tag has been added to various ventures, including the brilliant energy area. One of the most basic parts of an energy framework is the energy matrix. Utilizing

[1]Government Engineering College, Hassan, Karnataka, India
[2]Department of CS&E, Government Engineering College, Hassan, Karnataka, India

progressed data correspondence advances in various sections of the energy grid brings about an SG. The European Commission characterized SG as "a power network that can keenly coordinate the activities of all clients associated with it generators, buyers and those that do both to productively convey reasonable, prudent and secure power supplies" [6]. In correlation with the conventional energy lattices planned primarily for petroleum products, SG has a more prominent interest in sustainable power sources, such as breeze- and sun-based energy to help natural endeavors (Moretti *et al.*, 2017).

9.1   Introduction

SG, an advanced electrical grid, consists of technologies such as communication, automation, and information technology systems, which control flow of power starting at the place of generation and ending at the place of consumption (even till to the level of appliances) and in order to control flow of power or to balance any load that matches the generation of power in or near real time. The SG consists of IT applications that will allow a user-friendly integration along with higher sources of renewable energy. It is essential for rapid acceleration of the development and widespread usage of hybrid electric vehicles and their potential to use them as storage for the grid. This SG has million parts and pieces of controllers, computers, lines of power, and upcoming technologies and equipment. This will take some time to get fully functional as many of the technologies need to be perfected, equipment needs to be installed, and systems must be tested beforehand.

Electricity is a basic need of every human as it is required in day-to-day household chores but not limited although. Electricity is one among the best inventions that science has given to the mankind. This has become a part of our modern-day life, and we are into it as we cannot think of our lives without this. It has a number of uses in our daily life. Like it is used for lighting purposes, working of fans and also day-to-day appliances, such as induction stoves and air conditioners. All of these appliances will provide some comfortness to people. Apart from houses, electricity is also a major part in factories to run huge machineries and for transportation inside a factory and also for the public transports such as metro. Essential items such as foods, clothing, paper, and many other things are also the products that are produced with the help of the electricity.

The three major sources of energy used for electricity generation are as follows:

- fossil fuels
- nuclear energy
- renewable energy sources
- other sources

As generation of electricity is specific to location, i.e., in centralized fashion it has to be distributed across different locations, which are far from electricity generation points. To achieve this long transmission distance, electric wires and poles were used and also transformers were used to change the voltage output from a power station as voltage is increased with the help of step-up transformers at generation points and distributed through high power lines and at the user's end the step-down

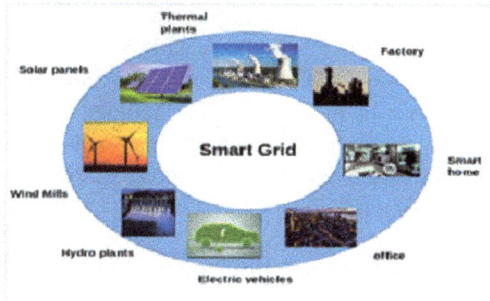

Figure 9.1   Smart grid

transformers are used in order to reduce the voltage, which is compatible for the usage in homes, factories, and offices. Even though we are delivering the electricity from the generation point to required locations, we are unable to meet the consumer or customer requirements of electricity and are unable to monitor the usage of electricity due to electricity theft and natural calamities, these power lines easily get damaged resulting in power cut for the consumers from the affected point.

To control and measure the usage of electricity, meters were introduced; these meters are placed at the end point of the consumers. Even though electricity meters were introduced one has to visit each meter personally to get the readings and also we are unable to cope up with the requirements of the consumer. Sometimes we end up in situations where electricity demand increases and its supply decreases, and in order to overcome this, electricity generation using non-renewable resources like nuclear power plants was setup, but these are affecting in different manner as the setup of nuclear power plants resulted in pollution so we need to think of their alternatives. We also have renewable sources of electricity generation points such as wind turbines and hydroelectric projects, but these could not produce electricity constantly as they are seasonal, for example, hydroelectric projects are productive at rainy seasons but not at other seasons and the quantity produced is also less. To overcome all these problems, SG were introduced.

9.1.1   SG introduction

• **What is SG**

> A smart grid is an electricity network based on digital technology that is used to supply electricity to consumers via two-way digital communication. This system allows for monitoring, analysis, control and communication within the supply chain to help improve efficiency, reduce energy consumption and cost, and maximize the transparency and reliability of the energy supply chain. The smart grid was introduced with the aim of overcoming the weaknesses of conventional electrical grids by using smart meters.

This SG is going to provide an opportunity to revolutionize energy industry to an advanced era of efficiency, availability, and reliability, which is more economic as well as environment friendly. With this transformation, time carrying out these tests is going to be critical, educating the consumer, technology enhancements, standards for development, regulations, and sharing of information in-between the projects to make sure that the benefits from SG are turned into a reality.

Some of the benefits of the SG includes:

1. Efficient manner of electricity transmission.
2. Robust transmission lines, even if one path fails transmission will be made through other paths.
3. Reduced maintenance costs resulting in lower prices of electricity.
4. Reduced fluctuations as production and consumption are always monitored and can be predicted using technology.
5. Interconnection of electricity generation points.
6. Monitoring can be done remotely, resulting in reduced in-person visits to end user's meters.
7. Opportunity for common people to earn by generating power in their own using renewable resources.
8. Reduced electricity thefts.
9. Improved management of billing and increase in revenue.

* **Why SG is smart?**

The SG consists of dual-way communication that is data transfer between consumer to grid and also transfer of electricity to consumer from grid and also two-way transfer of electricity from grid to prosumers (both consumer and producer), and also the sensing of transmission lines is one makes the SG avoiding electricity theft or leakage of electricity. SG consists of computers, automation, advanced controls, network, and advanced technology equipment will be working with each other. In SG, these technologies are going to work with the electrical grid and will respond dynamically and digitally for our ever-changing demands of electricity.

An electricity disruption may cause number of issues in different fields resulting in huge losses. Disruptions such as blackout can have adverse effects such as series of failures that can affect working of various sectors such as hospitals, banking, communications, education, traffic, and security.

SG are strong and odds of transmission disappointments are less as they can withstand substantial rains and diminish line blackouts as it is decentralized as every one of the grids are interconnected regardless of one electrical cable it will not influence the power supply. Two-way intelligence of the SG will take rerouting into consideration when power blackouts or lines failure happens. This will reduce the number of blackouts and the impacts from the blackouts will be limited when they do occur. When a power outage occurs, keen network innovations will identify the outage and separate them before they become enormous scope blackouts.

The new age advancements in SG will assist with guaranteeing that power recovery proceeds quickly and deliberately after a crisis. For the instance directing the power to crisis beneficiaries as priority. Furthermore, the brilliant network will search for the possessed power generators to generate electricity when it is not free from utilities. By including all these "appropriated age" assets, a local area could keep its well-being place, police division, traffic signals, telephone framework, hospitals, and supermarket working during crises. Likewise, the brilliant lattice is an approach to address a maturing energy foundation that should be redesigned or supplanted. It is a method to address energy productivity and to carry expanded attention to purchasers about the association between power use and the climate. What is more, it is a method to carry expanded public safety to our energy framework drawing on more prominent measures of local power that is more impervious to cataclysmic events and assault.

- **Concept of prosumer**

Prosumer can be easily understood as a person who performs a combined role of both the producer and a consumer. These prosumers will have a bi-directional flow of data and power enabled by advanced information, control, and communication. Prosumers play an important role in SG as they serve their purpose by producing a sum of energy and will play the role of producer by generating more than they require and contribute to the SG. They will act as producers by selling that produced electricity to the grid, and when needed, will be purchasing electricity from the grid [10].

- **Consumer controls of SG**

Consumer does not need to wait till month end to get the statement of electricity being used. With an SG, one can get a clear, accurate, and timely picture of readings using smart meters. "Smart meters," and other mechanisms, will also allow us to see how much electricity we use, at what time we use the most, and its cost. With ongoing evaluating, this will assist us with setting aside cash by utilizing less force when power is generally costly. While the likely advantages of the keen network are normally examined as far as financial matters, public safety, and sustainable power objectives, the brilliant lattice can possibly assist us with setting aside cash by assisting us with dealing with our power utilize and pick the best occasions to buy power as the cost will be high during the hour of pinnacle interest. Also, we can save more by generating our own power by installing renewable energy source-based electricity generators such as solar panels, gobar gas plants, and domestic wind mills.

- **Smart home**

Key role behind the great working of the smart home is the intuitive two-way connection among the framework operators, utilities, and buyers. Keen controls of home apparatuses with the assistance of Internet of Things (IoT) to control machines distantly and perform planned work can be set up. Furthermore to react to signals from energy suppliers to limit the energy utilization times when the grid is confronting popularity or busy time and furthermore can plan smart home appliances to work on occasions when the demand is less.

- **Energy management systems**

Transmission and consumption data generated at the smart meters of our homes can run through energy management systems (EMS), which will allow us to track our usage. An EMS will allow us to get real-time information regarding price details from the utility and can schedule power usage when prices are low or when demand is low.

- **Smart appliances**

Smart appliances are appliances that are capable of understanding signals from energy provider to reduce power usage during peak hours. This is not as simple as turning off the appliance for that time. Let us take an example of smart air conditioner where the frequency of vapor compression cycle is reduced so that the power used by the condenser to reduce temperature is reduced and the air intake cycles are also reduced; if all the conditioners start reducing the frequency under an SG at peak demand, then it results in a large amount of difference in the demand. Not only air conditioners but if air coolers, refrigerator, washing machines, etc. also act in a smart way then electricity crisis at peak hours can be reduced.

- **Home power generation**

Consumers can generate power using home energy generation systems such as domestic wind turbines, small hydropower system, gobar gas plants, and solar panels on roof tops. The SG, with its several controls and smart meters, helps in effective connectivity of all these mini-grids to the main grid, to provide data about their operation to utilities and owners, and to know what surplus energy is feeding back into the grid versus being used on site.

Islanding is the condition in which distributor continues to supply power even though external grid power is no longer provided due to some cause, but power will be supplied from the prosumers to the consumers even though the power is no longer available from the grid.

- **Renewable energy resources handling in SG**

The SG is capable to make better use of renewable energy resources by storing power from them even though demand is not high and after making use of the stored power to balance peak demand. For this, grid needs to have more energy storage capabilities.

- **Grid control centers**

SG technologies have come up with a new solution for monitoring and controlling the grid's transmission system with new technology called phasor measurement units (PMU). PMUs provide a new monitoring tool for the SG.

PMU: "A phasor measurement unit (PMU) is a device used to estimate the magnitude and phase angle of an electrical phasor quantity (such as voltage or current) in the electricity grid using a common time source for synchronization."

- **Distribution systems**

The dispersion framework is the blend of wires, switches and transformers that interface the utility substation to the consumer. The electrical cables that go through individuals' lawns are one piece of the force circulation system. A key part of dissemination knowledge is blackout discovery and reaction. By having sensors that can show when portions of the dispersion system have lost force, and by joining computerized exchange with a clever system that decides how best to react to a blackout, force can be rerouted to most clients very quickly, or maybe even milliseconds.

Distribution systems make use of various types of clustering routing algorithms to select the optimal route from the available route at the time of blackout. Clustering routing algorithms will select a cluster in available clusters as the whole network is divided as set of clusters based on a rule and the selected cluster is responsible for the activities such as data forwarding, collecting as well as integration of the information.

9.2 Use of blockchain in SG

9.2.1 Blockchain

A blockchain is a collaborative, tamper-resistant ledger that maintains transactional records. The transactional records (data) are grouped into blocks. A block is connected to the previous one by including a unique identifier that is based on the previous block's data. As a result, if the data is changed in one block, it's unique identifier changes, which can be seen in every subsequent block (providing tamper evidence). This domino effect allows all users within the blockchain to know if a previous block's data has been tampered with. Since a blockchain network is difficult to alter or destroy, it provides a resilient method of collaborative record keeping.

Blockchain as the name proposes is a chain comprising a few data blocks. Blockchain can screen all progressions made to any of its squares and as such it is absurd to expect to alter or erase any of its squares. This component of blockchain makes it entirely reasonable for use in decentralized exchanges without requiring the oversight of monetary organizations or governments.

All the more briefly, blockchain is a protected and alter mindful framework. Each square chain starts with an underlying square known as a beginning square, and every

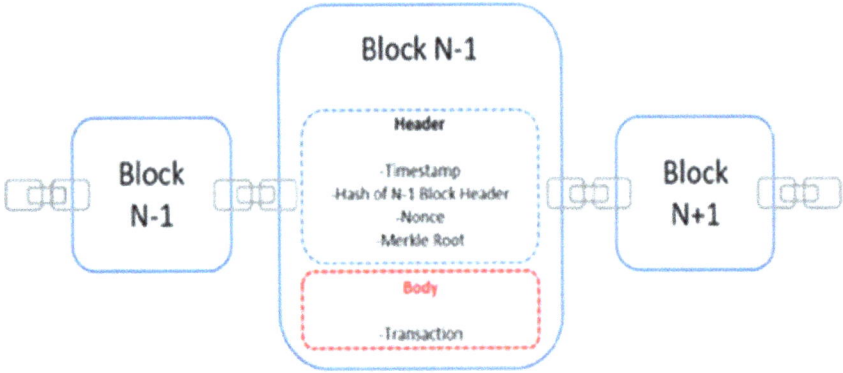

Figure 9.2   Basic structure of blockchain

other square is associated with the former squares and henceforth the beginning square. Each square contains data just as a hash that interestingly distinguishes it close by with its substance. Each square keeps its hash just as the hashes of past blocks. This makes block-chain a serious secure and invulnerable innovation. In the occasion where the substance of a square is changed, its hash changes since its hash relies upon its substance. The hash in ensuing squares does not change anyway and this nullifies the block. Before an exchange is added to the chain, a solicitation must be started by the client. After this is done, a square is made to address the exchange. The square is then communicated to each hub on the chain. Every hub in the chain approves and confirms the exchange block and the situation with the client by utilizing one out of a potential number of agreement calculations. When the exchange has been verified and approved, it is connected to every one of the past blocks in the square chain. Figure 9.3 shows how blockchain functions [25].

We have four different types blockchain models:

1. public blockchains
2. private blockchains
3. consortium blockchains
4. hybrid blockchains

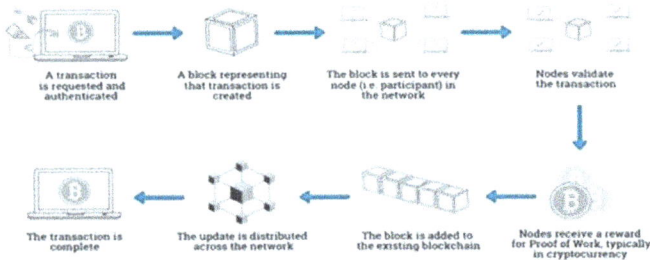

Figure 9.3   Blockchain functions

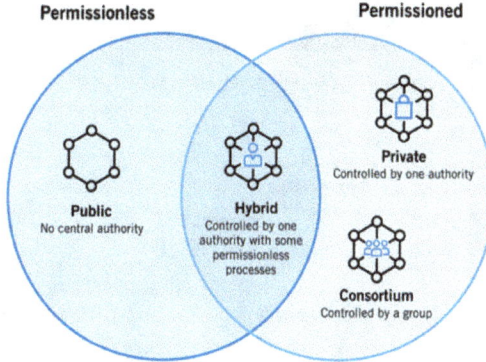

Figure 9.4   Blockchain types

What is a public blockchain?

A public blockchain does not have restrictions. Anyone with an internet connection can get access to the network and start validating blocks and sending transactions. Typically, such networks tend to offer some kind of incentive for users who validate the blocks [26]. Anyhow, this network tends to use proof of work or proof of stake (PoS) consensus algorithms for validating the transactions. It is a "public" network in a true sense.

Characteristics of public blockchain

There are certain characteristics of public blockchain architecture. Obviously, these features are different from other types of blockchain. Let us see what these are:

- Every node has access to read and write on the ledger.
- Anyone can download and add nodes to the system.
- The technology is fully decentralized in nature.
- It offers anonymity, which means no one can track your transactions back to you.
- It is a bit slower compared to the private blockchain.

What is a private blockchain?

Private blockchains are often referred to as "permissioned" blockchains. Unlike public blockchains, where anyone can download the software, form a node, view the ledger, and interact with the blockchain, private blockchains are often run and operated by an entity (the "trusted intermediary"). In this type of blockchain, only the entity participating in the transaction has knowledge about the transaction performed, whereas others will not be able to access it, i.e., transactions are private.

THE COMPARISON TABLE	PUBLIC BLOCKCHAIN	PRIVATE BLOCKCHAIN
Access	Anyone	Single organization
Authority	Decentralized	Partially decentralized
Transaction Speed	Slow	Fast
Consensus	Permissionless	Permissioned
Efficiency	Low	High
Data Handling	Read and Write access for anyone	Read and write for a single organization
Immutability	Full	Partial

Figure 9.5 Blockchain comparison table

Characteristics of private blockchain

- It focuses on privacy concerns.
- Private blockchain is more centralized.
- High efficiency and faster transactions.
- Better scalability—being able to add nodes and services on demand can provide a great advantage to the enterprise.

What is consortium blockchains?

A consortium blockchain is a hybrid form of public and private blockchains. The consortium blockchain straddles the line between public and private chains, incorporating aspects of both. The most noticeable difference between the two systems may be found at the level of unanimity. A consortium blockchain would be most useful in a situation when several companies operate in the same industry and need a single platform to conduct transactions or transmit information. Consortium blockchains Quorum and Corda are two examples.

What is hybrid blockchains?

Sometimes, organizations will want the best of both worlds, and they will use hybrid blockchain, a type of blockchain technology that combines elements of both private and public blockchain. It lets organizations set up a private, permission-based system alongside a public permissionless system, allowing them to control who can access specific data stored in the blockchain, and what data will be opened publicly.

9.2.2 Approach for blockchain-based SG

In the current power grids, blockchain likewise is found useful in expanding the security of the frameworks and the straightforwardness of smart meters (Shen and Pena-Mora, 2018). Information security is an extremely basic piece of SG, as it can seriously affect both energy organizations and clients in numerous ways. Broken data put away in the framework can make the clients experience the ill effects of additional charges and can likewise misdirect the energy choices of organizations that will hurt the entire framework (Xie *et al.*, 2019). Blockchain is known for its undeniable level information straightforwardness, trustworthiness, and dependability, which assist with meeting the necessities of a brilliant energy framework successfully.

A gathering of specialists presented a framework targeting the upgrading of the proportional fairness control in microgrids. This is an upgrade to the microgrid and oversees dispersed energy assets in the microgrid (Danzi, Angjelichinoski, Stefanovic, and Popovski, 2017). The assessment of the framework was finished through a reproduction, and the outcome showed that the blockchain model can possibly implement the security and trust of the control units. Then again, the analysts suggested further examination of more intricate situations so the plan can be improved. The creators likewise recognized costs that should be thought of, particularly mining and correspondence costs.

The security issue has been one of the basic worries in keen networks. Gao *et al.* (2018) presented a sealed framework that uses a blockchain to expand the security of customer information put away on the network. While the creators contrasted their plan and some current methodologies, their work stayed untested and required further tests for Contemporary Administration Exploration 211 execution assessment.

One more plan with regard to the security area was introduced by Liang, Weller, Luo, Zhao, and Dong (2019), which centers around expanding the obstruction of the force frameworks against digital assaults. This is accomplished by making a conveyed network utilizing meters as hubs and meter estimations as squares in the blockchain. The creators tried their model by running reproduction tests utilizing the IEEE-118 benchmark framework. The outcome upheld the end that blockchain can be applied to upgrade information security in current energy frameworks. Notwithstanding, the creators likewise showed that blockchain is a serious asset and enhancements in cost and speed are fundamental for the framework to turn out to be more dependable and capable.

Dong, Luo, and Liang (2018) explored the improvement of a brilliant lattice from a somewhat alternate point of view. They gave a calculated model of the total framework of a brilliant lattice, including both digital and actual foundations. On the digital side of the model, IoT, blockchain, and distributed computing were consolidated to establish a processing climate that can satisfy various necessities for future energy frameworks. The difficulties recognized by the creators included repetition, versatility, programming security, coordination, and the joining of blockchain and actual foundations. When of fruition of their report, the specialists were all the while dealing with an execution of the application, and the pragmatic unwavering quality

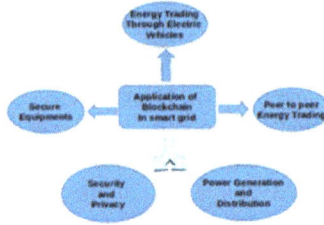

Figure 9.6   Applications of blockchain in cloud

was at this point to be affirmed. Notwithstanding industry, the cloud and blockchain are interwoven and blockchain applications will be regularly carried out and tried on cloud-based stages (Purdon and Erturk, 2017).

9.2.3   Applications of blockchain in SG

With the incorporation of remote sensor organizations and the IoT, the SG is being projected as an answer for the difficulties with regard to power supply later on. Notwithstanding, security and protection issues in the utilization and exchanging of power information present genuine difficulties in the reception of the SG. To address these challenges, blockchain innovation is being investigated for pertinence in the SG. The blockchain can be utilized for shared energy exchanging, where a credit-based installment plan can upgrade the energy exchanging measure. Productive information collection plans dependent on the blockchain innovation can be utilized to conquer the difficulties identified with protection and security in the network. Energy conveyance frameworks can likewise utilize blockchain to distantly control energy stream to a specific region by observing the utilization insights of that space. Furthermore, blockchain-based systems can likewise help in the determination and support of SG hardware.

9.2.4   Proposed architecture

Figure 9.7 shows the proposed architecture.

Let us consider a situation where Group A comprises power generation plant, Clusters B, C, and E comprise private structures and some of the structures are fit for creating their own power, which can be called as prosumers, Cluster D comprises exclusive force plants, and Group F comprises a public plug. Lattice stockpiles store overabundance power on the network.

A little figuring gadget is put in every element and every element can be called as a light hub which will equipped for creating a location and keeping up with address with its related equilibrium, this light hub will supplant electric meter and will likewise play outs crafted by the electric meter, this light hubs are remarkable addresses to distinguish the each end point and furthermore used to pass the message just as to get the data from the endpoints.

A full hub is a moderate gadget that acts in the middle of the light nodes (customers) and the generation point (producers) and comprises a registering gadget

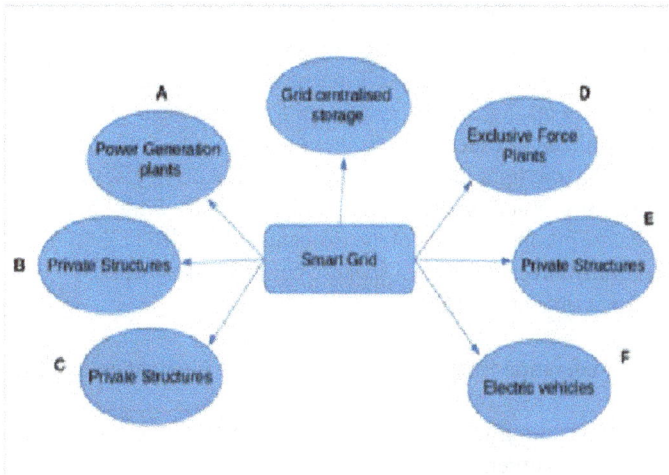

Figure 9.7   Proposed architecture

close to step down transformers. Full hubs approve exchanges, keeping up with agreement, likewise perform directing to the network (intelligence routing) determining a proficient course to convey the power, particularly when a specific power generator is unavailable. This will give the security of power supply. In the cryptographic money model, this hub additionally like the master node under which light hubs will come. Each producer and consumer is a light hub to decrease intricacy and costly equipment prerequisite.

- Possible nodes and their actions in SG.
- Producers: generate electricity and sell tokens.
- Consumers: consume electricity and purchase tokens.
- Prosumers: perform both the production and consumption of electricity, and tokens too.
- Distribution point: these perform as a full hub, keeping up with agreement, confirming exchanges, and perform routing.

Example: A house with a sunlight-powered charger has essential power from the dissemination line, through a bidirectional meter. The force is disseminated throughout the house. Sunlight-based charger creates power and stores the power in the batteries. The batteries are associated with an inverter and the bidirectional meter. The owner concludes whether to utilize the previously stored power, it is coordinated back to within lines, or chose to sell it, the electricity is sent back to the framework. Clients can screen and keep up with the interaction utilizing gadgets associated through the internet.

As per the research conducted by Anak Agung GdeAgung [16], the blockchain activity of the proposed engineering is given by computational hubs (PCs). The blockchain stage for the examination is Ethereum. The blockchain and keen agreements for this exploration run on the test organization. Sending a shrewd agreement in a genuine organization requires genuine Ether to spend as a charge. This will expect engineers to purchase Ether with fiat cash. The blockchain utilizes the verification of work agreement system and gives a stage to the shrewd agreement, which is utilized to execute exchanges among producers and consumers.

Each node (i.e. the light hub or the shopper) has a private key and address created. The key is novel and is utilized as a hub identifier, while the location is utilized for exchange purposes. Rules of the exchange are coded in shrewd agreements, which are naturally executed when a specific condition is satisfied.

Processes possible are as follows:

- Electricity generation from the producer, and as it is a light node it will generate a token.
- If producer is willing to sell its electricity then a selling order will be placed before placing an order, it will check for the availability and it will lock it before processing the order.
- If consumer is willing to purchase electricity, then they can buy a token with required quantity and amount will get deducted from the consumer account.
- If account does not hold required balance, transaction will fail.
- Once everything is done and clear token will be consumer, it will get used as the consumer utilizes electricity.

After completion of the above processes, transaction verification will be done by the miners

- Miners will look for the producers and consumers addresses.
- Tokens will be traced down to find its owner and authorization will be checked with the token. Previous owner history will also be made available as blockchain maintains transactions history.
- Miners will check with producer to make sure enough electricity is available to sell in the market.
- Once transaction verification is done then candidate will be added to the blockchain and will broadcast the same to all the nodes in the network then the smart contract will get executed.

The keen agreement itself goes about as an advanced convention, which consequently performs predefined activities once the prerequisites are satisfied by the recently concurred parties. For this situation, the keen agreement is utilized to guarantee the electricity is conveyed to the buyer once the customer follows through on the recently concurred cost to the producer. Since the keen agreements are composed as code and focused on the blockchain, it likewise acquires the changelessness of the blockchain, which implies nobody could adjust the agreed contract.

Contract logic will perform these checks:

- price, quantity of electricity, source of transfer, and destination of transfer

- Requirements will be checked, once the requirements are met and verified, access will be granted to the source, and consumption will get recorded once the measurement is reached then the access to the source will be revoked.

9.2.5 Transactions

In the energy market, electricity is an important and tradeable ware. Think about electricity as a resource, which is addressed by a token in a blockchain. Then, at that point, think about the three elements in the keen framework idea (producers, consumers, and prosumers, which can do both) as nodes. In digital money, a node joins and adds to a blockchain organization to get a prize. In our methodology, it is compulsory for a node to join and add to the organization, to devour and additionally sell the electricity. A node might be compensated for its positive commitments to the organization, as will be clarified in the prize system part beneath. To keep up with the framework, or when a node utilizes extra assistance, like utilizing public SG stockpiling, exchange charge can be added to the organization.

Consuming a token is to move it to a hopeless location, so it cannot be utilized more than once. In the proposed engineering, this location is predefined and coded into the convention. The symbolic sent there would be hopeless; however, everybody can perceive the amount they have spent. Consumers pay for the power they need ahead of time. Makers give power to the matrix. They create tokens, which they can sell at the organization at a specific cost. Shoppers purchase power available and get a symbolic worth the cash. The token, then, at that point, deducted when they burn-through electric from the network. Rather than purchasing from a solitary maker, they can see every one of the proposals from every maker. At the point when the measure of power is devoured, the addressed token is signed. Token consumed shows that the power addressed by it has been devoured.

To guarantee all gatherings perform activities as they should, each exchange in the lattice is put away as a shrewd agreement. The resource locking instrument goes about as an escrow component. All exchanges are done effectively. The savvy contract is proficient to deal with exchanges between a generator and customer and effectively consumed the token after the buyer has devoured the comparing power.

The joining of blockchain into the shrewd lattice permits the local area to keep up with exchanges in the framework in an agreement way. Exchanges are performed with keen agreements. Exchange history is put away in the blockchain and copied to every single full hub. The blockchain gives changelessness to the brilliant agreements and exchange information, by confining a record to be changed or deleted. In this manner, a savvy contract between a generator and a customer will consistently be executed, giving assurance that a maker will consistently convey the power when a buyer has paid.

The unchanging nature additionally gives recognizability, which is useful for review or addressing an exchange question. The Ethereum blockchain runs on verification of work agreement system, so exchanges are checked by excavators throughout the planet, including hubs running the full hub.

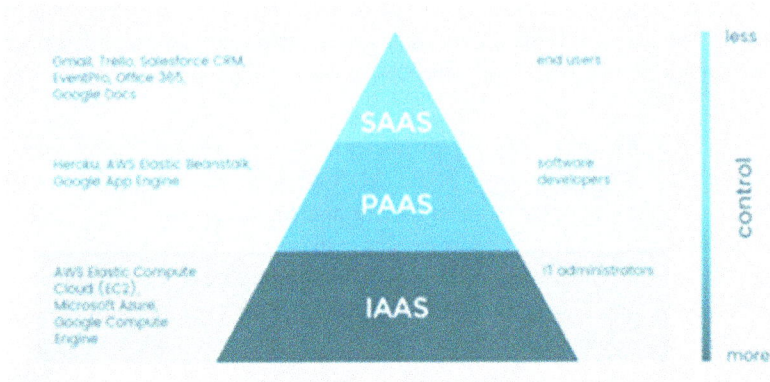

Figure 9.8   IaaS, PaaS, and SaaS module

9.3   Cloud computing uses in SG

9.3.1   Cloud computing

Cloud computing is now a days an important emerging computation model which is mainly used because that provides on-demand facilities, and shared resources over the Internet. [16]

Most prominent cloud computing services are as follows:

- platform as a service (PaaS)
- software as a service (Saas)
- infrastructure as a service (Iaas)

Platform as a service

Clients can access such programming models through cloud and execute their projects. PaaS is answerable for the turn of events and conveyance of programming models to IaaS [18].

Software as a service

The SaaS administration gives the displaying of programming arrangement where clients can run their applications without introducing it on his/her own PC. SaaS upholds all the applications in the cloud climate. This component of distributed computing is open through Internet browsers [16].

Infrastructure as a service

IaaS is the foundational administration concept that incorporates virtual machines and capacity. Clients can introduce admittance to required programming through virtual machines. Burden adjusting in distributed computing is performed utilizing IaaS. These virtual gadgets give on-request office to the clients [18].

Let us take a look at Table 9.1 that gives some information of different types of cloud-based applications along with the features they are providing, some of the applications like demand response, which are responsible for handling demands of consumers, peak demand, and dynamic pricing application, which is responsible for the dynamic pricing of the electricity based on the peak demand in the grid, and many more types of applications with various features can be found.

By absorbing ideas of cutting edge correspondences and future control advancements, SG has turned into the cutting edge power framework. Because of its more prominent vigor, productivity and adaptability over traditional electricity framework, its acquiring significance. As in present-day electrical force framework, need of assets and capacity is expanding, which can be managed by distributed computing. It is a promising innovation with usefulness of utilizing registering assets in adaptable and virtualized way. Distributed computing incorporates the electrical force framework assets through internal networks [11].

Problems encountered in non-cloud-based SG:

- The master–slave design (without mists) could trigger digital assaults (dispersed refusal of administration).
- Any master–slave plan disappointment might bring about framework disappointment, which does not happen in distributed computing.
- Just few (clients) can be upheld because of restricted worker limit. Because of its restricted limit and force, overhauling an immense number of clients will be difficult.
- The executives just as dependability issues are normal.

9.4 Integration of blockchain and cloud in SG

The SG offers greater security in terms of power supply. In SG network, information cloud foundation, cost, and information of the board are being shared. Because of the adaptability of distributed computing, data are recovered from the information cloud more advantageously. Adaptable assets and administrations partook in network, equal preparing and inescapable access are a few elements of distributed computing that are utilized for brilliant grid applications. Yet, it has a few limitations, for example, security and unwavering quality, which is tackled utilizing blockchain. Blockchain is an innovation, which empowers a local area to keep up with trust. At the point when we use blockchain in keen framework, it gives greater security and protection to the information exchange. Blockchain likewise builds straightforwardness of exchanges.

Table 9.1 Cloud computing applications for smart grid management comparison

Cloud applications	Smart grid features			
	Demand side management	Microgrid management	Load shifting	Dynamic pricing
Demand response	Available	Available	Available	Available
Peak demand and dynamic pricing	Available	Not available	Available	Available
Microgrid management	Not available	Available	Available	Not available
Real-time monitoring	Available	Not available	Not available	Not available
Power monitoring and early warning system	Available	Not available	Available	Not available
Information interaction using mobile agent	Available	Not available	Available	Not available
Dynamic demand response	Available	Available	Not available	Not available

Microgrids are profoundly decentralized frameworks. With the presence or non-appearance of a utility framework; age, markets what is more, different parts of a microgrid activity are inherently decentralized. Thus, blockchains are pertinent to microgrids. The issues of trust and motivating forces relevant to energy exchanging, metering, and request reaction in SG can be reached out to microgrids.

9.4.1   Benefits of using blockchain in SG

SG 2.0 discussed in drives the electricity network towards an automated, decentralized, self-healing, distributed, and democratic architecture where each node participating is equally important (Yapa *et al.*, 2021). On the other hand, SG 2.0 requires the following [27]:

- Distribution of generation sources to integrate more renewable energy
- Encourage individuals to contribute in power production and trade surplus electricity among peer nodes
- Automate energy trading and billing in a transparent mechanism with real-time pricing
- Fast decision-making protocols for efficient grid management
- Efficient energy data analysis and secure, privacy-preserving data storage
- Autonomous, self-healing grid operation

Furthermore, as the number of stakeholders increases, the management of bulk data by a single point would become a challenge. This will require fast computational capabilities to process a large amount of data at the central node (i.e. DSO), increasing its vulnerability to failures.

The involvement of an intermediary would result in additional costs that will limit the minimum amount of transactions executable in order to maximize the benefits of SG 2.0 (Blom and Farahmand, 2018). Decentralization, however, compromises the trust and will be vulnerable towards grid security (Miglani *et al.*, 2020). The successful implementation of SG 2.0 thus demands new tools, which facilitate trust management while decentralizing the grid topology.

9.4.2   Layers of SG and blockchain applications in different layers

Blockchain platforms enable decentralized participation of distributed electricity prosumers in the energy market. This is managed through a decentralized approach with minimal third-party interference, and facilitated through communication and computation-enabled devices (Baggio and Grimaccia, 2020). This eliminates the latency, inefficiency and embedded costs associated with the intermediary.

Energy information related to an instantaneous imbalance between supply and demand will be acquired through the smart meters and broadcasted to all authorized nodes (Mylrea and Gourisetti, 2017). The power producers would place their bids to trade the renewable energy generated. Blockchain will record all these transaction information in the form of blocks and add in a chronological order of occurrence, upon reaching consensus of the participating nodes (Parag and Sovacool, 2016).

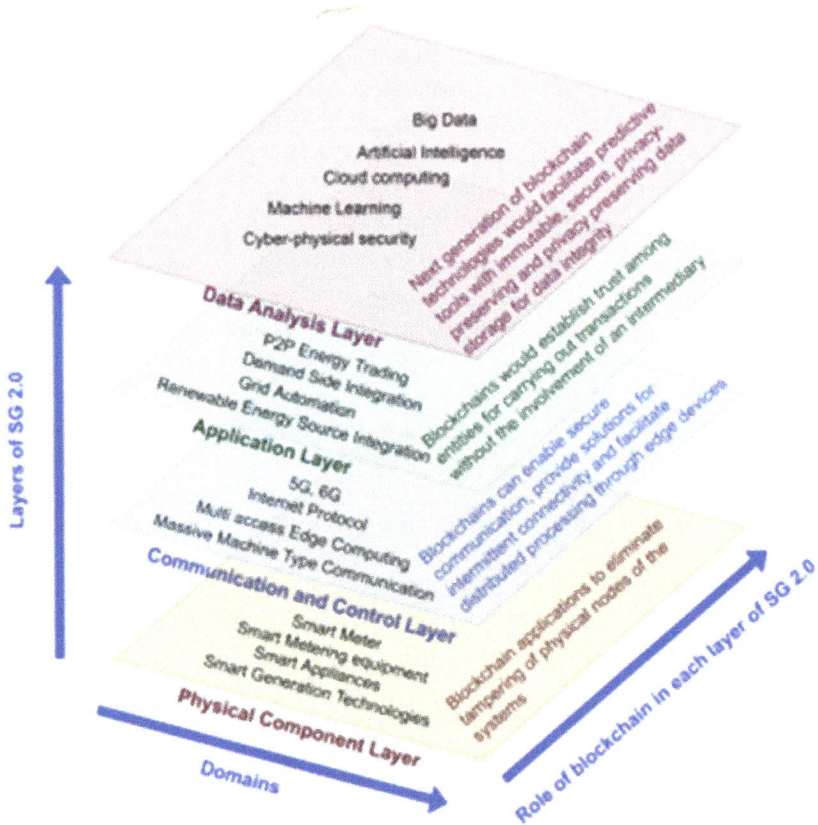

Figure 9.9   Layers of smart grid

This can be utilized to reduce the gap between scheduling and real-time dispatch and further facilitate decentralized coordination of millions of distributed devices.

9.5   How is blockchain run on cloud?

Blockchain as a service (BaaS) is the outsider creation and the executives of cloud-based organizations for organizations occupied with building blockchain applications. These outsider administrations are somewhat new improvement in the developing field of blockchain innovation. The utilization of blockchain innovation has moved well past its most popular use in digital money exchanges and has expanded to address secure exchanges, everything being equal. Subsequently, there is an interest for facilitating administrations.

BaaS offers an outside specialist co-op to set up all the fundamental blockchain innovation and foundation for free. When created, the supplier keeps on taking care of the complex back-end tasks for the client. The BaaS administrator regularly offers support exercises, such as transmission capacity, reasonable allotment of assets, facilitating prerequisites,

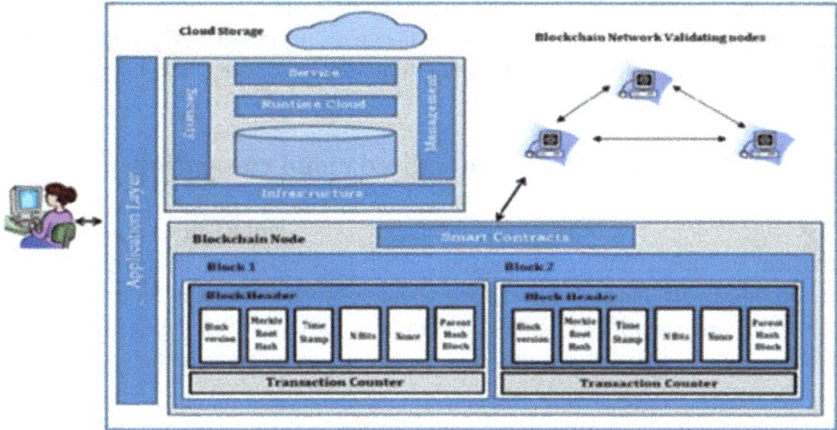

Figure 9.10 Blockchain integration with cloud

and information security highlights. The BaaS administrator liberates the customer to focus on his core work that is to focus on the functionality rather than administration.

A BaaS supplier's job is like that of a web facilitating supplier. The site designers make and run all the site content on their very own PCs. They might enlist support staff or sign up with an outer facilitating supplier like Amazon. These third-party organizations deal with the framework and upkeep issues. BaaS might be the impetus that prompts a more extensive and more profound entrance of blockchain innovation across different industry areas and organizations. Rather than making and running their own blockchains, a business, enormous or little, can now basically re-evaluate the in fact complex work and spotlight on its center exercises, instead of

Figure 9.11 Blockchain on cloud

developing everything from scratch and maintaining infrastructure one can opt for the cloud-based BaaS as it can be used on the go results in rapid set up of blockchain where we need to focus only on the main functionality of the blockchain.

9.6   Blockchain-based decentralized cloud computing

Distributed computing by arrangement may not be useful to little organizations or transient activities. Conversely, for some, enormous endeavors, distributed computing becomes valuable as it could bring about economies of scale. Since distributed computing frameworks are unified, administration blackouts are generally a disastrous chance. Network down time is frequently referred to as perhaps the greatest inconvenience of cloud innovation.

Edge registering further develops information pressure, moves information in the availability layer of the innovation stack, and diminishes network data transfer capacity. Haze registering could, moreover, see change of the populaces' PCs, workers, and cell phones in hubs ready to store information and/or perform calculations for other people, along these lines making a more extensive scope of IoT applications conceivable.

The blockchain-based edge-registering administrations empower decentralized association of P2P IoT gadgets across the SG framework. Furthermore, blockchain works with decentralized information assortment and gives decentralized information of the board and capacity administrations close to the information sources. The usage of blockchain on the edge-processing network benefits as far as consensus-based information detailing by consumer gadgets. Furthermore, it brings down the information correspondence cost inside the edge cloud network [20].

The advantages of using decentralized cloud along with blockchain

- Decentralized cloud storage with blockchain innovation stores just hashes of its information blocks. Furthermore, the encoded and disseminated hashes are sufficient to check these information blocks.
- Blockchain does not simply store information in an appropriated and encoded structure, yet additionally accommodates a successive chain where each square contains a cryptographic hash of the square. This connects the squares and, accordingly, makes a decentralized exchange record.
- A very much planned and freely available blockchain can supplant a significant number of the capacities that we right now depend on cloud mediators for giving a reliable exchanging climate, guarding against misrepresentation and dealing with, guaranteeing contract consistence, and monetary exchanges.

Challenges in integrating blockchain with cloud computing

At present, the research of trust-based approaches in cloud computing still faces huge challenges in theory and implementation.

- Most trust models are centralized, and even those that claim to be decentralized models still need a third-party trust or certification center, which may result in many security risks such as single point of failure, overload, and credibility loss.
- Trust evidence is not open to all participants and not traceable, so trust evaluation results are not convincing nor are they fully trusted.
- Inaccuracy of trust evaluation results. The existing trust models lack a sufficient description capability (trust data mostly in the form of numerical scoring), which is insufficient in real applications, such as e-commerce, where people's feedback often includes multiple data types such as numeric and characters.
- Less adaptive. Trust decision-making uses subjective methods, such as expert scoring and the averaging method, which makes the models subjective and lacks scientific and adaptability. Trust models are not robust enough to deal with malicious attacks (collusion), especially malicious recommendations.
- Huge management overhead. It limits trust solutions in a large-scale network applications.
- Lack prototype and platform. Performance tests of trust models are mostly achieved by some simulation experiments, needing further evaluation.

9.7   Blockchain-based decentralized cloud solutions

- Ankr
- Dfinity
- Solana

9.7.1   Ankr

The Ankr project is a decentralized cloud arrangement, a work in progress with a point of offering customers the foundation to run applications at less expensive costs contrasted with conventional cloud specialist organizations, and server farms the framework to make new income streams from their underutilized limit. This will be acquired by guaranteeing an undeniable degree of administration accessibility, simple incorporation, and secure correspondence. This degree will be reached by utilizing containerization, bunch organization, and confided in execution conditions (TEE).

9.7.2   Dfinity

The Dfinity project is a decentralized cloud arrangement whose point is to furnish a world supercomputer with "limitless" limit and computational force. It presented the idea of "The artificial intelligence is law" in which everything is dependent upon a mediator-free algorithmic administration framework that consolidates swarm astuteness and customary artificial intelligence innovations to freeze reprobate savvy contracts, which hurt the interests of those utilizing the stage. This successfully implies that a few exchanges can be changed and returned whenever supported by the algorithmic administration framework, a contrary methodology contrasted with that of activities like Bitcoin or Ethereum, where the standard "The Code is law" orders—a client indeed cannot return an exchange when prepared.

9.7.3 Solana

The Solana project intends to make a new blockchain design dependent on a clever idea, proof of history (PoH). PoH utilizes an obvious defer capacity to give the organization a trustless feeling of shared time. Also, exchange preparing on Solana is taken care of by GPUs, a methodology that scales with Moore's law. Moore's law is the perception that the quantity of semiconductors in a thick incorporated circuit pairs about like clockwork, in this manner considering substantially more computational force. This component, utilized in blend of a PoS agreement calculation, can make the foundation profoundly adaptable.

9.8   Cloud-based techniques for blockchain-based SG

9.8.1 Cloud-based demand response

In this EMS and brilliant meters will be the slaves, while expert will be the utility. To adjust the ongoing interest and supply bends, quick joining and investigation of data that stream from numerous smart meters at the same time is required that requires the versatile programming platform. Cloud stages fill in as fundamental parts because of the different advantages they offer, as referenced underneath:

- Cloud acts flexibly to keep away from expensive capital venture by the utility during the pinnacle hours.
- Clients can be profited from the ongoing data by sharing the constant energy utilization and evaluating data.
- A few information can be imparted to an outsider by utilizing cloud administrations, in the wake of meeting the information security strategies for creating insightful applications to tweak customer needs.

To take choices at various occasions, execution of particular information reflection for information streams produced from the various parts is needed for constant checking. Then again, outsider merchants are permitted to take an interest in such constant checking framework that requires characterizing a viable protection strategy as a security instrument. Virtualization is one of the most productive strategies for cost decrease, asset streamlining, and worker the board. Distributed computing can be carried out as various methodologies of the miniature frameworks.

9.8.2 Cloud-based economic power dispatching

The great obligation of the electric utilities is to meet the clients' prerequisites consistently with quality and amount as concurred. It coordinates with the force age by the utilities against the clients' force interest consistently. As the buyers' interest changes at each moment, the force age by the utilities should coordinate with the shoppers' interest. Truly, the force age cannot be changed at each occurrence; consequently, the age is changed ordinarily at 20 minutes span. The coordinating of force age against the shoppers' interest is known as force dispatching. In cloud-based financial force dispatching model, the utility and clients associate through the cloud, and the capacities for cost improvement are acted in the cloud. According to utility's point of view, cloud seems, by all accounts, to be a data framework, which takes a contribution from utility (e.g. power demand, climate information, fuel cost), measures the data, and gives a yield to utility and clients (e.g. ages of the singular plants, absolute creation cost). It merits bringing up that the cloud network plays out the force dispatching position according to the guidance of the electric utilities since the choice authority is the utility. For its administrations, the cloud gets its administration charges from the utilities. This plan is efficient for the utilities since they need not contribute on correspondence and figuring offices.

9.9 Challenges while merging blockchain with SG

9.9.1 Issues of scalability

Scalability issues are being faced when we try to coordinate blockchain with SG. This arises because in blockchain number of nosed and number of transactions are more. To validate each and every transaction, every time every node has to store and execute the corresponding computational task. And this leads to scalability issues. To overcome this scalability issue, we have to increase the information storing capacity of blockchain.

9.9.2 Chances for centralization

Blockchain is inherently centralized. And decentralized means almost distributed. Centralization should be done so that the whole control can be done by one single grid and no third party gets involved.

9.10 Conclusion

The SG is a very emerging technology in the energy sector, and it needs a reliable and secure framework for operations. Blockchain can be merged into the SG to open the doors to a wide range of possibilities. SG uses intelligent transmission and distribution networks in order to efficiently deliver the energy. We can conclude that using blockchain and cloud computing SG works in a more smarter way by giving more security and data management facilities to the users.

The coordination of blockchain and the SG framework permits the local area to keep up with exchanges in the framework in an agreement way. Exchanges are performed with smart contracts. Exchange history is put away in the blockchain and copied to every single full hub. The blockchain gives unchanging nature to the smart agreements and exchange information, by confining a record to be changed or deleted. Subsequently, a brilliant agreement between a generator and a purchaser will consistently be executed, giving assurance that a maker will consistently convey the power when a buyer has paid. The permanence additionally gives recognizability, which is useful for review or tackling an exchange debate.

Distributed computing is getting famous which is a framework with helpful, on request office to get to organize alongside the different coordinated figuring assets, for example, workers and capacity that can quickly be delivered with least administration exertion or specialist cooperation. At this point, electric grid with various use has a predetermined processor and capacity assets, subsequently distributed computing helps in most extreme usage of the capacity assets. With the assistance of distributed computing, different control calculations can be created to further develop power and burden adjusting.

Utilizing distributed computing applications, energy of the board strategies in SG network can be assessed inside the cloud, rather than between the end client's gadgets. This design gives more memory and capacity to assess processing system for energy the board, and cost-improvement.

So, we can finally conclude that the coordination of SG with cloud computing and blockchain has made the SG functionally more smarter.

References

[1]   Momoh J. *Smart Grid: Fundamentals of Design and Analysis*. 63. John Wiley & Sons; 2012.

[2]   Ye F., Qian Y., Hu R.Q. *Smart Grid Communication Infrastructures: Big Data, Cloud Computing, and Security*. Wiley-IEEE Press; 2018.

[3]   Crosby M., Pattanayak P., Verma S., Kalyanaraman V. 'Bitcoin—Upsides, downsides and bone of contention—A deep dive'. *Theoretical Economics Letters*. 2016, vol. 9(5), pp. 1384–92.

[5]   Ren Y.J., Leng Y., Cheng Y.P., Wang J. 'Secure data storage based on blockchain and coding in edge computing'. *Mathematical biosciences and engineering : MBE*. 2019, vol. 16(4), pp. 1874–92.

[6]   Jenkins N., Long C., Wu J. 'An overview of the smart grid in Great Britain'. *Engineering*. 2015, vol. 1(4), pp. 413–21.

[7]   Verma V.K. Blockchain technology: Systematic review of security and privacy problems and its scope with cloud computing. 2019. Available from https://www.semanticscholar.org/paper/Blockchain-Technology%3A-Systematic-Review-of-and-and-Verma/90a726327436def5797716c87d865c39648c621c reply Reply.

[8]   Pareek N.K., Patidar V., Sud K.K. 'The International Journal of Network Security & Its Applications (IJNSA)'. *Image Encryption with Chaos*, vol. 7(1), pp. 1–6.

[9]   Zhu L., Wu Y., Gai K., Choo K.-K.R. 'Controllable and trustworthy blockchain-based cloud data management'. *Future Generation Computer Systems*. 2019, vol. 91(99), pp. 527–35.

[10] Zafar R., Mahmood A., Razzaq S., Ali W., Naeem U., Shehzad K. 'Prosumer based energy management and sharing in smart grid'. *Renewable and Sustainable Energy Reviews*. 2018, vol. 82, pp. 1675–84.

[11] Mishra N., Kumar V., Bhardwaj G. 'Role of cloud computing in smart grid'. International Conference on Automation, Computational and Technology Management (ICACTM); 2019.

[12] Hu Y., Li W., Chen X. A Probabilistic Routing Protocol for Heterogeneous Sensor Networks. *Proceedings of the 2010 IEEE Fifth International Conference on Networking, Architecture, and Storage*; IEEE Computer Society; 2010. pp. 19–27.

[13] Jiaguangle. *Research on Routing Algorithms Based on Wireless Mesh Networks*. Beijing University of Posts and Telecommunications; 2007.

[14] Li Z., Kang J., Yu R., Ye D., Deng Q., Zhang Y. 'Consortium blockchain for secure energy trading in industrial Internet of things'. *IEEE Transactions on Industrial Informatics*. 2018, vol. 14(8), pp. 3690–700.

[15] Xiong Z., Zhang Y., Niyato D., Wang P., Han Z. 'When mobile blockchain meets edge computing: Challenges and applications'. 2017.

[16] Handayani R., GdeAgung A.A. 'Journal of King Saud University - Computer and information sciences'. *Blockchain for Smart Grid*. 2020, vol. 34(3), pp. 666–75.

[17] Bera S., Misra S., Rodrigues J.J.P.C. 'Cloud computing applications for smart grid: A survey'. *IEEE Transactions on Parallel and Distributed Systems*, vol. 26(5) 1477–94.

[18] Prodan R., Ostermann S. 'A survey and taxonomy of infrastructure as a service and web hosting cloud providers'. *IEEE*. 2009.

[19] Tchao E.T., Quansah D.A., Klogo G.S., *et al.* 'On cloud-based systems and distributed platforms for smart grid integration: Challenges and prospects for Ghana's grid network'. *Scientific African*. 2021, vol. 12(1), p. e00796.

[20] Aderibole A., Aljarwan A., Rehman H.M., *et al.* '*IEEE Access*'. *Blockchain technology for smart grids: decentralized NIST conceptual model*. 8; 2020. pp. 43177–90.

[21] Dielmann K., van der Velden A. Virtual power plants (VPP)—A new perspective for energy generation? Proceedings of the 9th International Scientific and Practical Conference on Modern Techniques and Technologies, Apr 2003; 2003. pp. 18–20.

[22] Du D., Li X., Li W., Chen R., Fei M., Wu L. 'ADMM-Based distributed state estimation of smart grid under data deception and denial of service attacks'. *IEEE Transactions on Systems, Man, and Cybernetics: Systems*. 2019, vol. 49(8), pp. 1698–711.

[23] Mengelkamp E., Notheisen B., Beer C., Dauer D., Weinhardt C. 'A blockchain-based smart grid: Towards sustainable local energy markets'. *Computer Science - Research and Development*. 2018, vol. 33(1-2), pp. 207–14.

[24] Yli-Huumo J., Ko D., Choi S., Park S., Smolander K. 'Where is current research on blockchain technology? - A systematic review'. *Plos One*. 2016, vol. 11(10),e0163477.

[25]   Tchao E.T., Quansah D.A., Klogo G.S., *et al.* 'On cloud-based systems and distributed platforms for smart grid integration: Challenges and prospects for Ghana's Grid Network'. *Scientific African*. 2021, vol. 12(1), p. e00796.

[26]   *What is a Public Blockchain? Beginner's Guide - 101 Blockchains* [online]. Available from https://101blockchains.com/what-is-a-public-blockchain.

[27]   'Survey on blockchain for future smart grids: Technical aspects, applications, integration challenges and future research'. *Science Direct*.

Chapter 10

Smart grid: energy storage and transaction

B Goutham[1], H L Gururaj[1], B R Sunil Kumar[1], and
V Ravikumar[1]

Energy storage units (ESUs) and transactions are becoming effective features for improved grid resilience, for effective demand response, and to lower bills of modern smart grids. This chapter gives an insight about smart grids and ESUs employed. The method could aid in the resolution of a number of complex issues relating to the integrity and reliability of fast, dispersed, and complicated energy transactions and data transfers. Employment of blockchain could lower transactive energy prices while also improving the security and long-term viability of distributed energy resource integration, removing hurdles to a more decentralized and resilient power system. This chapter explores more on the basic issues regarding this.

10.1 Introduction

In the present scenario, the electrical power grids are seeing a radical change in effective utilization of highly emerging information and communication technologies (ICTs). This development leads to smartness of power grids and coined as smart grid. Smart grid can be described as transparent two-way communicative power grid, as an evolved traditional power grid. Smart grid is an alternative approach to many traditional power grid problems. In smart grid, the operations pertaining to delivering, processing, and monitoring are under centralized control. This method of control has a few advantages such as easy setup, flexible management of the system, and so on, but this centralized control has a few serious issues such as privacy and security challenges. Smart grid should address the following issues, as shown in Figure 10.1.

- **Standardization:** Smart grids' multi-sectoral characters, the requirement for multiple technologies to be integrated, the large number of stakeholders, the

[1]Vidyavardhaka College of Engineering, Mysore, India

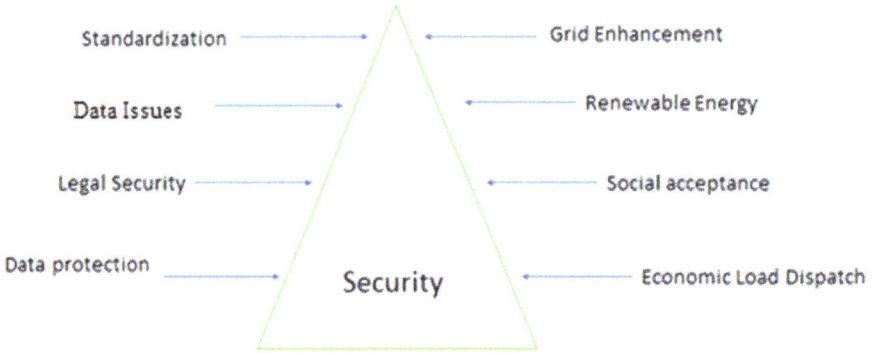

Figure 10.1   Security issues addressed by smart grid

required speed, the multiple global activities, and the ever-changing technical solutions make it a difficult assignment for standardization organizations around the world.

- **Data issues:** Duplicate data, unstructured data, missing data, multiple data formats, and trouble accessing the data can all cause data quality issues.
- **Legal security:** Smart grid is a system that is developed on the existing grid structure; these units will be handling many data with two-way communications. The smart grids developed should be in such a way that it should overcome security threats.
- **Data Protection:** Real-time network monitoring, consumption forecasting, and in particular, the introduction of dynamic pricing schemes based on the current availability of electricity from variable sources such as wind and solar are just three of the many smart grid functions that necessitate the use of high-resolution data. And it is utmost important for protection of these data.
- **Grid enhancement:** Grid enhancement has become an important consideration for the improvements required in the power grid to handle all the rapid technical advances occurring in transmission, electric power generation, and distribution [1].
- **Renewable energy:** Renewable energy sources (RESs) and energy storage systems are significant technologies for smart grid system, and they offer great prospects to decarbonize cities, regulate frequency and voltage aberrations, and react to critical time when the load exceeds the generation.
- **Social acceptance:** Energy technologies, particularly smart grids, have long been believed to require social acceptance in order to prosper [2]. That is, stakeholders (such as the general public or consumers) must be willing to utilize or allow the use of these technologies by others.
- **Economic load dispatch:** Economic load dispatch is an online procedure for allocating generation among available generating units in order to reduce total generation costs while also meeting equality and inequality restrictions.

The smart grid is defined as the two-way communication of energy information that is transparent, smooth, and immediate, allowing the electrical business to better manage energy distribution and transmission while also giving customers more control over their energy choices.

Essentially, the smart grid's objective is to provide far better visibility to lower-voltage networks while also allowing users to participate in the operation of the electrical system, primarily through smart meters and smart homes. A smart grid combines modern ICT elements to deliver real-time information and enable near-instantaneous supply and demand stability on the electrical grid. Smart grid and its subsystems will collect operational data that will allow system operators to immediately identify the appropriate line of attack to protect against attacks, susceptibility, and other threats posed by a variety of incidents. Smart grid, on the other hand, relies on understanding and researching important performance components, as well as designing an appropriate education program to equip current and future workers with the information and skills needed to utilize this highly advanced system [3].

Smart grid technology is a parallel approach for many typical problems. It creates a command center where energy transactions are governed, delivered, and handled. This technology has a few advantages, such as ease of setup and manageability; nonetheless, it faces a number of security and privacy issues, similar to other centralized computing solutions. Electrical power networks are undergoing drastic changes all across the world, spurred by the pressing need to decarbonize electricity supply, replace old resources, and make effective use of rapidly expanding ICTs. All these aspects are achieved by a common goal of smart grid [4].

The block diagram of smart grid is shown in Figure 10.2, which comprises the following components.

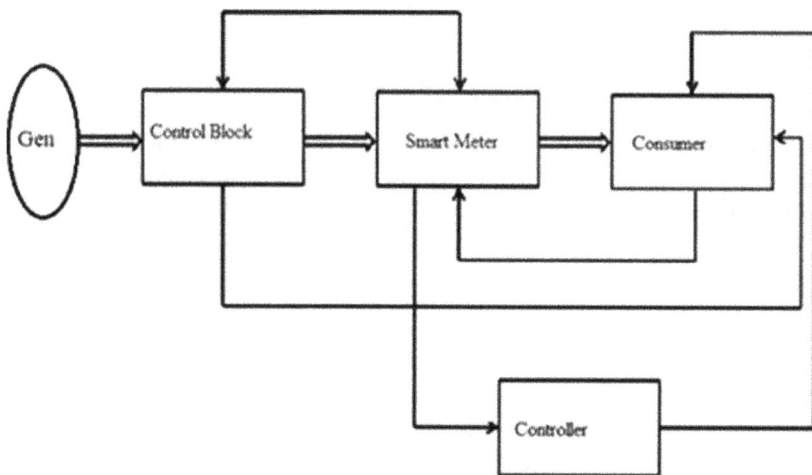

Figure 10.2   Block diagram of smart grid

- **Generation:** The conversion of a kind of primary energy into electricity at power stations or generating units is known as generation. In smart grid systems, generation is done by both conventional and nonconventional sources.
- **Control block:** It will be made of power electronic components for control of the conversation of electrical power. This module will be connected with different other modules for providing secure operations.
- **Smart meters:** A smart meter is an electronic device that keeps track of data such as electric energy consumption, voltage levels, current, and power factor. Smart meters transmit data to consumers for better understanding of consumption patterns, as well as to power suppliers for system monitoring and customer invoicing. It involves two-way communication.
- **Consumer:** Consumers are the end users for whom electricity has to be supplied.
- **Controller:** It is a hardware unit that manages the communication system with the smart meters [5].

Despite the fact that the smart grid idea has not yet been fully defined, the chapter will be useful in describing the essential enabling technologies and allowing the reader to participate in the debate over the smart grid's future. The improvement in the field of networking technology and energy makes a smart grid to share, transfer, and store energies. In the present market, there is an increasing demand for energy and its trading because the smart grid is a centralized network where the dealings of energy are processed, delivered, and governed.

At present, demand for energy is drastically increasing, because of many societal and other developments. The need for generation from renewable energy has become inevitable. Compared with all the renewables, solar and wind power generation play a vital role in curbing the rising energy demand. Figure 10.3 shows the estimated growth of RES in India by 2022 [6]. In comparison to generation from traditional schemes, DERs face challenges of intermittency issues of solar and wind generation. The main objective of electrical power system should be a stable system. Making the grid smarter gives the solution for this issue and even certain security issues that rise. Traditional power grids when implemented with modern communication and technology tools lead to effective utilization. Energy consumption is increasing all the time, and the demand for electricity as a source of energy is increasing even faster. Power generation from RESs will unavoidably play a larger part in meeting energy demand as a result of societal and technological advances.

10.2 Background

History and background of the fundamental concepts behind blockchain technology evolved in the late 1980s and early 1990s. Leslie Lampert wrote a the research article in 1989 and submitted it to the Parliament of ACM Transactions on Computer Systems in 1990 [7]. The paper was eventually published in a 1998 issue. The paper offers a proposed model for achieving a network where the computers or network itself could be operated safely. In the year 1991, a digital document was signed in an

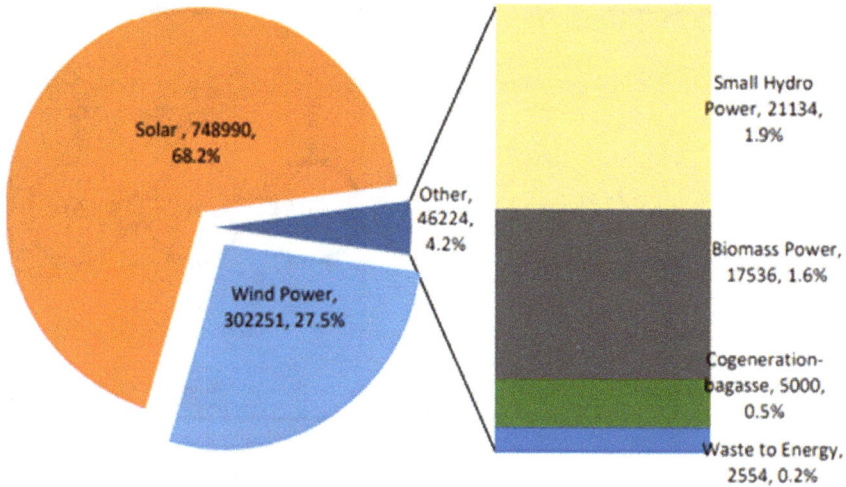

Figure 10.3   *Estimated growth of RES in India by 2022 (Department of Energy Statistics of India)*

electronic ledger in such a way that modification cannot be done easily. These concepts were integrated and used as electronic cash in 2008. Then Satoshi Nakamoto's pseudonymous article stated the same as Bitcoin. After all these developments peer-to-peer (P2P) electronic cash system was introduced in 2009, with the establishment of the Bitcoin cryptocurrency blockchain network [7]. The blueprint that most modern cryptocurrency schemes follow is outlined in Nakamoto's paper (although with variations and modifications). Bitcoin was only the first in a long line of blockchain applications. Prior to Bitcoin, there were many electronic cash schemes (e.g. ecash and NetCash), but none of them were used now [3].

Trading transactions were recorded in blocks so that all users could track and manage their transactions. Even if privacy protection was not properly addressed, this method mostly handles the mechanism design of employing blockchains in energy trade when considering efficiency. Another initiative related with electricity trade used consortium blockchain to enable P2P trading for hybrid electric vehicles (EVs). According to current blockchain-enabled research, adopting a pure blockchain system in a specific application situation could only provide minimal privacy protection [4].

10.2.1 Blockchain

Blockchains are distributed decentralized digital ledgers. At their most basic level, they allow a group of users to record transactions in a shared ledger within that group, with the result that none of the transaction can be modified once it has been published under standard blockchain network functioning. The blockchain concept was integrated with various other technologies and different computer concepts in 2008 to create cryptocurrencies [8]. These electronic cash are protected by cryptographic processes rather than a central repository or authority. With the creation of

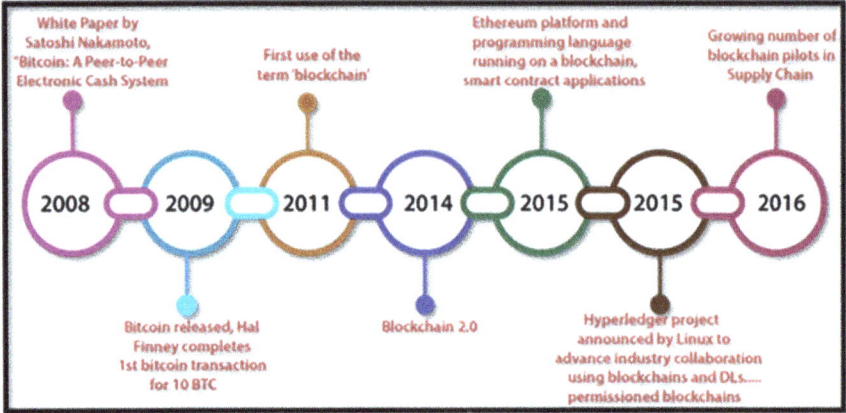

Figure 10.4   Development structure of blockchain

the Bitcoin network in 2009, the first of several modern cryptocurrencies, this tech-nology became well known.

The transfer of digital information that represents electronic cash takes happen in a distributed system like Bitcoin and comparable systems. The users of Bitcoin can sign digitally and transfer their rights and information to another user, and the Bitcoin blockchain publicly records this transfer, allowing all network participants to independently verify the transactions' legitimacy. A distributed set of people inde-pendently maintains and manages the Bitcoin blockchain, as shown in Figure 10.4. This, in conjunction with cryptographic mechanisms, provides the blockchain resis-tant to later attempts to alter the ledger (modifying blocks or forging transactions).

Blockchain technology has allowed the creation of numerous cryptocurrency systems, including Bitcoin and Ethereum1. As a result, blockchain technology is frequently regarded as being closely linked with Bitcoin or even cryptocurrency solutions in general [2]. Moreover, the technology exists for a greater range of applications and is now being researched in a number of industries. The many components of blockchains, as well as its dependency on cryptographic primitives and distributed networks, can make it difficult to comprehend. Each component, however, may be stated simply and utilized as a building brick to comprehend the broader complicated system. Informally, blockchains are distributed digital ledgers of cryptographically signed transactions that are organized into blocks. Following validation and a consensus decision, each block is cryptographically connected to the previous one (making it tamper obvious) [9]. Older blocks become progres-sively difficult to change as new blocks are added (creating tamper resistance). New blocks are duplicated across network ledger copies, and any conflicts are addressed automatically using predefined criteria. The flow of bitcoin is shown in Figure 10.5.

Bitcoin was just the first of many blockchain applications. Many electronic cash schemes existed prior to Bitcoin (e.g. ecash and NetCash), but none of them achieved widespread use. The use of a blockchain enabled Bitcoin to be implemented in a

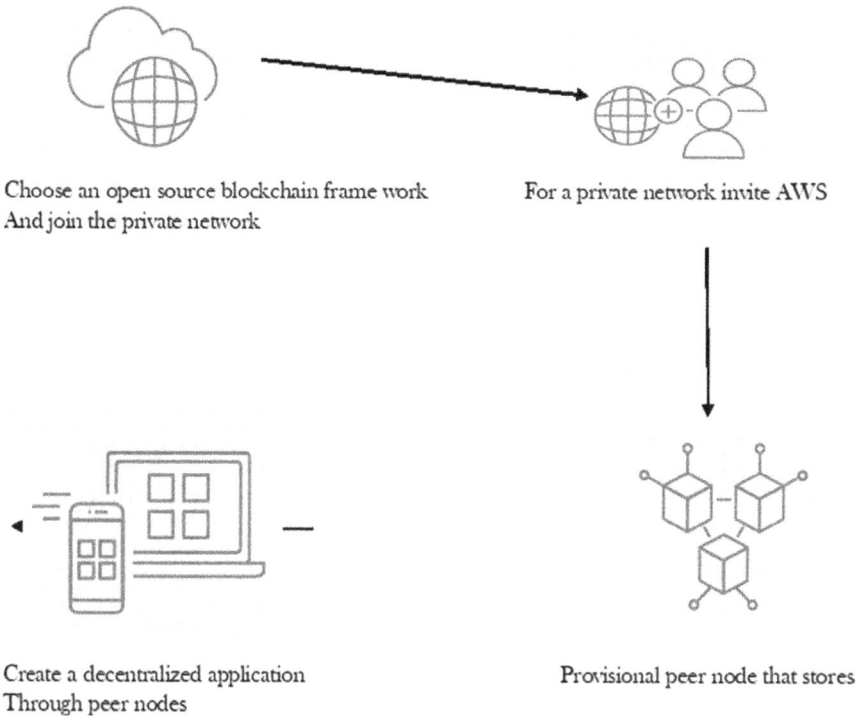

Choose an open source blockchain frame work
And join the private network

For a private network invite AWS

Create a decentralized application
Through peer nodes

Provisional peer node that stores

Figure 10.5   Bitcoin structure flow

distributed fashion such that no single user controlled the electronic cash and no single point of failure. Its primary benefit was to enable direct transactions between users without the need of a third party. It also enabled the issuance of new cryptocurrency in a defined manner to those users who manage to publish new blocks and maintain copies of the ledger, and such users are called miners in Bitcoin. The automated payment of the miners enabled distributed administration of the system without the need to organize. By using a blockchain and regular based, a self-maintaining mechanism was created to ensure that only valid transactions and blocks were added to the blockchain [10].

10.2.2   Smart grid

Recent technological advancements and the ever-increasing demand for energy demand a crucial role for smart grids. The smart grid idea has been around for a while, and there are just a few countable smart grid installations, mostly in academic institutions and far-flung areas. Smart grid terminology was a niche technology until the Internet of Things arrived. The advancement of communication technology has transformed the generation and distribution of electrical energy. Smart grids with a higher percentage of RESs are on the verge of becoming an integral part of today's power system. Consumers and service providers will benefit from new technological advancements. Reliability and energy sustainability will also promote active engagement of all stakeholders. The entire structure will be built on top of an intelligent communications infrastructure, the smart grid is a combination of hardware, management, and reporting software. Consumers and utility corporations alike have capabilities to manage, monitor, and respond to energy challenges in the smart grid era. The amount of energy from the utility to the consumer will become a two-way communication, reducing consumers money and energy while also increasing transparency and lowering carbon emissions.

Scientists agree that greenhouse gases (GHGs) are contributing to severe climate change. Emissions from the power industry must be lowered to near-zero levels by 2050. The predicted future decarbonized electrical power system in the world is likely to rely on a mix of renewables, nuclear generators, and fossil-fueled facilities with carbon capture and storage [11]. Because its intermittency of renewable power generation, the huge central generators are made to run at a consistent manner. Meanwhile, with the development in the electrification of heating and transportation, electricity consumption will become more variable. As a response, the electrical grid will have to become more intelligent in terms of balancing demand and supply. The smart grid is commonly seen as the way of the future for modern power systems. It is an updated existing network, depends on how they operate, promoting changes in energy customers' behavior, providing new services, and assisting the transition to a low-carbon economy that is sustainable [12].

The developed smart grid should comprise the following parameters, as shown in Figure 10.6.

- Microgrid and RES: A microgrid is a small power system that serves a specific geographic area, such as a university campus, a medical complex, a commercial center, or a residential neighborhood and renewable energy system.
- Advance management and distribution: The software platform that covers the full range of distribution management and optimization is known as an advanced distribution management system (ADMS). An ADMS offers services that automate outage restoration and improve distribution grid performance.
- Smart meter and phasor measurement: A phasor measurement unit, also known as a Phase Measurement Unit (PMU) or a synchrophasor, is a vital tool in use on power grids to help operators see what's going on across the huge grid network.
- Intelligent transportation system: It is a technology that includes state-of-the-art wireless, automated technologies, and electronics.
- Efficient carbon management: Smart grid should be made efficient enough to manage carbon emission.

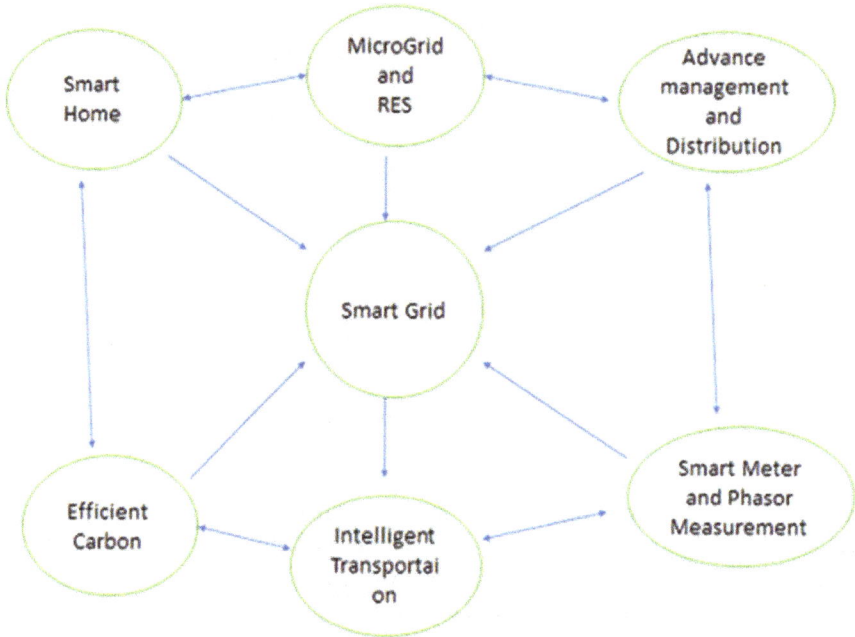

Figure 10.6   Parameters of smart grid

10.2.3   Need for smart grids

In today's society, reduced carbon emissions are mandated due to the implementation of the COP21 accord (Paris Climate Accord) around the world. Under these conditions, the progress in the energy industry is critical. For achieving super economic growth, the develop in the energy sector is mandate. Locally managed and extremely reliable power generation will be possible with a smart grid system that includes an intensive technology and superior communication network. As fossil fuel supplies deplete, smart networks with more renewable energy penetration will become more cost-effective. Peak hour demand can be well-handled using a combination of renewable and storage systems, lowering energy costs during peak hours of use. The smart grid will use storage and high-cost instantaneous power sources at the local level to control demand and supply in order to meet creatively at all times. This will lower the capital cost of installed capacity at the neighborhood, district, state, and national levels. According to a recent research by *The Wall Street Journal*, an attack on just nine vital substations out of a total of 55 000 could bring the whole US power grid to a halt for weeks or months [13]. Localized generation, distribution, and consumption would be the best choice to avoid such an occurrence and improve the reliability of the power system. In a word, the smart grid solution will meet the needs of the environment, reliability, sustainable energy, and economic growth. The smart grid has been rapidly evolving over the last few years but in an inconsistent

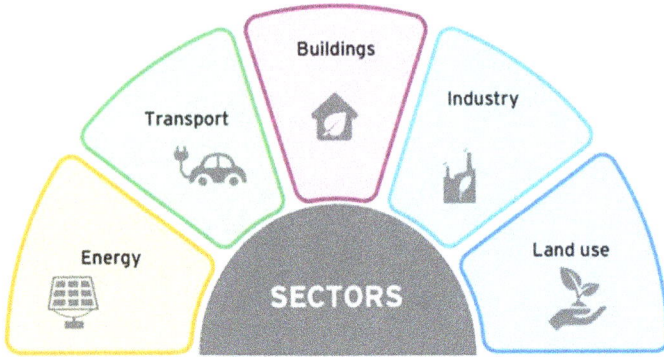

Figure 10.7   Sectors of smart grids influence

manner. Recent events in Japan, such as the nuclear reactor meltdown at Fukushima, have made the energy equation even more complicated, driving the need for a smart grid with higher renewable penetration, as shown in Figure 10.7. The need of smart grids will provide the following social benefits:

- provides high-quality power on a local level,
- reasonably adaptable to local conditions,
- secured and reliable operation,
- optimize consumption pattern based on cost of energy,
- energy buying and selling with utility with real-time data analysis,
- helps to improve environment through green energy,
- reduced line losses, thereby reduction in cost,

- **Energy**
  Most of systems get their energy from the sun. Solar energy can help you save money on your energy bills over time if you use it to power your house or company. Tracking the energy users online and for maintenances of online maintenance makes the energy sector to develop.
- **Transport**
  Smart grid technology helps in transportation sectors, i.e. smart transportation relies on smart grids for power consumptions. To sustain important transportation arteries and modes, transportation and traffic systems would work in tandem with energy systems.
- **Buildings**
  The city intends to employ IT and automation systems to instrument and integrate all infrastructure and services in order to maximize the assets, as well as analytic tools to forecast demand, maintenance patterns, reliability,

emergencies, and other occurrences. If the powered automation is supplied through smart grids great benefits may be established.

- **Industry**

   Industries smart cities, like smart grids, will develop gradually but steadily. They will effectively harness, combine, and use data that will be exchanged among departments, infrastructure operators, and citizens.

- **Land Use**

   Implementation of smart grids will reduce the land use when compared with traditional power grids. Hence more utilization of land will be available for other purposes.

10.3   Blockchain applications in smart grid

Researchers know that the improvement in blockchain technology will help in the development of the traditional electric grids and smooth shift toward a smart grid. The decentralization of technology is considered the foundation of technology of the smart grid. With the integration of RESs, energy storage devices, and EVs into the electrical grid, a major study area on novel control techniques to solve these difficulties has begun. The various and desired benefits of blockchain technology have sparked a lot of interest in researching and implementing it in smart grids. The application of blockchain technology to smart grid is shown as follows:

- Power generation: Blockchain technology provides dispatching agencies with real-time information on the entire operation status of a power system. This allows them to devise profit-maximizing dispatching strategies.
- Power transmission and distribution: The use of blockchain technologies helps in automation and allows for decentralized solutions that overcome the major drawbacks of traditional centralized systems.
- Power consumptions: As in transmission and generation, the use of blockchain can be found efficient in managing the trading of energy between prosumers and different energy storage in Figure 10.8.

10.3.1   Blockchain applications in energy transaction

Energy trading has become a hot topic for numerous researchers and industry executives as microgrid and distributed generation grow more prevalent. The implementation of blockchain technology in trading process removes barriers. In P2P trading systems, the blocks in the chain keep track of the units of generated electrical energy, allowing sellers and purchasers to transact quickly and independently. Instead, then relying on an intermediary agency, users (owners and buyers) have complete control over their preferences, choices, and prices.

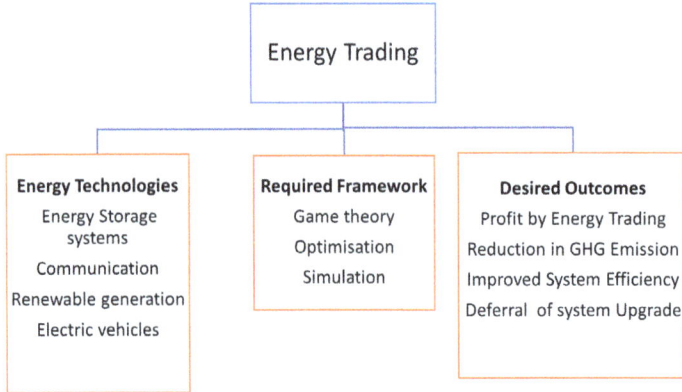

Figure 10.8   Energy trading

10.3.1.1   Benefits

- Improved system efficiency

  The involvement of Distributed Generations (DG) with storage provides additional services to the unit and hence, bus voltages and power quality would be improved by injecting active power from distributed generating choices, which is primarily achieved by eliminating voltage sags. Furthermore, with energy trading, customer demand will be satisfied locally, reducing the need for far-flung high-capacity generators. Congestion on transmission lines will be reduced as a result, and line losses (in the form of heat) will be reduced as a result.

- Reduced system operation cost

  For meeting up with the customers demand, the generation portfolio is dispatched according to the operating cost. To meet base load demand, large-scale, low-cost generating units are frequently favored. In contrast, with increased demand from customers, system operators send more generators to meet the varied requirements of customers minute by minute. During peak hours, which is around 10% of the day, utilities use fast start-ups and high cost of operation to meet a high demand for energy, mainly gas power generators. One of the key reasons for energy trading is to reduce local energy trading in peak hours to peak-to-average demand [14].

- Reduction in GHG emissions

  Targets are set by the countries across the globe for reduction in GHG emission. For example, EU countries are aiming to reduce GHG emissions by 20% by 2020. Renewable and distributed energy production is an integral part of the reduction of GHG emissions because one-fourth of the world's GHG emissions are accounted for by the power grid operations.

- Energy profiling

  When smart grid generators are networked along with smart virtual to profile users' needs and govern the supply, then the generated energy is most efficient.

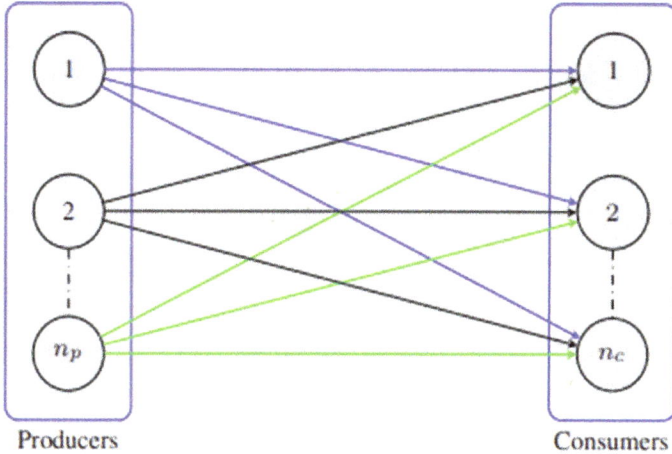

Figure 10.9   P2P energy trading

Smart grids generate, distribute, and regulate the flow of electricity energy to users in the same way as the main power grid does but on a smaller scale. They have substantially lower line losses since they are local, as opposed to the larger line losses associated with long-distance energy transmission. Being network-ing of grids is an excellent approach to incorporate renewable resources on a local level and to enable customers to participate in the power business, balance supply and demand near the point of use. Developing energy trading mecha-nisms are the foundations of the future's ideal power grid. One of the most sig-nificant advantages of a networked smart grid is that it can be better adjusted to suit future energy consumption and CO_2 reduction objectives, as well as enable faster reaction to demand and supply balancing at smaller granularity than the central grid can [15].

Figure 10.9 shows P2P energy trading has the following advantages:

a.   One of the most promising techniques for establishing decentralized electricity market concepts is P2P energy trading [16].
b.   P2P trade involves each actor negotiating directly with a group of trading part-ners without the use of a traditional intermediary.
c.   In the electrical network, a P2P market platform permits direct energy transac-tions between producers and consumers.
d.   Because P2P energy trading in smart grids is a new concept, a robust market clearing mechanism for P2P energy trading between diverse producers and con-sumers is necessary [17].

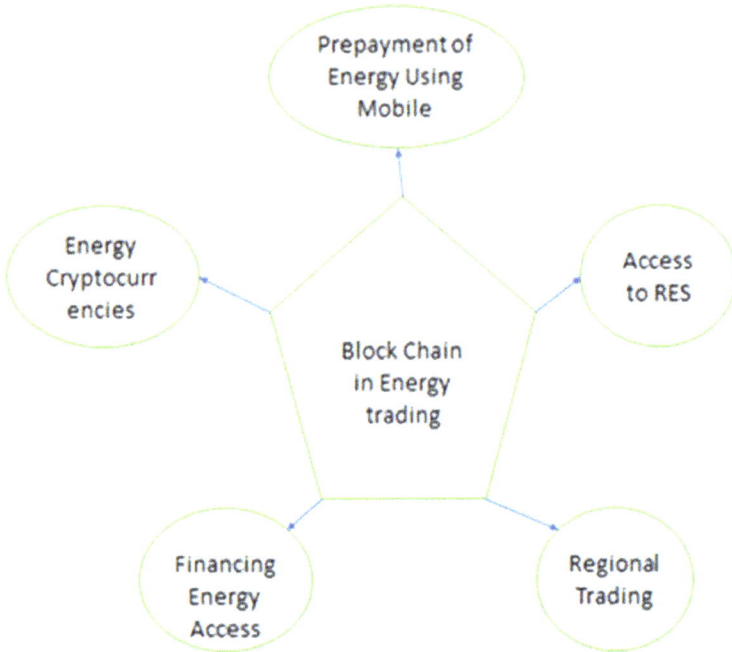

Figure 10.10   Blockchain in energy trading

10.3.2   Energy storage in blockchain

ESUs, which include both household batteries and EVs, are a viable way to improve the grid's functionality [18]. ESUs, in particular, are a powerful emergency backup that may be deployed in the case of a power outage, improving grid resilience [19]. Furthermore, ESUs offer a solution to the intermittent nature of RESs, allowing for a high level of integration of environmentally beneficial energy sources [13]. Additionally, the stored energy in such units can be utilized during peak load periods, reducing the strain on the power grid during these times and allowing for effective demand response. Also, ESUs provide financial benefits to users by lowering their electricity bills because ESU owners can buy electricity from the grid during low-tariff periods and can use it during high tariff periods, as shown in Figure 10.10.

Energy trading using blockchain has many advantages, as shown in Figure 10.10. It makes the use of energy cryptocurrencies. Access to RESs and even provision will be made for payment using mobile phones. The usage of bitcoins and energy cryptocurrencies facility can also be used [20].

There are many benefits from ESUs and even posses many challenges during the period of integration. More particularly during simultaneous mass-scale uncontrolled charging of ESUs may result in a mismatch between charging demands and energy supply, compromising the grid's overall resiliency [18]. In many cases, this may even lead to blackouts. A charging coordinating mechanism is necessary to reduce the pressure on the distribution system and prevent power outages in order to reduce such impacts. ESUs should report data like time-to-complete-charging (TCC), battery state-of-charge (SoC), and the amount of required charging in a charging coordination mechanism. Then, a trustworthy third party, such as a charging controller, determines which ESUs should be charged first and which should be deferred to a later time slot. Even with these features, the present mechanisms suffer from many limitations.

The existing charge coordination systems [19, 21, 22] have a number of drawbacks. To begin, they rely solely on entity to handle charging requests, called the charging coordinator (CC). As a result, if a successful denial of service assault is conducted against the CC, it will be down, and as a result, a large number of charging requests will be unable to be coordinated. Second, most existing works regard the CC as a reliable partner who is entirely honest when it comes to charging requests. As a consequence, ESU owners have no way of knowing if the charging plans are valid. Third, current charging coordination techniques demand ESUs to transmit certain data to the CC, such as whether or not an ESU requires charging, the TCC, the battery SoC, and the amount of charging required. This discloses private information about the ESU owners, such as the placement of an EV and the activities of the inhabitants of a dwelling [14, 23]. The restrictions outlined above emphasize the necessity for a decentralized, transparent, and confidentiality of charging coordination system. The restrictions outlined above emphasize the necessity for a decentralized, accessible, and privacy-preserving charging coordination system. Blockchains have recently attracted the attention of academics and industry in a variety of disciplines [24, 25]. With the use of blockchain in place, applications can be operated in a decentralized and transparent manner, eliminating the need for a central authority while attaining the same functionality as before [17, 26]. Furthermore, blockchains allow for the creation of smart contracts, which are pieces of code on the blockchain that perform a specific action when certain conditions are met. Smart contracts can be self-executed without the use of third parties because they are stored on the blockchain [6, 27].

10.3.3   Blockchain applications in energy trading

In the contemporary power system, the existence of more proactive actors or agents has prompted the creation and adaption of a more decentralized paradigm to power systems and electricity market operation. One of the most promising techniques for establishing decentralized electricity market models is P2P energy trading.

Energy trading has become a hot topic for numerous researchers and industry executives as smart grid, microgrid and distributed generation grow more prevalent. Blockchain is playing a major role in this field. By removing intermediaries from the market, this usage of blockchain in the energy trading process helps to reduce the time and effort required. P2P energy trade becomes a very promising future procedure thanks to blockchain. In P2P trading systems, the blocks in the chain record the units of generated electrical energy, allowing sellers and purchasers to transact quickly and independently. Instead, then relying on an intermediary agency,

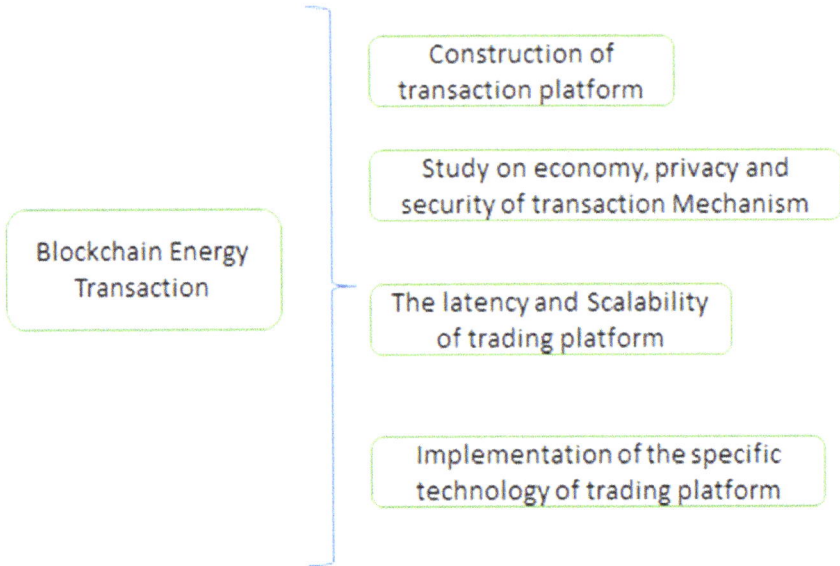

Figure 10.11   Blockchain energy transaction

users (owners and buyers) have complete control over their preferences, choices, and prices [17].

- Motivation of blockchain usage in smart grid systems

According to a market survey performed by Energy Charter Secretariats on energy trading, one of the key issues in local energy trading between providers and consumers is a lack of openness [2, 5].

Local sellers have difficulty selling their energy to a centralized authority, so a safe transaction method is also required [6]. The authority (energy supplier) buys energy at a lower price and sells it at a higher price, thereby creating a demand-supply mismatch. This is also bad for users who buy energy from local suppliers at a cheaper price and then sell it for a higher price. There will be energy losses at different levels at user end and supplier. As a result, a dependable mechanism is essential to provide a seamless local energy exchange between users and service providers [27].

- Application of blockchain in energy trading

Most studies on energy trade using blockchain for smart grid are still in the early stages, according to several research focuses, four aspects are considered, as shown in Figure 10.11.

10.4   Construction of transaction platform

On the demand side, constructing an integrated demand response (IDR) has been used to improve system operating flexibility and energy consumption efficiency by optimizing the operations of flexible loads, energy conversion, and storage equipment [3]. For complete use of integrated scattered resource (IDR), [7] have presented an energy transaction mechanism. Furthermore, [4] has created the Crypto-Trading Project, emphasizing the importance of blockchain technology and smart contracts in the management and control of the energy market's innovative model.

10.4.1   Handling of transaction request

The process of handling the transactions has two types such as:

1. Transaction access
2. Store transaction

To form a consensus agreement for safe communication, these transactions are relayed through various nodes in the network. The requester key and the requestee key are the two types of keys utilized in blockchain technology. Through a requester key, the requester sends a request to the cloud storage for data access. The cloud storage uses the requestee public key to offer access permissions to the requesting smart grid. On the smart grid, the transactions that flow via the network are essentially stored in a local blockchain. The brief explanation of the types of transaction is explained below [27].

10.4.2   Store transaction

This type of transaction is used for storing the data in cloud. For example, in the case of storing energy meter reading, usage data on a utility server for billing purposes, etc. In this situation, the data will be stored in the cloud storage using different block IDs and a hash. There cannot be a same block and same hash tag two data. So, the user is authentic with the specified block ID and hash value when the data are stored in the cloud successfully. Once the data are saved in the cloud storage, the data owner can access it using a public key. The following are the steps followed for storing the data in the cloud storage [17]:

*   The request is sent by the normal node for miner node to data in the cloud.
*   The miner node verifies the normal node using the local blockchain. Then the local blockchain generates a block ID number with a hash value after node verification, which is modified at the miner node [26].
*   The block ID number (together with the hash value) and the data to be stored are used by the mining node to communicate with the cloud storage. The cloud storage calculates a unique hash value for the data it receives and compares it to the hash value given by the miner node. When this value is found to be

identical, the cloud storage generates a new block identification number with only a shared key and sends it back to the miner node [15].
- The data received is stored in the local block by the miner node, which then sends the blocks ID and shared key to the regular node that requested the storage.
- These entities are used by the normal node to communicate its data to the cloud for storage [27].

10.4.3   Access transaction

The data owner or the service provider uses the access transaction to obtain information from the cloud storage (known as requestee node). The following are the steps involved in the access transaction:

- The transaction data, which includes the service provider's signature, requested data, and the requestee's shared key, is sent from the requestee node to the miner node.
- After authentication, the miner node submits the transaction data to the local blockchain, which generates a random block ID number and hash value.
- The requested data are sent to the cloud storage with the block ID and a hash value that matches the hash values. Only if the data are same the miner node will be given access for the requested data by the cloud storage.
- The data are obtained from cloud storage and delivered to the requestee node as an output after being stored in the local blockchain by the miner node.

10.5   Conclusion

DERs, grid edge devices, and associated energy infrastructure are currently vulnerable to cyberattacks since the power grid lacks the essential security and resilience. Behind the meter, cyber vulnerabilities and functionality affect building automation and controls systems (in smart grids). The use of blockchain technology could improve the fidelity and security of building-to-grid communications. Furthermore, numerous consumers can use the same widely seen blockchain to cryptographically validate the data of other entities as necessary, resulting in a distributed trust mechanism. Blockchain could also aid in the resolution of various optimization and reliability issues that have arisen as a result of grid modernization.

References

[1]   Nikolaidis A.I., Charalambous C.A., Mancarella P. 'A graph-based loss allocation framework for transactive energy markets in unbalanced radial distribution networks'. *IEEE Transactions on Power Systems*. 2019, vol. 34(5), pp. 4109–18.

[2]   Secretariat EC. *Market Trading Mechanisms for Delivering Energy Efficiency*. Brussels, Belgium; 2010.

[3]   Wang M., Ismail M., Zhang M., Shen M., Serpedin M., Qaraqe M. 'A semi-distributed V2V fast charging strategy based on price control'. *Proceedings Of Global Communications Conference (GLOBECOM), IEEE*; 2014. pp. 4550–5.

[4]   Sortomme E., Hindi M.M., MacPherson S.J., Venkata S. 'Coordinated charging of plug-in hybrid electric vehicles to minimise distribution system losses'. *IEEE Transactions on Smart Grid*. 2011, vol. 1(1), pp. 198–205.

[5]   Tushar W., Saha T.K., Yuen C. 'IEEE transactions on smart grid'. *Grid Influenced Peer-to-Peer Energy Trading*. 2019, pp. 1–1.

[6]   'Estimated growth of Res in India by 2022'. *Department of Energy Statistics of India*. 2022.

[7]   Kumar N., Parizi R.M., ChooA K.K.R., Kumari S., Tanwar S., Tyagi S. 'A taxonomy and process model for FOG data analytics'. *Journal of Network and Computer Applications*. 2019, vol. 128, pp. 90–104.

[8]   Wang M., Ismail M., Zhang R., Shen X., Serpedin E., Qaraqe K. 'Spatio-temporal coordinated V2V energy swapping strategy for mobile PEVs'. *IEEE Transactions on Smart Grid. 2016*, vol. 9(3), pp. 1566–79.

[9]   Guerrero J., Chapman A.C., Verbic G. 'Trading arrangements and cost allocation in P2P energy markets on low-voltage networks'. *IEEE Power Energy Society General Meeting*. 2019, pp. 1–5.

[10]   Soliman S.A.-H., Mantawy A.-A.H. *Modern Optimization Techniques with Applications in eEectric Power Systems*. Springer Science & Business Media; 2011.

[11]   Sun Y., Wu X., Wang J., Hou D., Wang S. 'Power compensation of network losses in a microgrid with BESS by distributed consensus algorithm'. *IEEE Transactions on Systems, Man, and Cybernetics Systems*. 2020, pp. 1–10.

[12]   Kim J., Dvorkin Y. 'A p2p-dominant distribution system architecture'. *IEEE Transactions on Power Systems*. 2019, pp. 1–1.

[13]   Shafee A., Baza A., Talbert A., Fouda A., Nabil A., Mahmoud A. 'Mimic learning to generate a shareable network intrusion detection model'. 2019.

[14]   Callaway D.S., Hiskens I.A. 'Achieving controllability of electric loads'. *Proceedings of the IEEE*. 2011, vol. 99(1), pp. 184–99.

[15]   Kumar N., Vasilakos A., Rodrigues J.J.P. 'A cloud-based DC nano grid with multiple tenants for self-sustaining smart buildings in smart cities'. *IEEE Communications Magazine*. 2017, vol. 55(3), pp. 14–21.

[16]   Paudel A., Khorasany A., Gooi A. 'Decentralized local energy trading in microgrids with voltage management'. *IEEE Transactions on Industrial Informatics*. 2020, pp. 1–1.

[17]   'Benefits of demand response in energy markets and proposals for achieving them'. *US Department of Energy*. 2006.

[18]   Baza M., Pazos-Revilla M., Nabil M., Sherif A., Mahmoud M., Alasmary W. 'Privacy-preserving and collusion-resistant charging coordination schemes for smart grid'. *arXiv preprint arXiv:1905.04666*. 2019.

[19] Baza M.I., Fouda M.I., Eldien M.I., Mansour M.I. 'An efficient distributed strategy for key management in microgrids'. *11th International Computer Engineering Conference (ICENCO). IEEE*; 2015. pp. 19–24.

[20] Amer R., Saad W., ElSawy H., Butt M., Marchetti N. 'Caching to the sky: Performance analysis of cache-assisted COMP for cellular connected uavs'. *arXiv preprint arXiv:1811.11098*. 2018.

[21] Li M., Weng J., Yang A., *et al.* 'Crowdbc: A blockchain-based decentralised framework for crowdsourcing'. *IEEE Transactions on Parallel and Distributed Systems*. 2018.

[22] Weng J., Zhang J., Li M., Zhang Y., Luo W. *Deepchain: Auditable and Privacy-Preserving Deep Learning with Blockchain-Based Incentive*. 2018/679. Cryptology ePrint Archive; 2018.

[23] Liu D., Alahmadi A., Ni J., Lin X., Shen X. 'Anonymous reputation system for IIoT-enabled retail marketing atop PoS blockchain'. *IEEE Transactions on Industrial Informatics*. 2019, vol. 15(6), pp. 3527–37.

[24] Paschalidis I.C., Li B., Caramanis M.C. 'Demand-side management for regulation service provisioning through internal pricing'. *IEEE Transactions on Power Systems*. 2012, vol. 27(3), pp. 1531–9.

[25] Onlocation INC. *RELOAD database documentation and evaluation and use in NEMS*. 2001. LoadShapesReload2001.pdf [online]. Available from http://www.onlocationinc.com.

[26] Rahimi F., Ipakchi A. 'Demand response as a market resource under the smart grid paradigm'. *IEEE Transactions on Smart Grid*. 2010, vol. 1(1), pp. 82–8.

[27] Zhao C., He J., Cheng P., Chen J. 'Consensus-based energy management in smart grid with transmission losses and directed communication'. *IEEE Transactions on Smart Grid*. 2017, vol. 8(5), pp. 2049–61.

Chapter 11

Development of novel cryptocurrencies for IoT—blockchain-enabled smart grid platforms

Anil Kumar[1], Gururaj H L[2], Sunil Kumar B R[2],
Goutham B[2], and Ravikumar V[2]

The concept of the Internet of Things (IoT)-blockchain-enabled smart grid has been introduced as a current vision of the usual power grid to solve a systematic way of incorporating renewable and green energy technologies. To secure energy from anywhere at any time, there is an approach called energy internet (EI), where the smart grid is connected through the internet an innovative and emerging technology. The main goal of this approach is to improve the energy sector more efficient and available to improve the environmental and economic conditions of our society. There are many challenging issues in the existing centralized energy grid system like coordinating and integrating a huge number of growing grid connections. Accordingly, the IoT-blockchain-enabled smart grid is going through an alteration from a centralized topology to a decentralized topology. Precisely, more development has been introduced as cryptocurrency initiatives are emerging in trials of products and projects. Furthermore, the IoT-blockchain-based model has a few excellent features which cause a favorable application for the smart grid model. Here we aim to provide an IoT-blockchain-based platform for smart grid and identified some significant security issues of smart grid scenarios.

11.1 Introduction

The traditional centralized energy generation systems which include fossil fuel-fired power plants in the past few decades have been facing major challenges like air pollutant emissions, power crisis, pollution, and long-distance transmission. To build a efficient energy system and reliable system, there should be some efficient energy generation system should be used other than centralized system for enhancing the efficiency usage of energy. It is an advanced energy generation, delivery, and

[1]Department of ISE, CIT, Gubbi, India
[2]Vidyavardhaka College of Engineering, Mysore, India

consumption system that includes smart metering, interconnected energy systems, and communication technology. Energy monitoring and management are correctly managed in this smart grid technology breakthrough. Multi-directional data communication, in particular, aids in connecting many nodes and energy exchange for more efficient use of green and renewable energy technology. Regrettably, when it comes to upgrading and improving access to distributed and scalable energy resources, the smart grid is struggling to keep up on a wide scale. Several researches have been undertaken to overcome current challenges, constraints, and develop the system by merging smart grid 2.0 with energy technology. By helping in the migration of various forms of storage, lodes, and energy resources, the EI provides end-to-end energy distribution on a huge scale. EI is also known as the internet of energy (IoE), which is a current energy infrastructure that is combined with renewable energy power generation technologies and modern internet technology, in contrast to the smart grid.

In a conventional centralized smart grid system if connection expands, it will also increase several concerns, such as managing energy providers, their customers, electric vehicles management etc., must integrate and coordinate. Managing such networks in centralized systems, which are constantly growing, necessitates a sophisticated and costly information and communication infrastructure. The aforementioned challenges can be overcome using smart grid components that can work in a decentralized and dynamic manner, which are the basic requirements of the energy sector. The development of the internet in the smart grid is a growing trend. However, in a decentralized smart grid system with complicated linkages and a high number of components, privacy, security, and trust concerns may arise, necessitating advanced and inventive new technologies to handle [1–6]. On the other side, the blockchain which is a promising and emerging technology provides various new and advanced opportunities to build a decentralized system. Blockchain is a decentralized technology that requires no central trusted authority to manage the system; instead, each entities has the authority of creating, maintaining, and storing a chain of blocks. Every entity in the network has the access to verify that data in blocks have not been tempered and the chain order is correct. This decentralized system solves many of the issues in a centralized system and makes any system redundant and withstands cyberattacks and system failures. Although it was first presented and populated as digital money, blockchain has a variety of good qualities that are attracting interest in a variety of non-monetary applications [7, 8]. Simultaneously, it encourages the development of secure, trusted, and privacy-preserving smart grid innovations in the direction of decentralization.

Our contributions and related surveys: Though it is a new area of research, blockchain in the smart grid has already gotten a lot of attention. Many studies have recently been conducted to use blockchain to address smart grid issues. Table 11.1 summarizes the contributions of our research to others.

Table 11.1 mentioned works point to some issues or survey topics only, but there is no complete work that covers the full aspects of IoT smart grid domain in blockchain research. This state drives us to present this work with a complete literature survey on all the activities of blockchain in IoT smart grid research. Here, the paper contribution and summary of survey work are summarized as follows:

Table 11.1   Blockchain contribution

Paperwork	Contribution
EI blockchain technology	• EI—initiative related to blockchain
Blockchain purposes in smart grid—review and frameworks	• New framework of blockchain and grid test beds
Blockchain technology, challenges, and opportunities in the energy sector	• Future EI • Energy trading • Challenges and opportunities in blockchain
Blockchain technologies—challenges and solutions for smart energy systems	• Fundamentals and solutions of smart energy systems
Current projects and concepts of blockchain in microgrids—overview	• Microgrids use blockchain technologies
A survey on distributed ledger technology (DLT) for peer-to-peer—potentials and energy exchange in regional energy markets	• Transactive energy trading using peer-to-peer
Present work	• Motivation—implementing blockchain in smart grid • Blockchain usage—energy CPS, microgrid, EV management, decentralized energy market, and trading • Smart grid research oriented—future work • Cryptocurrency initiatives

• introduction to the blockchain (blockchain categories, smart contract, DLT, and consensus mechanism)
• major requirement outline (smart grid privacy, security, and trust objective)
• discussion on key challenges like scenarios and different parts of smart grid domain, blockchain contribution to the domain challenges
• cryptocurrency initiative for IoT-blockchain research in smart grid
• we further mentioned challenges that remain to be discussed based on our study and future research directions.

The following is a list of the paper's organization. The background of blockchain technology is explored in Section 11.2. The smart grid's transformation into a decentralized system and its contribution are discussed in Section 11.3. Section 11.4 discusses the role of blockchain in smart grids. Section 11.5 provides a quick overview of cryptocurrency initiatives. The conclusion and future research directions are presented in Section 11.6. Table 11.2 foci a list of abbreviations and their meanings used in this proposal.

Table 11.2   Summary of blockchain-based solutions in smart grid

Work	Major contribution	Focused application
[9]	Facilitating secure and fast energy exchange of DERs by means of blockchain-based AMI	Transactive energy applications
[10]	A blockchain-enabled distributed ledger for storing smart meter data (considered as energy transactions) to utilize in making a balance between energy demand and production	Decentralized demand response programs management
[11]	A permissioned blockchain to ensure privacy and energy security (traceable and transparent energy usage) in a smart grid	Traceable energy usage and transparent energy usage
[12]	A blockchain-based privacy-preserving energy scheduling model for energy service companies	Energy demand and supply information
[13]	A consortium-based energy blockchain	Peer-to-peer energy trading

11.2   Blockchain background

This section gives a brief description of blockchain fundamentals with blockchain structure and the definition of blockchain technology, DLT, and smart controls. Later discussion is on blockchain classifications and consensus mechanisms.

11.2.1   Blockchain fundamentals

A blockchain is a chained series of blocks which is used to store data. Blockchain gives trust, transparency, and security to data in all the applications, specifically in critical systems using the distributed ledger. Here the information and data are distributed in the network rather than copied material. The traditional architecture employs a client-server model, in which data are stored on the server and easily modified because the server has access to the data. However, with blockchain, all nodes have the same authorization and privileges to maintain, change, and approve new blocks.

The block structure is depicted in Figure 11.1. One block consists of three elements: a header, a data blob, and metadata. The blockchain storey always begins with block B0, which represents state S0 of the state machine architecture that was used to model it. At time t and state S0, nodes begin sending the transaction id embedded in block B1. Each node will have a unique address called a public key, which is used to perform some secure operations within a digital application. A safe mathematical relation is put into the message to specify the address identity. This is a digital signature that can only be formed using a component that is encrypted and kept secret by the node, namely its private key. Receivers' nodes could validate the sender's authenticity by verifying this signature with the visible public key of the

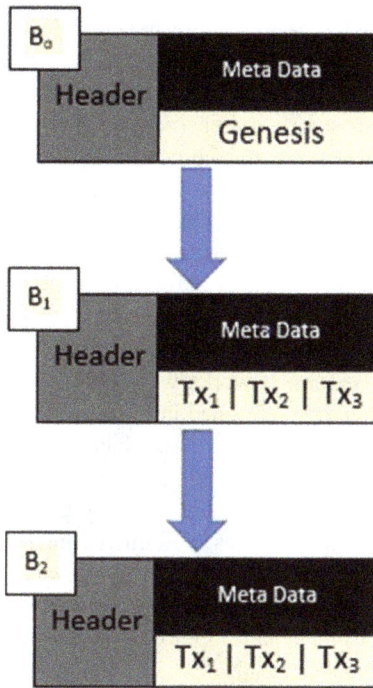

Figure 11.1   Blockchain block structure

incoming transaction. Asymmetric cryptography refers to the use of a public/private key pair in this way.

11.2.2 DLT, blockchain, and smart contract

- DLT: It is a protocol that works for the secure processing of decentralized digital repositories. It also refers to the technical infrastructure and rules that enable concurrent validation, access, and updating the record in an unchangeable manner over the network that is laid out across multiple locations or entities. To keep track and check against manipulation, no authority is required in the distributed network. Here nodes updated themselves with corrected and new replicas of the ledger and for giving security to the nodes always signatures and cryptographic keys are used.

- Blockchain technology: Blockchain technology is made up of a chain of blocks in which information is recorded, making it tough or difficult to manage, manipulate, or circumvent the system. Because it is a digital record of operations, blockchain is replicated and duplicated over a network of computer systems. Each block contains data such as the number of transactions, and new transactions are added to the blockchain at regular intervals. Each operation is always documented in the ledgers of all members. Each transaction on the blockchain is kept and documented with a cryptographic signature known as a hash, with each block referring to the hash of the earlier data block. The timestamp, nonce, smart contract strips [14], a Merkle tree hash tree [15], and so on are all included in each block of the blockchain. Cryptographic keys, hash, and Merkle tree are used to verify block integrity and ensure the block content is not updated. If intruders want to change or corrupt a blockchain system, they must change each block in the chain across the entire chain's distributed network. Because each block has the hash of the previous block, deliberately corrupting the blockchain is impossible. Aside from linear structures, there is a type of structure known as a directed acyclic graph, in which each block refers to several preceding blocks.

- Smart contract: It [16, 17] is a system software package or a protocol which automatically controls, executes, or documents with specific events and actions according to the agreement. It works with simple conditions statements like if, if/when, if/then, that are mentioned in the blockchain. When predetermined conditions have been met within the blockchain, then a computer executes the action part. The blockchain is updated after each transaction, which means that the transaction cannot be modified and that only those who have been granted permission can access the outcomes. The smart contract is placed on the block-chain and can run without any centralized mechanism. Ethereum, a customized blockchain platform, is one of the most prominent smart contract platforms. Figure 11.2 depicts a blockchain topology, the operating basis of smart con-tracts, and a typical block structure.

11.2.3   Blockchain categories

- **Permissioned vs Permissionless:** While creating new blocks, there will be two types of restrictions: permissioned and permissionless. Access to blockchain

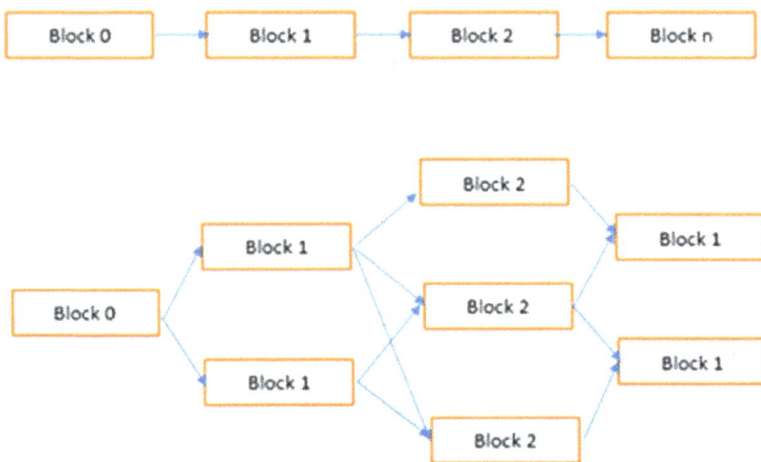

Figure 11.2   Smart contract: a typical block structure and blockchain topology

content is based on these two restrictions. Before using any block there should be a prior approval for a permissioned blockchain whereas a permissionless blockchain block can participate at any time within the system. Both systems behave the same because they cannot be used for the same things. Many of us don't know how to use a permissioned currency and no one has control over how it works within the system as mentioned one of the major drawbacks of cryptosystem.

- **Private vs Public:** One more important category of blockchain is private and public. If blockchain is public then it is belonging to the category of permissionless. Anyone within the blockchain system can connect with the network and write, read, or participate. No single entity is controlling the network in the public blockchain because it is decentralized. In a public blockchain, data are secured and it is not possible to alter or modify data once it is validated (Figure 11.3). If a blockchain is private, then it is belonging to the category of

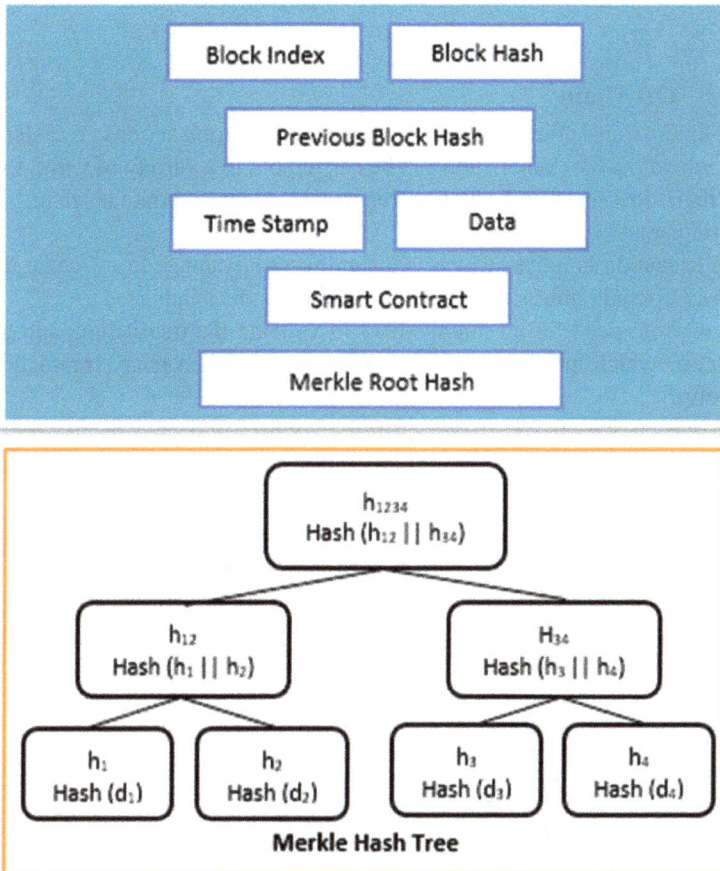

Figure 11.3   A typical block structure and Merkle hash tree

permissioned. Here in this type of blockchain, the access controls are restricted between the people who are participating in the network. Most of the blocks which control the network help the third-party dependency. Participating entities will have the access to the entire blockchain in a private system and others will not be able to access it (Figure 11.4).

- **On-chain vs Off-chain:** There are two ways of transactions that we can do within blockchain; they are on-chain and off-chain which differ in many ways.

11.2.3.1 On-chain

1. These transactions are available to all users on the network.
2. Transactions are validated by a huge number of blocks and provide all necessary information to the network of blockchain.
3. Each transaction is recorded on the appropriate block, making it irreversible.

11.2.3.2 Off-chain

1. Off-chain transactions, in contrast to on-chain transactions, transfer value outside of the blockchain and can be carried out in a variety of ways. Because of their low cost, off-chain categories are more popular among major participants.
2. The execution is quick, which is a significant advantage in off-chain transactions. Off-chain transactions have no transaction fees indicated.
3. There is no need for an intermediary to validate the transaction, and there is no cost, which makes this an appealing choice when many transactions are involved.

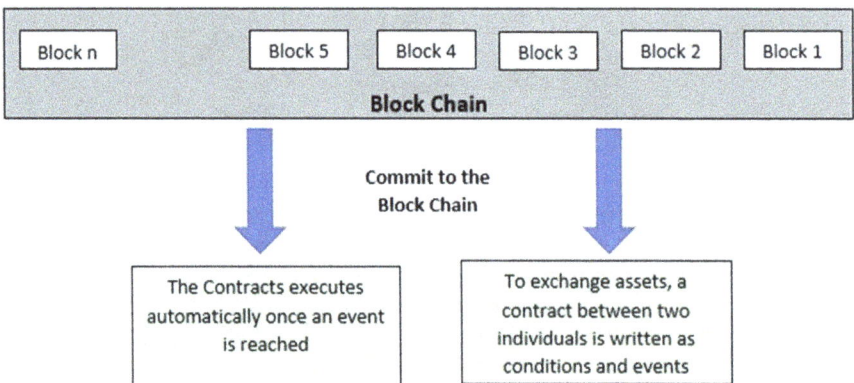

Figure 11.4   The basic working principle of smart contract

11.2.4   Consensus mechanisms

Finding flaws in the system is one of the most important features of blockchain technology. This method is a fault-lenient mechanism used to accomplish the necessary bond between distributed systems on a single state of the network in the blockchain. This is a term used in the cryptocurrency world. In both private and public blockchains, the network's trust is ensured by a consensus mechanism in which a group of validators/miners agrees on whether a block is valid or not. Decentralized, self-regulating public blockchains operate on a global scale without relying on a single authority. Many kinds of consensus technique algorithms are available with different working principles are explained below:

- **Proof of Work (PoW):** One of the highly common consensus algorithms used by many cryptocurrencies like Litecoin and Bitcoin is PoW. Here adding a node or a block must prove about work done and must qualify themself for receiving a new block in the blockchain. The complete mechanism of cryptocurrencies needs a longer processing time and energy consumption. The key idea behind this method is to solve the consensus nodes for creating new blocks, and an expensive puzzle is given to solve the node known as PoW, which is easy to verify but hard to solve. After solving, the new block is attached to the solution, and it is broadcasted across the network. With the help of this connection, any additional nodes can confirm the rightness of a new block released by the node. PoW is not suitable for many blockchain applications because it has some limitations like high energy consumption, high latency, and inefficient throughput involved in the mining process.
- **Proof of Stake (PoS):** PoS [18, 19] is one more consensus mechanism that consumes low energy and low cost compared to the PoW algorithm. The main responsibility of this algorithm is to keep up the public ledger to a new node in proportion to the number of virtual currency tokens held by it. PoS helps to promote cryptocurrency for saving instead of spending. PoS is the most preferred alternative to PoW and its aim is to improve the limitations of the PoW. In PoS-based blockchain, the concentrating will be on validating the block. In this PoS consensus mechanism, the main concern is validating, and validation is replaced by the mining process, which means blocks are validated rather than mined. The validation process for new blocks is random and related to the assets of its own.
- **Delegated Proof of Stake:** It is another consensus algorithm designed to secure the blockchain, mentioning the transaction within the blockchain. To protect the blockchain from malicious usage, centralized system voting process is used to implement a technology-based system. One of the variants of PoS is Delegated Proof of Stake (DPoS), but the only difference is generating and validating blocks, and this can be done by some selected blocks.
- **Leased Proof of Stake:** It is a variant of PoS. Leased Proof of Stake (LPoS) is a consensus algorithm which is mainly used to allow block holders to lease their tokens to the full node of the network and earn payment as a stake. With

the regular process of this platform, each node adds new node to the blockchain. Normally, nodes are shared proportionally. Here, the main agenda is to lease the nodes is to increase the portability of being validators and reduce the portability of the blockchain network being ruled by a single group of node.

- **Proof of Burn:** An alternative to PoS and PoW, an alternative method of consensus mechanism is introduced called Proof of Burn (PoB). PoB is mainly used for bootstrapping one cryptocurrency to another cryptocurrency. Here, the miner should burn some coins and must show proofs, and later send those coins to a verifiably unspendable address. Spending coins is mainly considered an investment. PoB consumes no resources rather than burning underlying resources. Hence a user can become an authorized validator by making their stakes on the chain after investing.

- **Proof of Authority:** For permissioned blockchain, Proof of Authority (PoA) [20] is designed. In this consensus mechanism, any block prior to becoming an authority to bring out a block must confirm its identity within the network. Here participant's identity is considered as a stake instead of coins or assets. PoA always informs the malicious activities of other nodes.

11.3   IoT blockchain in smart grid

In this section, the discussion is mostly on how we can apply blockchain in smart grid technology. As such, we first will discuss on detail what the future smart grid system will be. Here we are showing how blockchain provides protection, security, and trust to participants for forming a good network.

11.3.1   Moving toward decentralized smart grid system

IoTs for some of the digital computations, and communication technologies, for creating new grid infrastructure for some of the computations, to modernize the grid into more effective. In response to the demand for new fuel sources and the rapid change in the environment, the smart grid is being upgraded and modernized. Changing the energy landscape by coordinating and employing more sustainable and widely distributed energy assets, rather than relying on petroleum-based ages, is one of the most essential components of these transformations and modernization. The intelligent network perspective brings producers and buyers closer together by transporting autonomously appropriated sustainable power makers, whilst the old heritage lattice feeds buyers via long-distance transmission lines.

Of late, the EI idea [21–28] is presented, which is characterized as the overhauled adaptation of a shrewd network framework. The EI is portrayed by Internet advances to foster the up-and-coming age of brilliant framework by coordinating data, energy, and financial aspects. The EI is intended to give an incredible chance to work with the consistent combination of assorted perfect and sustainable power sources with the matrix and, additionally, give more collaborations among different components of the force matrix to create a completely independent and insightful energy organization. The EI's main idea is to efficiently distribute both energy and data, similar to how information is transferred on the internet.

However, while the smart network and EI are designed to adjust both dispersed and concentrated energy ages, one major drawback of the current plan is that it has brought together geography where energy ages, transportation and conveyance organization, and markets are all reliant on a unified or go-between elements in some way. In this unified framework, the components of the keen lattice connect and speak with unified elements that can screen, gather, and cycle information, backing all components with suitable control signals. Surprisingly, the current plan of shrewd lattice framework raises a few concerns due to the infiltration of sustainable power sources as well as the continually growing quantity of components. Adaptability, expandability, heavy computational and communicational weights, accessibility assaults, and the inability to regulate future force frameworks with a large number of parts are all issues.

Changing to a decentralized framework is a tendency in keen networks to bring more strong, clever, and proactive features, all things considered. The network foundation itself is likewise going through a variation and advancing toward a completely robotized network having decentralized geographies to increment the connections among all segments of keen matrix frameworks in a unique manner. The network and openness that EI offers also arrive at a more significant level of conservative, proficient, furthermore, dependable activity of the savvy matrix framework. Table 11.3 addresses a concise correlation between the shrewd network and the EI-empowered future decentralized smart lattice.

11.3.2 Motivations of applying blockchain in smart grid paradigm

The key concerns of any framework are security, protection, and trust. Similarly, the future brilliant frameworks should have some level of security [14, 35, 36], including (i) fostering an issue open-minded organization having opposition against accessibility assaults, (ii) making more productive checking, (iii) utilizing progressed security safeguarding strategies to ensure data exposure, and (iv) expanding the trust, straightforwardness, and majority-rule government among every one of the substances.

In [35], Nakamato tackles the issue of building up faith in a disseminated framework by presenting a narrative agreement system that creates Bitcoin the best blockchain function up until this point. This is because of utilizing this agreement component as well as using different methods such as cryptographically ensured information structure, computerized signature procedure, timestamp, and remunerating plan. In blockchain applications, the agreement systems are generally used to build up trust as it were. Then again, the various types of cryptographic methods are additionally used to tackle essentially the fundamental security prerequisites including secrecy, uprightness, validation, approval, and nonrepudiation just as protection.

Right now, a large portion of the arrangements are based on unified models where keen framework parts are reliant upon all things considered concentrated stages or go-betweens to get administrations like charging, checking, offering,

Table 11.3   Summary of a few cryptocurrencies

Name	Ref.	Organization	Description
Charg Coin	[29]	WeCharg	Charg Coin is mainly used for EVs that need a huge source of energy distribution through charging stations using IoE
CyClean coin	[30]	CyClean Pte Ltd.	CyClean coins provide the pre-extracted coin for the users as rewards based on the usage of the electric motorbike, electric bicycle, and electric car
Pylon network blockchain	[31]	Pylon Network	In this Pylon network blockchain platform, energy usage and creators' information are stored in a repository. Here, data owners are having all the access to data and make decision on who can retrieve the data
EXERGY	[32]	LO3 Energy	EXERGY is an authenticated data platform which deploys data and energy in localized markets for transforming energy
Energy Web Chain	[33]	Energy Web Foundation (EWF)	Web chain is an open-source, decentralized platform. This blockchain platform is mostly used in the energy industry to speed up user-centric and low-carbon energy transactions.
Powerledger	[33]	Power Ledger Pty Ltd	Powerledger is a blockchain-assisted trading platform that provides a user-friendly interface for energy markets. It also gives detailed information on all markets' pricing mechanisms and power units.
Sunchain	[34]	GreenTech Verte	The Sunchain platform manages solar energy sharing and pooling using IoT and blockchain.

energy exchanging, and so forth. Though these arrangements are developed and working appropriately, a few testing issues are related to the present smart lattice framework. Also, we earlier referenced beforehand that the shrewd lattice is working with the incorporation of an enormous number of EVs, Distributed energy resources (DERs), prosumers, and digital actual frameworks. Along these lines, the framework geography itself is adjusting and moving from brought-together geography to a decentralized and completely robotized organization to permit more noteworthy connections among the segments.

In relation to decentralized systems, using blockchain gives an opportunity to provide this transformation due to its following features which makes it suitable to apply.

- **Decentralization:** The blockchain system is normally kept up with various decentralized hubs through agreement conventions. This organization can typically run in distributed way without confiding in a brought together and confide in an expert for approval and support.

- **Permanence:** Since the blockchain innovation uses cryptographic methods and keeps a worldwide record which is synchronized among the hubs, the substance inside the squares can't be changed except if the dominant parts become vindictive.
- **Straightforwardness and auditability:** The blockchain network's design is extremely simple, as the organization's hubs can authenticate the validity of the records and confirm that the squares have not been modified. Furthermore, because of this simplicity, any hub in the network can audit the squares because all of the records are exposed to everyone.
- **Versatility:** Blockchain innovation allows a strong and flaw-lenient organization in which any flaw activity may be associated and remedied without difficulty. This robustness stems from the design's decentralization, which leaves no one weak spot, as well as the ability to store the entire tea.
- **Secure script deployment**: Along with the changelessness and decentralization comes the ability to send secure scripts inside the blockchain. It's also known as a brilliant agreement in the blockchain context. The agreements for clever contracts are usually stored on the blockchain. Without human intervention, middleman, or focal approval, these agreements can operate freely and intuitively based on a few set principles.

As a result of these benefits, as well as the state-of-the-art cryptographic safety benefits, blockchain can be a hopeful substitute to traditional brought together frameworks for furthering security, protection, and trust while assisting in the removal of obstacles to become a more decentralized and versatile framework.

11.4 Blockchain contribution in smart grid

Here, we listed out several recent contributions of blockchain toward the smart grid. Later, will oversee the blockchain outcome on IoT smart grid integrity, privacy, and security. Here, more focus is on some of the smart grid applications, and those details are explained below. Some of the blockchain components are showcased in Figure 11.5 and each section will give a specific description of the blockchain application and address the blockchain contributions.

11.4.1 Blockchain for advanced metering infrastructure

In this advanced metering infrastructure, companies, contractors, and buyers in the smart grid network will be able to communicate with one another via two-way communication and automatic smart meters. These advanced metering infrastructures are used to collect data on energy production and consumption, as well as energy status and detailed statistics. We don't obtain this level of detail from regular meters. Monitoring, billing, troubleshooting, and user application control are all common uses for the data. These unique data transmissions are carried out over a wide area

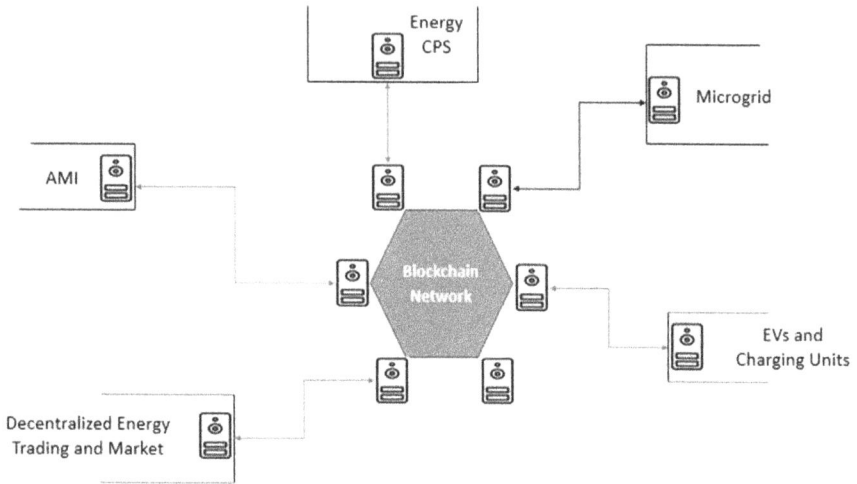

Figure 11.5   Blockchain applications in the smart grid domain covered in this work

network (WAN) using standard centralized repositories like the cloud. Centralized computing systems may have a few concerns like single system failure, privacy crack, and risk of alteration. In addition to this, a centralized system leads to many problems like availability, scalability, and response.

Along with this, the IoT smart grid system's EVs and smart meters generate many digital payment records and power usage figures, those details are regularly distributed with others for trading, billing, and monitoring. Here, in centralized computing system information processing and sharing creates different confidential threats since information such as identities, energy profiles, energy usage and generation trends, intermediaries, trusted third parties, and locations may be exposed. As a result, consumers and producers may find it difficult to accept centralized parties' openness and fairness. The distributed system can be used to overcome these problems.

11.4.2   Blockchain in decentralized energy trading and market

An IoT smart grid is required for decentralized trading and market, and to increase the number of producers and consumers. The bidirectional energy stream enables customers to perform as producers and producers as customers. With this decentralized method, we can reduce load peaks and power loss in transmission, enable green energy frameworks, and provide energy supply with distributed sources like EVs, microgrids, and energy storage units. As a result, it is critical to coordinate energy trading, as well as its fundamental norms, in order to provide bid processing, exchange, and contract execution among the participants. Producers and consumers can also trade energy with one another in a transparent and seamless manner. This direct energy exchange without the need for intermediaries can also boost the benefits, all else being equal, and is advantageous for long-term power deals. However, under traditional methods, customers exchange with one another in a circuitous way through various go-between parties and merchants, which will reveal some potential concerns and difficulties. As a result, it has significant operational and administrative costs, which are passed on to customers, producers, and prosumers in the end. Failures contaminated or pernicious intermediaries, and imposing business model motivational elements, prizes, and punishments result in an uncompetitive market, a lack of openness and rationality, and imposing business model motivating factors, awards, and punishments. The unique characteristics of blockchain make it an excellent tool for planning a more decentralized and open energy market.

11.4.3   Use of blockchain to monitor, measure, and control

The supervisory control and data acquisition (SCADA) framework is commonly used to control power grid systems. The SCADA architecture, which is linked to the internet, will enable massive, large-scale distributed checking, estimation, and control in a more evolved manner. Smart devices, sensors, and Phasor Measurement Unit (PMUs) in the IoT collect power gadget status data and share it with Maximum Transmission Unit (MTUs) via Remote Terminal Unit (RTUs), with MTUs serving as central archives and control points. These fine-grained estimations among CPS components, various grid administrators, providers, and purchasers are used by the smart grid network framework to enable wise control, wide-area monitoring, and administration to more effectively deal with grid safety, dependability, and reliability quality, as well as preventing power theft and misfortune. Be that as it may, cyber assaults can be dispatched in various manners such as middleman attack or insider attack, for example, changing information inside central controllers, dispatching accessibility assaults, and infusing bad information box sensors and PMUs. Subsequently, the attacker can assume control over control channels and furthermore, produce vindictive orders. In a decentralized framework, the blockchain provides a new process to measure, monitor, and control.

11.4.4   Blockchain on EVs and charging units management

In IoT smart grid, EVs are more often used in charging stations, energy storage devices, and trade energy with the power grid in a peer-to-peer manner. Grid-to-vehicle, vehicle-to-grid, and vehicle-to-vehicle situations will all be possible. EVs can play a critical role in delivering a significant interest reaction, boosting grid flexibility, and reducing load tops by engaging in energy charging and releasing activities. In contrast to the current situation, EI is used to transporting and dealing with many EVs. Regardless, in the three situations described above, constant two-way power and information interactions, short communication distances, and the portability of EVs can all offer significant security and protection concerns. Creating decentralized and plain EVs and charging the executives' systems are critical in this regard, as shown in Figure 11.6.

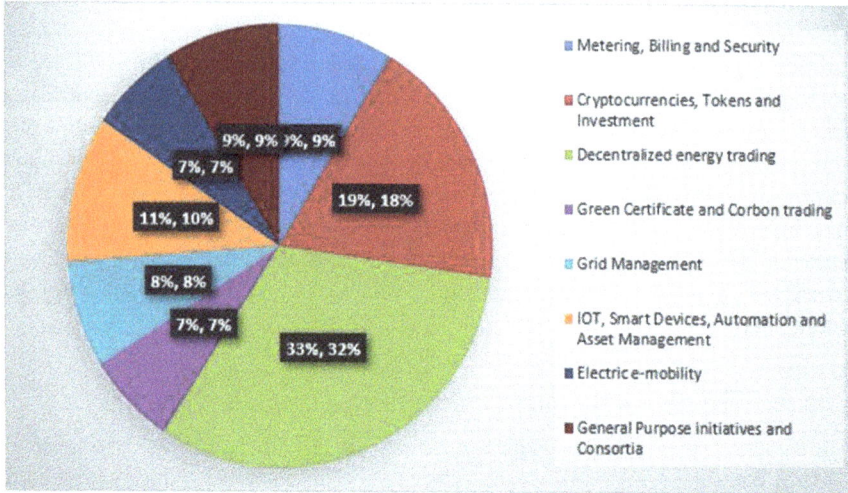

Figure 11.6   Blockchain usage in smart grid

11.4.5   Use of blockchain in microgrid

Energy management is an important part of the smart grid since it allows for the usage of DERs such as fuel, wind, and solar cells. Smart appliances and loads, EVs, distributed energy resources, and battery storage units make up the microgrid. It is always near the loads, usually near the generation units. Compared to fossil-based generation system, it produces large scale of electricity at a centralized facility. The microgrid produces small-scale generation of electricity. The main aim is to provide a dependable electricity supply, utilize renewable sources, and reduce transmission losses. Microgrid networks have many dispersed energy resources, and microgrids are interconnected to power grids in a distributed manner to provide necessary electricity to the grid. EI-enabled smart grid helps to construct a plug-and-play microgrid for faster connectivity with the other grids.

In any case, the top stage of DERs insertion into the grid from the microgrid system may pose significant power worries, potentially making the power grid unstable. In addition, the structure of the microgrid, paired with the smaller grid, provides certain open testing issues in energy exchange and management. Dispatch optimization, congestion pricing, and control are some of the energy management difficulties. The blockchain can be applied to added energy trading issues like centralized monopoly markets and improper mechanisms to prompt their adoption. By constructing decentralized microgrids, the blockchain can be used to address these issues.

11.5 Cryptocurrency initiatives

11.5.1 SolarCoin

SolarCoin can be called digital currency, which is fully based on blockchain technology. It is an inventiveness to produce offers and rewards for solar energy creators. The main aim of this currency is to provide rewards for the solar planet. The rate of one SolarCoin per one MWh is maintained with this digital token in the solar power plant. This enhances the usage of SolarCoin across the globe. Solar energy producers will grant this currency freely for transaction purposes. For any cost of electricity, the SolarCoin currency helps to compensate for the cost. Producers will give SolarCoin to anyone. Initially, participants must register for solar installations, and solar producers must submit a claim to SolarCoin currency affiliates. Then the claim participants can download some free coins wallet for creating receiving address. This address acts as a bank account, and this will be shared with the currency affiliates for data documentation purposes. Claimants will receive SolarCoins from the SolarCoin Foundation to the wallet at the rate of one SolarCoin per one MWh of validated electricity. Claimants can exchange, save, and spend the SolarCoins whenever necessary. SolarCoins are equated to government coins on crypto exchange and can be spent for any business purpose or for foreign exchange. VeriCoin [37] uses the PoS time mechanism to maintain the blockchain of SolarCoin.

11.5.2 NRGcoin

NRGcoin is developed at Vrije University, Brussel; this NRGcoin project is jointly carried out by industry and academia. Currently, Enervalis has taken up the project and carries forward it to the industry level. It uses green energy content in the local grid and makes it beneficial economically to consumers and governments. Here energy is not for trading, and we can sell and buy it by means of a smart contract. NRGcoin not only is a cryptocurrency but also supports every renewable resource. NRGcoin contains three important components; they are gateway devices, smart contracts, and currency. Traditional risky renewable-supported policies are replaced by blockchain-based smart contracts. The Ethereum platform supports the smart contract. Here, the customers are offering produced green energy for 1 NRGcoin using a smart contract directly for every 1 KWH. For creating new NRGcoins, every time reports are audited followed by the smart contract; it's all up to the producers whether they want to sell on a current market or use them to pay for green resources. Based on the consumptions, customers can buy NRGcoins from the cryptocurrency market. NRGcoins have some popular and non-popular cryptocurrencies such as Dollar, Euro, and Bitcoin. This helps to develop digital currency transactions and new currency exchange platforms for the centralized or decentralized exchange markets. One NRGcoin value is always 1 KWH produced energy, and it is not dependent on the price of electricity.

11.5.3   Electronic energy coin

Blockchain and smart contracts are using the green energy platform for developing electronic energy coin (E2C) [38] project. By using this coin, fair, secure, and proper energy distribution take place, and this coin is considered an energy revolution for energy distribution. For building this coin, Ethereum token standards are used; by this, we can compare other cryptocurrencies and the process is faster. Blockchain and smart contract help to interact with energy producers and consumers. The E2C platform helps the user to exchange and trade E2C coins for energy. Energy demand and supply dependency can easily be predicted using this E2C platform at each energy transaction. For future investments, better decisions can be made by the users to access the energy transactions.

11.5.4   KWHCoin

KWHCoin [39] is supported by clean energy units and it is a blockchain-based currency. It has a clear mission of expanding clean energy, where it reduces the cost of blockchain transactions. Using this, coin producers and consumers can set up and link power generation resources, and KWHCoin used as a token for decentralized application. With the use of blockchain-based energy platform, KWHCoin permits customers to sell and buy energy resources with smart grid. Smart grid makes use of programs and software for collaborating among producers and consumers. By using green energy and renewable resources, power grid can provide 100% clean energy. KWH token helps consumers to select their energy suppliers and producers to sell their produced clean energy to others with the help of power grid.

11.5.5   TerraGreen coin

One of the special blockchain initiative is the TerraGreen coin. This blockchain-assisted initiative is used to manage agriculture wastes and forest resources. Here, these resources are converted into useful energy. These products have good low-cost value for processing into energy. The meaning of terra is earth, and the main purpose of this coin is to build the earth greener. This coin constructs on a consensus network and provides a decentralized payment mechanism. Instead of a PoW mechanism, a firm PoS mechanism is used to make the blockchain most efficient. TerraGreen coin takes less amount of time for setting new transactions and for creating new blocks. Here users can trade securely and easily via TerraGreen user interface with the use of a secure wallet. We can create *n* number of sidechains using TerraGreen coin for cryptocurrencies and projects in a green energy platform (Figure 11.7).

11.6   Blockchain contributions in smart grid

While presenting numerous recent releases, we focus on blockchain contributions to the smart grid. This section looks at blockchains for common smart grid privacy and trust issues. The blockchain's services to smart grid communities such as improved metering architecture, distributed energy trading and market, energy

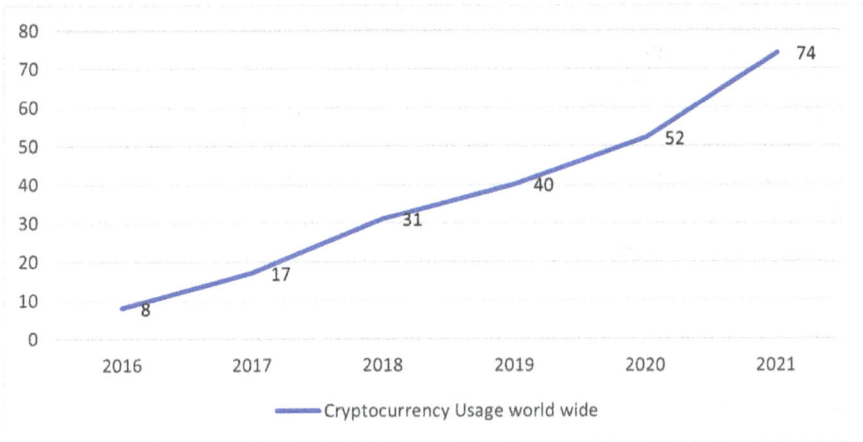

Figure 11.7   Cryptocurrency usage worldwide

malware system, administration of EVs and their fueling units, and ultimately microgrid are discussed. Figure 11.1 shows an example of a blockchain application. Furthermore, before discussing the blockchain advances, we explore the issues that can be addressed in each component.

11.6.1   Blockchain-based infrastructure for smart metering

With the development of high metering architecture, energy providers, users, and suppliers in the smart transmission networks will be able to communicate even more through smart meters that facilitate automatic and two-way communication. Smart meters, as opposed to standard meters, are modern ones that can collect detailed analysis of energy consumption and output, as well as status and diagnostic data. Payment, user device control, tracking, and debugging are all common uses for this information. These various data transfers, on the other hand, are carried out through a wide area network and kept in a conventional dedicated storage system or the cloud. The presence of a single network may come with its own set of problems, such as the danger of change, data leakage, and a sense of powerlessness. In [40], the researchers create a framework in which they investigate blockchain and smart contracts for smart grid robustness and safety. The contracts will operate as a middleman between energy client and vendor, lowering costs and increasing transaction rates while increasing transaction security. When a transaction occurs, the smart meter linked to the public blockchain will submit the record to the distributed ledger to produce a new block with a stamp for later verification. After that, the client can be billed depending on the information on the ledger. Nevertheless, one major criticism of this study is the absence of extensive technical discussion.

Reference [41] introduces a service delivery paradigm for smart power networks to achieve autonomy. The blockchain is used in this architecture to create a decentralized, safe, and automatic equipment setup in which all nodes operate independently with no need for central oversight and Distributed System Operator

(DSO) control. Furthermore, it is used to record consumption data in tamperproof blocks that can be obtained from smart meters.

Finally, using energy usage and generation traces from UK buildings' databases, this model is verified and validated by creating a prototype on the Ethereum blockchain platform. According to the findings, this model is capable of adjusting demand in near real-time by performing energy adaptability levels and validating all power storage agreements. However, how energy characteristic privacy has been guaranteed on this public blockchain is not precisely stated. It is possible to retrieve the user by studying publicly available transactions. Reference [42] proposes a blockchain-assisted, and edge computing-assisted strategy to improving the usability and energy supplies of smart grid networks (Figure 11.8).

The blockchain is mostly used for financial transactions. To prevent criminal operations within channels of communication and centralized data facilities, it secures the privacy of all participants and decentralizes data storage. This blockchain design is allowed to access three components like edge devices, super nodes, and smart contract servers (as shown in Figure 11.2) being established to secure the blockchain network's correctness and reliability. In this case, edge devices, super nodes, and smart contract servers are treated as standard nodes like in the modern blockchain system. Super nodes, on the other hand, are particular form of nodes that have the ability to pick certain devices from the edge to engage in the decision and polling process. Super nodes must confirm the identities of edge nodes before requesting to participate in the voting process (Figure 11.9).

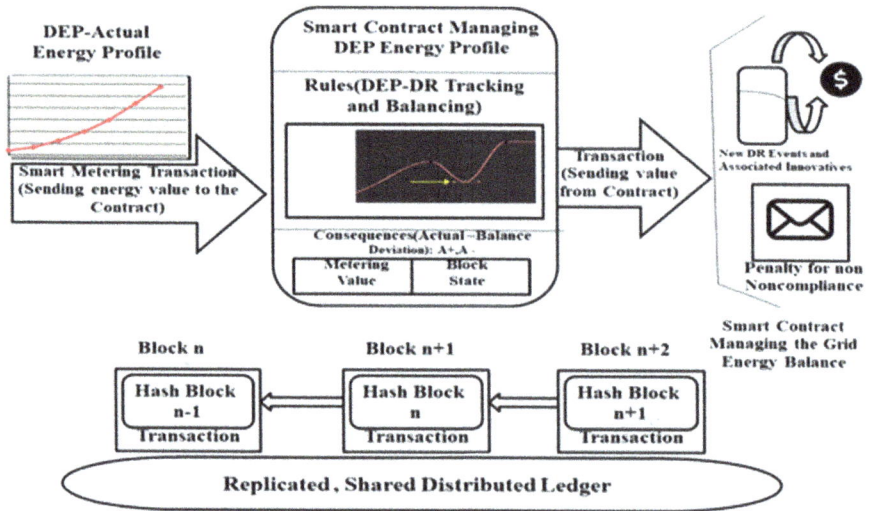

Figure 11.8   The self-enforcing contract structure

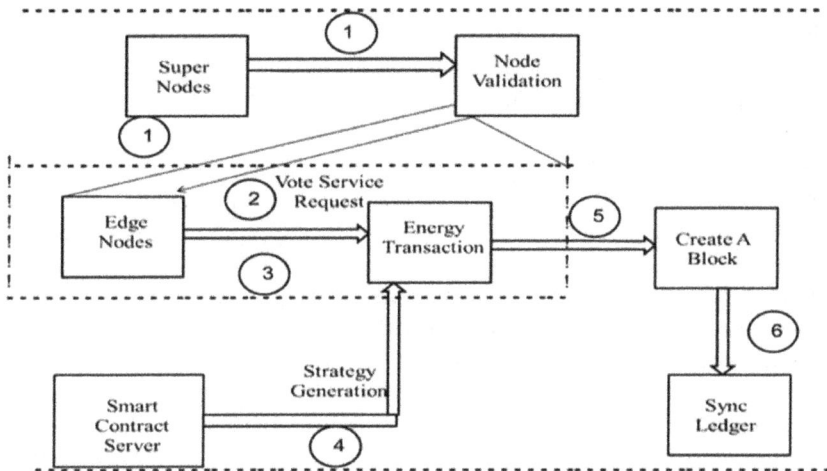

Figure 11.9   The three main entities of the proposed approach

11.6.2   Autonomous energy trading and market utilizing blockchain

Consumers can function as suppliers and vice versa because of the bidirectional energy and information transfer capability. In decentralized energy exchange scenarios, the smart grid is expected to accommodate a rising group of consumers, suppliers, and buyers (producer + consumer). As a result, they must be free to transact excess power from multiple resources such as microgrids, electric cars, and power storage units with one another in ways that generate benefits such as lowering load peaks, reducing power outage in the transmitter, easing the hardship on the power grid in order to promote green systems, and balancing energy supply and demand. As a result, it is important to integrate energy trading, as well as its necessary formalities, in order to provide bid processing, negotiation, and contractual arrangements among the parties.

Additionally, buyers and sellers are able to buy and sell the power generated smoothly with one another. This primary energy trade, which avoids the use of brokers, can benefit the people and is beneficial for sustainable energy installations. Legacy methods, on the other hand, allow buyers and sellers to engage in such trading procedures with one another only indirectly through a slew of middlemen, external parties, and merchants, all of whom will face problems and concerns. As a result, substantial legal and management costs are imposed, which are eventually passed on to consumers, manufacturers, as well as prosumers. Dysfunction, corrupt, or hostile middlemen also result in an unlikely to consider, a lack of openness and honesty, and hegemonic incentives, rewards, and penalties. Hence, the key characteristics of blockchain consider energy trade is a great instrument for creating more based on the distributed system.

11.6.3   Monitoring, measuring, and controlling using blockchain

The current smart grid CPS is primarily based on a single SCADA system that is hierarchically integrated with numerous elements such as MTUs, RTUs, PMUs, and a variety of sensors. To control and maintain electricity grids, the SCADA system is frequently used. The SCADA system, which is connected to the internet, will improve highly distributed surveillance, assessment, and management. IoT intelligent technologies, monitors, and PMUs typically collect power device status content and communicate it with MTUs via RTUs, with MTUs acting as primary concentration and control systems. These perfectly alright measurements between CPS components, different grid operators, providers, and users are used by the smart grid infrastructure to provide control algorithms and wide-area surveillance and oversight to improve grid security, consistency, and dependability, as well as track electricity theft and wastage. Cyber hackers or insiders can launch cyber-threats in a number of ways, including modifying data within core processors, launching uptime acts of violence, and introducing faulty information through detectors and PMUs. As a consequence, the intruder can take the power of modulation schemes and issue fraudulent orders. The blockchain provides additional surveillance, metering, and control capabilities in the distributed smart grid infrastructure.

Reference [43] concentrates on information security in equipment. To improve the safety of plant performance information, ICS-BlockOpS, a blockchain-based architecture, has been deployed. This architecture was created primarily to overcome two significant ICS issues.

Using blockchain technology, you can achieve preservation and durability. The preservation of data is ensured by the tamperproof nature of blockchain. On the other hand, to assure redundant information, a blockchain-based efficient duplication technique influenced by the Hadoop Distributed File System is proposed. Unfortunately, there really is no mention of when to deal with malevolent or hacked nodes leaking misleading information, or how asset-sensing devices will function in a blockchain system. Reference [44] describes a smart grid control system based on blockchain technology to assure power consumption protection and traceability. Electric utilities, agreement nodes, and utility companies are among the three types of nodes in this blockchain system, each with its own set of tasks. Smart people are usually the ones who make the best decisions.

11.7   Hash algorithm

As another important encryption algorithm in the Bitcoin blockchain and a hash algorithm is an important tool for modern information system encryption. Hash calculation is a digest calculation. Data with different lengths can be compressed and mapped to a unique value of fixed size by a hash algorithm. For different input X, it is possible to get the same hash value Y, but Y cannot be used to deduce the input X. By changing only one digit of X, the value of Y' is completely different from that of Y. The transaction data in the blockchain has the corresponding hash value through the encryption or hash algorithm [45]. Once the data is modified, the hash value will change completely. Forgery data needs to forge hash values together, and the cost of forging a single hash value is very high. The blockchain further improves the difficulty of hash value forgery through the Merkle tree.

11.7.1 Mathematical implementation of Merkle trees for creation and verification

1. Creation

Merkle tree is created by taking two nodes from each layer and hashing them to create the parent node by representing the tree in matrix form; we can mathematically write it as:

$$tree[i][j] = hash\ (\ tree[i+1][2j].\ hash\ +\ tree[i+1][2j+1]\ .\ hash)$$

where,

$$j \text{ is the index of node in } i^{th} \text{ layer}$$
$$0 \leqslant i < tree.\ length$$
$$0 \leqslant j < tree[i].\ length$$

This makes the root of the tree available at tree [0][0].

2. Verification

Verification is a bottom-up approach where we start from the data, find its hash, and calculate the parent and continue this until we find the root.

let position of the node to be verified be p
then,

$$tree[i]\left[\left\lfloor\frac{p}{2}\right\rfloor\right] = \begin{cases} hash(\ tree[i+1][p].\ hash\ +\ tree[i+1][p+1].\ hash, & p \text{ is even} \\ hash(\ tree[i+1][p].\ hash\ +\ tree[i+1][p-1].\ hash, & p \text{ is odd} \end{cases}$$

where,

$$0 \leqslant i < tree.\ length$$
$$0 \leqslant p < tree[i].\ length$$

3. Implementing a Merkel Tree in Node.js

```
const crypto = require("crypto");
function hash(data)
{
    return data != null
    ? crypto
    .createHash("sha256")
    .update(data.toString())
    .digest("hex")
    : "";
}
module.exports = hash;
```

11.8   Conclusion

As a new vision of conventional power grid, the smart grid concept has been introduced in which the energy is basically going to be monitored, metered, and managed properly, which involves bidirectional data communication and energy exchange to find an efficient way of delivering and integrating green and renewable energy technologies and managing them. In this chapter, a broad study has been made for future IoT smart grids. Initially, we have given a brief introduction to blockchain technology followed by a blockchain background. We looked at some decentralized IoT smart grids as well as the benefits of blockchain technology in smart grids. Some features of the smart grid, such as privacy, security, and trust, can be addressed with blockchain technology. We've discussed some of the blockchain's contributions to the smart grid and outlined a few recent Bitcoin smart grid initiatives. Finally, we identified a few research issues for the future. We anticipate that this work will provide a broad overview of blockchain technology's application in smart grids and will assist researchers in furthering their research in this field.

Blockchain technology furnishes many in and out applications and opportunities in the field of IoT smart grid. Without the support of a central authority, entities can communicate with each other by using some of the blockchain technologies, such as smart contracts, consensus mechanisms, and modern cryptosystem. Although blockchain system depends on some pre-established protocols, no intermediary is present during dynamic runtime and operation. Consequently, blocks are always vital, and they are secure, dependable, and precise. Furthermore, blockchain is yet at a growing stage and has many problems such as prevailing implementations, prevailing protocols, and diminished transaction loads are still demanding issues for users, practitioners, and researchers. There are some potential limitations in the smart grid and blockchain is also facing some challenges in smart grid implementation.

References

[1]   Gunduz M.Z., Das R. 'Cyber-security on smart grid: Threats and potential solutions'. *Computer Networks*. 2020, vol. 169(11), p. 107094.

[2]   Islam S.N., Baig Z., Zeadally S. 'Physical layer security for the smart grid: Vulnerabilities, threats, and countermeasures'. *IEEE Transactions on Industrial Informatics*. 2019, vol. 15(12), pp. 6522–30.

[3]   Kumar P., Lin Y., Bai G., Paverd A., Dong J.S., Martin A. 'Smart grid metering networks: A survey on security, privacy and open research issues'. *IEEE Communications Surveys & Tutorials*. 2019, vol. 21(3), pp. 2886–927.

[4]   Ghosal A., Conti M. 'Key management systems for smart grid advanced metering infrastructure: A survey'. *IEEE Communications Surveys & Tutorials*. 2019, vol. 21(3), pp. 2831–48.

[5]   De Dutta S., Prasad R. 'Security for smart grid in 5G and beyond networks'. *Wireless Personal Communications*. 2019, vol. 106(1), pp. 261–73.

[6]   Rastogi S.K., Sankar A., Manglik K., Mishra S.K., Mohanty S.P. 'Toward the vision of all-electric vehicles in a decade [energy and security]'. *IEEE Consumer Electronics Magazine.* 2019, vol. 8(2), pp. 103–7.

[7]   Shayeghi H., Shahryari E., Moradzadeh M., Siano P. 'A survey on microgrid energy management considering flexible energy sources'. *Energies.* 2019, vol. 12(11), p. 2156.

[8]   Nakamoto S. Bitcoin: A peer-to- peer electronic cash system. National seminar, US sentencing commission 2018.; 2008.

[9]   Hossain E., Khan I., Un-Noor F., Sikander S.S., Sunny M.S.H. 'Application of big data and machine learning in smart grid, and associated security concerns: A review'. *IEEE Access.* 2019, vol. 7, pp. 13960–88.

[10] Mylrea M., Gourisetti S.N.G. 'Blockchain for smart grid resilience: Exchanging distributed energy at speed, scale and security'. *Resilience Week (RWS)*; IEEE; 2017. pp. 18–23.

[11] Pop C., Cioara T., Antal M., Anghel I., Salomie I., Bertoncini M. 'Blockchain based decentralized management of demand response programs in smart energy grids'. *Sensors.* 2018, vol. 18(1), p. 162.

[12] Gai K., Wu Y., Zhu L., Xu L., Zhang Y. 'Permissioned blockchain and edge computing empowered privacy-preserving smart grid networks'. *IEEE Internet of Things Journal.* 2019, vol. 6(5), pp. 7992–8004.

[13] Tan S., Wang X., Jiang C. 'Privacy-preserving energy scheduling for ESCOs based on energy blockchain network'. *Energies.* 2019, vol. 12(8), p. 1530.

[14] Mrabet Z.E., Kaabouch N., Ghazi H.E., Ghazi H.E. 'Cyber-security in smart grid: survey and challenges'. *Computers & Electrical Engineering.* 2018, vol. 67(5), pp. 469–82.

[15] Wood G. 'Ethereum: A secure decentralised generalised transaction ledger'. *Ethereum Project Yellow Paper.* 2014, vol. 151, pp. 1–32.

[16] Merkle R.C. A digital signature based on a conventional encryption function. *Conference on the Theory and Application of Cryptographic Techniques.* Springer; 1987. pp. 369–78.

[17] Szabo N. Smart contracts: Building blocks for digital markets [online]. 1997. Available from http://www.fonhum.uva.nl/rob/Courses/InformationInSpeech/CDROM/Literature/LOTwinterschool2006/szabo.best.vwh.net/smartcontracts2.html [Accessed Nov 2019].

[18] Smart contracts [online]. 1994. Available from http://www.fon.hum.uva.nl/rob/Courses/InformationInSpeech/CDROM/Literature/LOTwinterschool2006/szabo.best.vwh.net/smart.contracts.html [Accessed Nov 2019].

[19] Proof of stack [online]. Available from https://en.bitcoin.it/wiki/Proof of Stake [Accessed Nov 2019].

[20] Nguyen C.T., Hoang D.T., Nguyen D.N., Niyato D., Nguyen H.T., Dutkiewicz E. 'Proof-of-Stake consensus mechanisms for future blockchain networks: Fundamentals, applications and opportunities'. *IEEE Access.* 2019, vol. 7, pp. 85727–45.

[21] Al-Turjman F., Abujubbeh M. 'IoT-enabled smart grid via Sm: An overview'. *Future Generation Computer Systems.* 2019, vol. 96(6), pp. 579–90.

[22] Saleem Y., Crespi N., Rehmani M.H., Copeland R. 'Internet of things-aided smart grid: Technologies, architectures, applications, prototypes, and future research directions'. *IEEE Access*. 2019, vol. 7, pp. 62962–3003.

[23] Kabalci Y., Kabalci E., Padmanaban S., Holm-Nielsen J.B., Blaabjerg F. 'Internet of things applications as energy Internet in smart grids and smart environments'. *Electronics*. 2019, vol. 8(9) 972.

[24] Wang K., Yu J., Yu Y., *et al.* 'A survey on energy Internet: Architecture, approach, and emerging technologies'. *IEEE Systems Journal*. 2017, vol. 12(3), pp. 2403–16.

[25] Yijia C., Qiang L., Yi T., *et al.* 'A comprehensive review of energy Internet: Basic concept, operation and planning methods, and research prospects'. *Journal of Modern Power Systems and Clean Energy*. 2018, vol. 6(3), pp. 399–411.

[26] Dong Z., Zhao J., Wen F., Xue Y. 'From smart grid to energy Internet: Basic concept and research framework'. *Automation of Electric Power Systems*. 2014, vol. 38(15), pp. 1–11.

[27] Chen S., Wen H., Wu J., *et al.* 'Internet of things based smart grids supported by intelligent edge computing'. *IEEE Access*. 2019, vol. 7, pp. 74089–102.

[28] Hussain S.M.S., Nadeem F., Aftab M.A., Ali I., Ustun T.S. 'The emerging energy Internet: Architecture, benefits, challenges, and future prospects'. *Electronics*. 2019, vol. 8(9), p. 1037.

[29] Proof-of-Stake-Time (PoST) by Vericoin whitepaper: A time accepted periodic proof factor in a nonlinear distributed consensus [online]. Available from https://www.vericoin.info/downloads/VeriCoinPoSTWhitePaper10 May2015.pdf [Accessed Nov 2019].

[30] Charg Coin (CHG) whitepaper [online]. Available from https://chgcoin.org/white-paper/ [Accessed Nov 2019].

[31] CyClean green blockchain ecosystem whitepaper. Available from https://www.adb.org/sites/default/files/publication/464821/adbi-wp883.pdf [Accessed Nov 2019].

[32] Pylon network blockchain whitepaper. Available from https://pylon-network.org/wp-content/uploads/2019/02/WhitePaper_PYLON_v2_ENGLISH-1.pdf [Accessed Nov 2019].

[33] Exergy an LO3 energy innovation whitepaper [online]. Available from https://exergy.energy/wp-content/uploads/2019/03/TransactiveEnergy-PolicyPaper-v2-2.pdf [Accessed Nov 2019].

[34] The Energy Web Chain: Accelerating the energy transition with an open-source, decentralized blockchain platform, whitepaper [online]. Available from https://github.com/energywebfoundation/paper/blob/master/EWF-Paper-v2.pdf [Accessed Nov 2019].

[35] De Angelis S., Aniello L., Baldoni R., Lombardi F., Margheri A., Sassone V. 'PBFT vs Proof-of-Authority: Applying the caps theorem to permissioned blockchain'. *Italian Conference on Cyber Security, Milan, Italy*. 2018.

[36] Weerakkody S., Sinopoli B. 'Challenges and opportunities: Cyberphysical security in the smart grid, *Smart Grid Control*'. *Berlin/Heidelberg: Springer*. 2019, pp. 257–73.

[37] Li Z., Kang J., Yu R., Ye D., Deng Q., Zhang Y. 'Consortium blockchain for secure energy trading in industrial Internet of Things'. *IEEE Transactions on Industrial Informatics*. 2017, vol. 14(8), pp. 1–3700.

[38] Power Ledger whitepaper [online]. Available from https://www.powerledger.io/wp-content/uploads/2019/05/power-ledgerwhitepaper.pdfnurl [Accessed Nov 2019].

[39] KWHCoin. Available from https://arxiv.org/pdf/1911.03298.pdf.

[40] Mylrea M., Gourisetti S.N.G. 'Blockchain for smart grid resilience: exchanging distributed energy at speed, scale and security'. *2017 Resilience Week (RWS). IEEE*. 2017–18–23.

[41] Pop C., Cioara T., Antal M., Anghel I., Salomie I., Bertoncini M. 'Blockchain based decentralized management of demand response programs in smart energy grids'. *Sensors*. 2018, vol. 18(1), p. 162.

[42] Gai K., Wu Y., Zhu L., Xu L., Zhang Y. 'Permissioned blockchain and edge computing empowered privacy-preserving smart grid'. *IEEE Internet of Things Journal*. 2019, vol. 6(5), pp. 7992–8004.

[43] Maw A., Adepu S., Mathur A.' ' 'ICS-BlockOpS: Blockchain for operational data security in industrial control system". *Pervasive and mobile computing*. 2019, vol. 59,p. 101048

[44] Gao J., Asamoah K.O., Sifah E.B., *et al.* 'Grid Monitoring: Secured sovereign Blockchain based monitoring on smart grid'. *IEEE Access*. 2018, vol. 6, pp. 9917–25.

[45] Wang X. 'Collisions for hash functions MD4s MD5, NAVAL-128, and RIPEMD[J'. *Cryptology Eprint Archive Deports*. 2004, vol. 2004.

Index